A History of Thermodynamics

Ingo Müller

A History of Thermodynamics

The Doctrine of Energy and Entropy

Springer

Professor Dr. Dr.h.c. Ingo Müller
Thermodynamik
Technische Universität Berlin
10623 Berlin
Germany
E-mail: ingo.mueller@alumni.tu-berlin.de

Library of Congress Control Number: 2006933419

ISBN-10 3-540-46226-0 Springer Berlin Heidelberg New York
ISBN-13 978-3-540-46226-2 Springer Berlin Heidelberg New York

Springer is a part of Springer Science+Business Media
springer.com

Typesetting: by the author and techbooks using a Springer LaTeX macro package
Cover design: *design & production* GmbH, Heidelberg

Printed on acid-free paper SPIN: 11553786 89/techbooks 5 4 3 2 1 0

Preface

The most exciting and significant episode of scientific progress is the development of thermodynamics and electrodynamics in the 19th century and early 20th century. The nature of heat and temperature was recognized, the conservation of energy was discovered, and the realization that mass and energy are equivalent provided a new fuel, – and unlimited power.

Much of this occurred in unison with the rapid technological advance provided by the steam engine, the electric motor, internal combustion engines, refrigeration and the rectification processes of the chemical industry. The availability of cheap power and cheap fuel has had its impact on society: Populations grew, the standard of living increased, the environment became clean, traffic became easy, and life expectancy was raised. Knowledge fairly exploded. The western countries, where all this happened, gained in power and influence, and western culture – scientific culture – spread across the globe, and is still spreading.

At the same time, thermodynamics recognized the stochastic and probabilistic aspect of natural processes. It turned out that the doctrine of energy and entropy rules the world; the first ingredient – energy – is deterministic, as it were, and the second – entropy – favours randomness. Both tendencies compete, and they find the precarious balance needed for stability and change alike.

Philosophy, – traditional philosophy – could not keep up with the grand expansion of knowledge. It gave up and let itself be pushed into insignificance. The word came up about *two cultures*: One, which is mostly loose words and subjective thinking – in the conventional style –, and scientific culture, which uses mathematics and achieves tangible results.

Indeed, the concepts of the scientific culture are most precisely expressed mathematically, and that circumstance makes them accessible to only a minority: Those who do not shy away from mathematics. The fact has forced me into a two-tiered presentation. One tier is narrative and largely devoid of formulae, the other one is mathematical and mostly relegated to *Inserts.* And while I do not recommend to skip over the inserts, I do believe that that is possible – at least for a first reading. In that way a person may acquire a quick appreciation of the exciting concepts and the colourful personages to whom we owe our prosperity and – in all probability – our lives.

Berlin, *Ingo Müller*
July 2006

Contents

1 Temperature

Temperature – also *temperament* in the early days – measures hot and cold and the word is, of course, Latin in origin: *temperare* - to mix. It was mostly used when liquids are mixed which cannot afterwards be separated, like wine and water. The passive voice is employed – the "*-tur*" of the present tense, third person singular – which indicates that some liquid *is being mixed* with another one.

For Hippokrates (460–370 B.C.), the eminent, half legendary Greek physician, proper mixing was important: An imbalance of the bodily fluids blood, phlegm, and black and yellow bile was supposed to lead to disease which made the body unusually hot or cold or dry or moist.

Klaudios Galenos (133–200 A.C.), vulgarly Galen, – another illustrious Greek physician, admirer of Hippokrates and polygraph on medical matters – took up the idea and elaborated on it. He assumed an influence of the climate on the mix of body fluids which would then determine the character, or temperament (sic), of a person. Thus body and soul of the inhabitants of the cold and wet north were wild and savage, while those of the people in the hot and dry south were meek and flaccid. And it was only in the well-mixed – *temperate* – zone that people lived with superior properties in regard to good judgement and intellect,[1] the Greeks naturally and, perhaps, the Romans.

Galen mixed equal amounts if ice and boiling water, which he considered the coldest and hottest bodies available. He called the mixture neutral,[2] and installed four degrees of cold below that neutral point, and four degrees of hot above it. That rough scale of nine degrees survived the dark age of science under the care of Arabian physicians, and it re-emerged in Europe during the time of the Renaissance.

Thus in the year 1578, when Johannis Hasler from Berne published his book "De logistica medica", he presented an elaborate table of body temperatures of people in relation to the latitude under which they live, cf. Fig. 1.1. Dwellers of the tropics were warm to the fourth degree while the

---

[1] Galen: "Daß die Vermögen der Seele eine Folge der Mischungen des Körpers sind." [That the faculties of the soul follow from the composition of the body] Abhandlungen zur Geschichte der Medizin und Naturwissenschaften. Heft 21. Kraus Reprint Liechtenstein (1977).

[2] It is not clear whether Galen mixed equal amounts by mass or volume; he does not say. In the first case his neutral temperature is 10°C in the latter it is 14°C; neither one is of any obvious relevance to medicine.

eskimos were cold to the fourth degree. Persons between latitudes 40° and 50°, where Hasler lived, were neither hot nor cold; they were given the neutral temperature zero.

One must admit that the idea has a certain plausibility and, indeed, the nine degrees of temperature fit in neatly with the 90 degrees of latitude between the equator and the pole. However, it was all quite wrong: All healthy human beings have the same body temperature, irrespective of where they live. That fact became soon established after the invention of the thermometer.

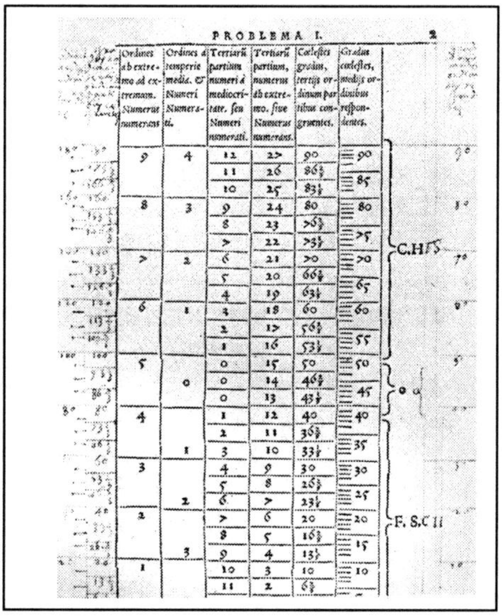
PROBLEMA I. 2

Ordines ab extremo ad extremum. Numerus numerans	Ordines à temperie media. & Numeri Numerati.	Tertiarũ partium numeri à mediocritate, seu Numeri numerati.	Tertiarũ partium, numerus ab extremo. siue Numerus numerans.	Cœlestes gradus, tertijs ordinum partibus congruentes.	Gradus cœlestes, medijs ordinibus respondentes.
9	4	12	27	90	90
		11	26	86⅔	85
		10	25	83⅓	
8	3	9	24	80	80
		8	23	76⅔	
		7	22	73⅓	75
7	2	6	21	70	70
		5	20	66⅔	
		4	19	63⅓	65
6	1	3	18	60	60
		2	17	56⅔	
		1	16	53⅓	55
5		0	15	50	50
	0	0	14	46⅔	
		0	13	43⅓	45
4		1	12	40	40
		2	11	36⅔	
	1	3	10	33⅓	35
3		4	9	30	30
		5	8	26⅔	
	2	6	7	23⅓	25
2		7	6	20	20
		8	5	16⅔	
	3	9	4	13⅓	15
1		10	3	10	10
		11	2	6⅔	

Fig . 1.1. Hasler's table of body temperatures in relation to latitude

The instrument was developed in the early part of the 17th century. The development is painstakingly researched and well-described – as much as it can be done – by W.E. Knowles Middleton in his book on the history of the thermometer.[3] Another excellent review may be found in a booklet by Ya.A. Smorodinsky.[4] It is not clear who invented the instrument. Middleton complains that *questions of priority are loaded with embarrassment for the historian of science...*, and he indicates that the answers are often biased by nationalistic instincts.

---

[3] W.E. Knowles Middleton: "The History of the Thermometer and its Use in Meteorology". The Johns Hopkins Press, Baltimore, Maryland (1966).
Hasler table of body temperatures, cf. Fig.1.1, is the frontispiece of that book.

[4] Ya.A. Smorodinsky: "Temperature". MIR Publishers, Moscow (1984).

So also in the case of the thermometer: According to Middleton there was some inconclusive bickering about priority across the Alps, between England and Italy. One thing is certain though: The eminent scientist Galileo Galilei (1564–1642) categorically claimed the priority for himself. And his pupil, the Venetian diplomat Gianfrancesco Sagredo accepted that claim after at first being unaware of it. Sagredo experimented with the thermometer and on May 9th, 1613 he wrote to the master [5]:

> The instrument for measuring heat, invented by your excellent self ...[has shown me] various marvellous things, as, for example, that in winter the air may be colder than ice or snow; ...

Another quaint observation on well-water is communicated by Sagredo to Galilei on February 7th, 1615, cf. Fig. 1.2[6]. It is clear what Sagredo means: If you bring water up in summer from a deep well and you stick your hand into it, it feels cool, while, if you do that in wintertime, the water feels warm.

GIOVANFRANCESCO SAGREDO a GALILEO
in Firenze
Venezia, 7 febbraio 1615
Molto Ill.re S.r Ecc.mo

... Con questi istrumenti ho chiaramente veduto, esser molto più freda l'aqua de' nostri pozzi il verno che l'estate; e per me credo che l'istesso avenga delle fontane vive et luochi soteranei, anchorchè il senso nostro giudichi diversamente.

Et per fine li baccio la mano

In Venetia, a 7 Febraro 1615
Di V.S. Ecc.ma
Tutto suo Il Sag.

Fig. 1.2. Galileo Galilei. A cut from a letter of Sagredo to Galilei with the remarkable sentence: *I have clearly seen that well-water is colder in winter than in summer ..., although our senses tell differently*

Misconceptions due to the subjective feeling of hot and cold were slowly eliminated during the course of the 17th century. A serious obstacle was that no two thermometers were quite alike so that, even when there were

---

[5] Middleton. loc.cit. p. 7.

[6] "Le Opere di Galileo Galilei", Vol. XII, Firenze, Tipografia di G. Barbera (1902) p.139. The letter, and other letters by Sagredo to Galilei are replete with flattering, even sycophantic remarks which the older man seems to have appreciated. Part of that may be attributed, perhaps, to the etiquette of the time. But, in fact, it may generally be observed – even in our time – that, the more eminent a scientist already is, the more he demands praise; and a diplomat knows that.

scales on them, it was difficult to communicate objective information from one place to another.

Scales were just as likely to run upwards as downwards at that time. Middleton[7] lists the scale on a surviving thermometer built by John Patrick in or around the year 1700; it runs downward with increasing heat from 90° to 0°, thus maintaining remnants of Galen's scale of 9 degrees, perhaps.

90°	Extream Cold	55°	Cold Air	15°	Sultry
85°	Great Frost	45°	Temperate Air	5°	Very Hott
75°	Hard Frost	35°	Warm Air	0°	Extream Hott
65°	Frost	25°	Hott		

Fix-points were needed to make readings on different thermometers comparable. From the beginning, melting ice played a certain role – either in water or in a water-salt-solution – and boiling water, of course. But alternatives were also proposed:

the temperature of melting butter,
the temperature in the cellar of the Paris observatory,
the temperature in the armpit of a healthy man.

The surviving Celsius scale uses melting ice and boiling water, and one hundred equal steps in-between. However, since Anders Celsius (1701–1744) wished to avoid negative numbers, he set the boiling water to 0°C and melting ice to 100°C, – for a pressure of 1atm. Thus he too counted *downwards*. That order was reversed after Celsius's death, and it is in that inverted form that we now know the Celsius scale, or centigrade scale.

Gabriel Daniel Fahrenheit (1686–1736) somehow thought that three fix-points were better than two. He picked

a freezing mixture of water and sea-salt (0°F),
melting ice in water alone (32°F),
human body temperature (96°F).

Later he adjusted that scale slightly, so as to have boiling water at 212°F, exactly 180 degrees above melting ice. One cannot help thinking that 180° is a neat number, at least when the degrees are degrees of arc. However Middleton, who describes the development of the Fahrenheit scale in some detail, does not mention that analogy so that it is probably fortuitous. Anyway, after the readjustment, the body temperature came to 98.6°F. That is where the body temperature stands today in those countries, where the Fahrenheit scale is still in use, notably in the United States of America.

From the above it is easy to calculate the transition formula between the Celsius and the Fahrenheit scales: $C = 5/9(F - 32)$.

[7] Middleton: loc.cit. p. 61.

There were numerous other scales, advertised at different times, in different places, and by different people. It was not uncommon in the 18th and early 19th century to place the thermometric tube in front of a wide board with several different scales, – up to eighteen of them. Middleton[8] exhibits a list of scales shown on a thermometer of 1841:

Old Florentine	Delisle	Amontons
New Florentine	Fahrenheit	Newton
Hales	Réaumur	Société Royale
Fowler	Bellani	De la Hire
Paris	Christin	Edinburg
H. M.Poleni	Michaelly	Cruquius

All of these scales were arbitrary and entirely subjective but, of course, perfectly usable, if only people could have agreed to use *one* of them, – which they could not.

A new *objective* aspect appeared in the field with the idea that there might be a lowest temperature, an *absolute minimum.* By the mid-nineteenth century, two hundred years of experimental research on ideal gases had jelled into the result that the pressure p and the volume V of gases were linear functions of the Celsius temperature (say), such that

$$pV = m\frac{k}{\mu}(273.15°\mathrm{C} + t)$$

m is the mass of the gas.[9] Therefore, upon lowering the temperature to $t = -273.15°\mathrm{C}$ at constant p, the volume had to decrease and eventually vanish, and surely further cooling was then absurd. At first people were unimpressed and unconvinced of the minimal temperature. After all, even then they suspected that all gases turn into liquids and solids at low temperatures, and the argument did not apply to either.

However, in the 19th century it was slowly – painfully slowly – recognized that matter consisted of atoms and molecules, and that temperature was a measure for the mean kinetic energy of those particles. This notion afforded an understanding of the minimal temperature, because

[8] Middleton: loc.cit. p. 66.

[9] In much of the 19th century literature this equation is called the *law of Mariotte and Gay-Lussac.* Nowadays we call it the *thermal equation of state* for an ideal gas. The pioneers of the equation were Robert Boyle (1627–1691), Edmé Mariotte (1620–1684), Guillaume Amontons (1663–1705), Jacques Alexandre César Charles (1746–1823), and Joseph Louis Gay-Lussac (1778–1850). Their work is now a favourite subject of high-school physics courses. Therefore I skip over its motivation and derivation. I only emphasize that the value 273.15 is the *same* for all gases. That value was established by Gay-Lussac when he measured the relative volume expansion by heating a gas of 0°C by 1°C. [The value 273.15 is the modern one; in fact it is 273.15 ± 0.02. Gay-Lussac and others at the time were up to 5% off.] [The factor k/μ is also modern. k is the Boltzmann constant and μ is the molecular mass. Both are quite anachronistic in the present context. However, I wish to avoid the ideal gas constant and the molar mass in this book.]

when temperature dropped, so did the kinetic energy of the particles – of gases, liquids, and solids – and finally, when all were at rest, there was no way to lower the temperature further.

Therefore William Thomson (1824–1907) (Lord Kelvin since 1892) suggested – in 1848 – to call the lowest temperature *absolute zero*, and to move upward from that point by the steps or degrees of Celsius. This new scale became known as the absolute scale or Kelvin scale, on which melting ice and boiling water at 1atm have the temperature values 273.15°K and 373.15°K respectively. K stands for Kelvin. It became common practice to denote temperature values on the Kelvin scale by T, so that we have

$$T = \left(273.15 + \frac{t}{°\mathrm{C}}\right)°\mathrm{K}.$$

Kelvin's absolute scale was quickly adopted and it is now used by scientists all over the world. However, the scale has subtly changed since its introduction. In 1954, by international agreement the temperatures of melting ice and boiling water were abolished as fix-points. They were replaced by a single fix-point:

$$T_{tr} = 273.16°\mathrm{K} \qquad \text{for the triple point of water.}$$

The triple point of water occurs when ice, liquid water and water vapour can coexist; its pressure is $p_{tr} = 6.1\mathrm{mbar}$, and its temperature is $t_{tr} = 0.01°\mathrm{C}$ on the Celsius scale. The modern *degree* is defined by choosing 1°K as $T_{tr}/273.16$. This unit step on the Kelvin scale was internationally agreed on in 1954 so as to coincide with the familiar 1°C. The 13th International Conference on Measures and Weights of 1967/68 even robbed temperature of its little decorative adornment "°" for degree. Ever since then we speak and write of temperature values prosaically as so many "K" instead of "degrees K", or "°K".[10]

The lowest temperatures reached in laboratories are a few μK – a few millionth of one Kelvin –, the highest may be 10MK – ten million Kelvin –, and we believe that the temperature in the centre of some stars are as high as 100 million K, cf. Chaps. 6 and 7.

For the early researchers there was no need to *define* temperature. They knew, or thought they knew, what temperature was when they stuck their thermometer into well-water, or into the armpit of a healthy man. They were unaware of the implicit assumption, – or considered it unimportant, or self-evident – that the temperature of the thermometric substance, gas or mercury, or alcohol, was equal to the temperature of the measured object.

[10] Temperature measurements at extremely low temperatures are still a problem. The interested reader is referred to the publication "Die SI-Basiseinheiten. Definition, Entwicklung, Realisierung." [The SI basic units. Definition, development and realization] Physikalisch Technische Bundesanstalt, Braunschweig & Berlin (1997) p. 31–35.

This in fact is the defining property of temperature: That the temperature field is continuous at the surface of the thermometer; hence temperature is measurable. Axiomatists call this the zeroth law of thermodynamics because, by the time when they recognized the need for a definition of temperature, the first and second laws were already firmly labelled.

2 Energy

The word energy is a technical term invented by Thomas Young (1773–1829) in 1807. Its origin is the Greek word ἐνεργεια which means *efficacy* or *effective force*. Young used it as a convenient abbreviation for the sum of kinetic energy and gravitational potential energy of a mass and the elastic energy of a spring to which the mass may be attached. That sum is conserved by Newton's laws and Hooke's law of elasticity, although the individual contributions might change.[1] The term *energy* was not fully accepted until the second half of the 19th century when it was extrapolated away from mechanics to include the internal energy of thermodynamics and the electro-magnetic energy. The *first law of thermodynamics* states that the *total* sum is conserved: the sum of mechanical, thermodynamic, electro-magnetic, and nuclear energies. We shall proceed to describe the difficult birth of that idea.

Eventually – in the early 20th century – energy was recognized as having mass, or being mass, in accord with Einstein's formula $E = mc^2$, where c is the speed of light.

Caloric Theory

In the early days of thermodynamics nobody spoke of energy; it was either heat or force. And nobody really knew what heat was. Francis Bacon (1561–1626) mentions heat in his book "Novum Organum" and – true to his conviction that the laws of science should be gleaned from a mass of specific observations – he tabulated sources of heat such as: flame, lightning, summer, will-o'-the-wisp, and aromatic herbs which produce the feeling of warmth when digested.[2]

A little later Pierre Gassendi (1592–1655), a convinced atomist, saw heat and cold as distinct species of matter. The atoms of cold he considered as tetrahedral, and when they penetrated a liquid that liquid would solidify, – somehow.

---

[1] The observation that mechanical energy is conserved is usually attributed to Gottfried Wilhelm Leibniz (1646–1716), who pronounced it as a law in 1693.

[2] Francis Bacon: "Novum Organum" (1620).

An important step away from such interesting notions was done by Joseph Black (1728–1799). Black melted ice by gently heating it and noticed that the temperature did not change. Thus he came to distinguish the *quantity* of heat and its *intensity*, of which the latter was measured by temperature. The former – absorbed by the ice in the process of melting – he called latent heat, a term that has survived to this day.

The next step – unfortunately a step in the wrong direction – came from Antoine Laurent Lavoisier (1743–1794), the pre-eminent chemist of the 18th century, sometimes called the father of modern chemistry. He insisted on accurate measurement and therefore people say that he did for chemistry what Galilei had done for physics one and a half century before. The true nature of heat, however, was beyond Lavoisier's powers of imagination and so he listed heat – along with light – among the elements,[3] and considered it a fluid which he called the *caloric*. Asimov[4] writes that ... *it was partly because of his* [Lavoisier's] *great influence that the caloric theory ... remained in existence in the minds of chemists for a half century.* The idea was that caloric would be liberated when chips were taken off a metal in a lathe (say) and thus the material became hot.

Benjamin Thompson (1753–1814), Graf von Rumford

Benjamin Thompson, later Graf von Rumford – ennobled by the Bavarian elector Karl Theodor – was first to seriously question the caloric theory. Thompson was born in Woburn, Massachusetts to poor parents, just like Benjamin Franklin (1706–1790), the other famous American scientist of the 18th century; their birthplaces are only two miles apart. Both, although congenial as scientists, subscribed to different political views. Indeed, Thompson supported the British in the war of independence; he spied for them and even led a loyalist regiment, – a *Tory regiment* for American patriots – the King's American Dragoons.[5]

Perforce, after the colonials had won their independence, Thompson left America and, by his intelligence and his captivating demeanour, he became a man of the world, welcome in courts and scientific circles. He proved to be an inventor of everything that needed inventing: a modern kitchen – complete with sink, overhead cupboards and trash slot –, a drip coffee pot,

[3] A.L. Lavoisier: "Elementary Treatise on Chemistry" (1789).

[4] I. Asimov: "Biographical Encyclopedia of Science and Technology''.- Pan Reference Books, London (1975).

[5] Kenneth Roberts: ''Oliver Wiswell." Fawcett Publications, Greenwich, Connecticut. (1940).

and the damper for chimneys.[6] Also he was a gifted organizer of anything that needed to be organized:

- The distribution of a cheap, nourishing and filling soup – the Rumford soup[7] – for the poor people of Munich,
- the transplanting of fully grown trees into the English garden of the elector of Bavaria,
- and a factory for military uniforms staffed by the beggars from the streets of Munich.

The grateful elector made him a count: Graf von Rumford, see Fig. 2.1. Rumford was a town in Massachusetts, where Thompson had lived; later it was renamed Concord – now in New Hampshire; it was a hotbed of the American revolution. Needless to say that the elector knew neither Rumford nor Concord. Actually, one cannot help feeling that the two of them, the elector and Thompson, may have had a good laugh together: The elector, who had no jurisdiction over Rumford county and Thompson, – the new Graf von Rumford – who could not show his face there without running the risk of being tarred and feathered and made to ride a fence.

Fig. 2.1. Lavoisier and Thompson (Graf Rumford), both married to the same woman, – at different times

Graf Rumford was put in charge of boring cannon barrels for the elector. He noticed that blunt drills liberated more *caloric* than sharp-edged ones, although no chips appeared. By letting the blunt drill grind away for some length of time he could liberate more caloric than was known to be needed to melt the whole barrel. Thus he came to the only possible conclusion that the caloric theory was bunk and that

---

[6] According to Varick Vanardy: "Gen. Benjamin Thompson, Count Rumford: Tinker, Tailer, Soldier, Spy." http://www.rumford.com.

[7] A variant of that soup was handed out in German prisons until well into the 20th century. It was then known as "Rumfutsch". According to Ernst von Salomon: "Der Fagebogen" [The Questionaire] Rowohlt Verlag Hamburg (1951).

> ...it was inconceivable to think anything else than that heat was just the same as what was supplied to the metal as continually as heat was appearing in it namely: *motion.*[8]

Considering the jargon of the time that was a direct hit. Even fifty years later Mayer could not express the 1st law more clearly than by saying: *motion is converted to heat*, – and Mayer did still shy away from saying: Heat *is* motion.

Rumford even made an attempt to give an idea of what was later called the *mechanical equivalent of heat.* His drill was operated by the work of two horses – *of which one would have been enough* – turning a capstan-bar, and Rumford notes that the heating of the barrel by the drill

> equals that of nine big wax candles.

Actually, he became more concrete than that when he said that *the total weight of ice-water that could be heated to* 180°F *in* 2 *hours and* 30 *minutes amounted to* 26.58 pounds.[9] Joule fifty years later[10] used that measurement to calculate Rumford's equivalent of heat to 1034 foot-pounds.[11] For the calculation Joule adopted Watt's measurement of one horsepower, namely 33000 foot-pounds per minute.

It is probably too much to suppose that Rumford thought about conservation of energy, but he did say this:

> One would obtain more heat [than from the drill], if one burned the fodder of the horses. Thus he gave the impression, perhaps, that he may have suspected those amounts of heat to be the same.

Rumford *through his arrogance and the general unpleasantness of his character* – so the American author Asimov[12] – *eventually outwore his welcome in Bavaria.* He went to England where he was admitted into the Royal Society. He founded the Royal Institution, an institute which may be regarded as the prototypical postgraduate school. Rumford engaged Thomas Young and Humphry Davy as lecturers, who both became eminent scientists in their own time. Jointly with Davy, Rumford continued his

---

[8] Rumford: "An inquiry concerning the source of the heat which is excited by friction". Philosophical Transactions. Vol. XVIII, p. 286.

[9] Rumford: loc.cit. p. 283.

[10] J.P. Joule: "On the mechanical equivalent of heat". Philosophical Transaction. (1850) p. 61ff.

[11] This means that a weight of 1 pound dropped from a height of 1034 feet would be able to heat 1 pound of water by 1°F. [Joule's best value in 1850 is 772 foot-pounds, see below.]

[12] I. Asimov: "Biographies...." loc.cit.
Americans do not like their countryman Graf Rumford because of his involvement in the war of independence on the side of the loyalists. They scorn him and revile him, and largely ignore him. This is punishment for a person who fought on the wrong side – the side that lost. *We* must realize though that the American revolutionary war was as much a civil war as it was a war against the British rule; and civil wars have a way of arousing strong feelings and long-lasting hatred.

experiments on heat: He carefully weighted water before and after freezing and found the weight unchanged, although it had given off heat in the process. Therefore he concluded that the caloric, if it existed, was *imponderable*. This observation should have disqualified the caloric, but it did not, not for another 40 year.

After England, Rumford went to Paris where, posthumously, he crossed the path of Lavoisier, because he married the chemist's widow. Asimov writes

> The marriage was unhappy. After four years they separated and Rumford was so ungallant as to hint that she was so hard to get along with that Lavoisier was lucky to have been guillotined[13]. However, it is quite obvious that Rumford was no daisy himself.

Rumford's insight into the nature of heat was largely ignored and the caloric theory of heat prevailed until the 1840s. At that time, however, in the short span of less than a decade three men independently – as far as one can tell[14] – came up with the first law of thermodynamics in one way or other. Basically this was the recognition that the gravitational potential energy of a mass at some height, or the kinetic energy of a moving mass, may be converted into heat by letting it hit the ground. The three men who realized that fact in the 1840s were Mayer, Joule and Helmholtz. All three of them are usually credited with the discovery. And although all three devote part of their works to the discussion of the weightless caloric – actually to its refutation – it is clear that that theory had run its course. Says Mayer in his usual florid style: *Let's declare it, the great truth. There are no immaterial materials.*

Robert Julius Mayer (1814–1878)

Mayer was first and he went further than either of his competitors, because he felt that energy *generally* was conserved. He included tidal waves in his considerations and conceived of falling meteors as a possible source of solar heat- and light-radiation. Nor did he stop at chemical energy, not even chemical energy connected with life functions.

Mayer was born and lived most of his life in Heilbronn, a town in the then kingdom of Württemberg. Württemberg was one of the several dozen independent states within the loose German federation, whose rulers

---

[13] Lavoisier was executed on May 8, 1794 because of his involvement in tax collection under the *ancien régime*. On the eve of his execution he wrote a letter to his wife. The chemist was being philosophical: "*It is to be expected*" the letter reads "*that the events in which I am involved will spare me the inconvenience of old age*."

[14] This is what is usually said. It is not entirely true, though. To be sure, it is likely that Joule and Helmholtz were unaware of Mayer's ideas, but Helmholtz was fully aware of Joule's measurements, he cites them, see below.

suppressed all activity to promote German unity. Unity, however, was vociferously clamoured for by the idealistic students in their fraternities; therefore fraternities were declared illegal. But in Tübingen, where Mayer studied medicine, he and some friends were indiscreet enough to found a new fraternity. He was arrested for that – and for *attending a ball indecently dressed* – and relegated from the university for one year.

Mayer made good use of the enforced inactivity by continuing his medical education in Munich and Paris and then took hire as a ship's physician – a *Scheeps Heelmeester* – on a Dutch merchantman for a round-trip to Java. This left him a lot of free time since, in his words, *on the high seas people tend to be healthy*. He learned about two important phenomena which he lists in his diaries:

- The navigator told him that during a storm the ocean water becomes warmer,[15] and
- while bleeding patients he observed that in the tropics venous blood is similar in colour to arterial blood.

The first observation could be interpreted as motion of the water waves being converted to heat and the second seemed to imply that the des-oxidization of blood is slower when less heat must be produced to maintain the body temperature.

The flash of insight, a kind of ecstatic vision, came to Mayer when his ship rode at anchor off Surabaja taking on board a consignment of sugar. Henceforth he was a changed man, a fanatic in the effort of spreading his gospel. And he hurried back home in order to let the world know about his discovery.[16]

The gospel, however, left something to be desired. At least nobody wanted to hear it. Right after his return from Java Mayer rushed out a paper: "Über die quantitative and qualitative Bestimmung der Kräfte."[17] Actually there was nothing quantitative in the paper and, moreover, it was totally and completely obscure. There was hapless talk in hapless mathematical and geometrical language which could not possibly mean anything to anybody. The only saving grace is the sentence: *Motion is converted to heat*, which Rumford had said 40 years before. The paper ends characteristically in one of the hyperbolic statements which are so typical for Mayer's style: *In stars the unsolvable task of explaining the continuous creation of force, i.e. the*

---

[15] This observation is also mentioned by J.P. Joule: "On the mechanical equivalent of heat". Philosophical Transaction (1850) p. 61 ff.

[16] Later, in 1848, Mayer was involved in a political squabble and he was ridiculed publicly as having travelled as far as East India without setting his foot on land. This, however, seems to be untrue, if Mayer's diary is to be believed. He did leave the ship for a short excursion; cf. H. Schmolz, H. Weckbach: "Robert Mayer, sein Leben und Werk in Dokumenten". Veröffentlichungen des Archivs der Stadt Heilbronn. Bd. 12. Verlag H. Konrad (1964) p. 86.

[17] "On the quantitative and qualitative determination of forces".

differentiation of 0 *to* MC – MC, *is solved by nature; the fruit of this is the most marvellous phenomenon of the material world, the eternal source of light.* And in unshared enthusiasm Mayer finishes the paper with the hopeful words

Fortsetzung folgt = to be continued.

Well, Poggendorff, to whose "Annalen der Physik and Chemie" Mayer had sent the paper on June 16th 1841, was unimpressed. Certainly and understandably he did not want to encourage the author. Despite several urgent reminders by Mayer – the first one on July 3rd 1841 (!) – Poggendorff never acknowledged receipt, nor did he publish the paper.[18] He must have thought of Mayer as of some queer physician in Heilbronn with an unrequited love of physics.

Mayer had started a practice in Heilbronn, and in May 1841 he was appointed *town surgeon* which gained him a regular salary of 150 florin. Later he changed to *Stadtarzt*, at the same salary, and in that capacity he had to treat the poor, – free of charge – and also the lower employees of the town, like the prison ward or the night watchman.[19]

Mayer's problem in physics was that he did not know mechanics. He took private instruction from his friend Carl Baur who was a professor of mathematics at the Technical High-School Stuttgart, but Mayer never graduated to the knowledge that the gravitational potential energy mgH of a mass m at height H is converted to the kinetic energy $\frac{m}{2}v^2$ when the mass falls and acquires the velocity v; specifically the factor ½ remained a mystery for him. To be sure, he never used the word energy in the above sense: gravitational potential energy was *falling force* for him and kinetic energy was *life force*.[20]

All he knew was, that motion, or the *life force* of motion could be converted into heat and he even came up with a reasonable number: the *mechanical equivalent of heat,* cf. Insert 2.1.

$$1^\circ\,\text{heat} = 1\,\text{gram at} \left\{ \begin{matrix} 365\,\text{m} \\ 1130\,\text{Parisian feet} \end{matrix} \right\}\ \text{height}\,.$$

---

[18] The manuscript did survive and, when Mayer's work was eventually recognized, the paper was published in journals and books on the history of science, e.g. P. Buck (ed): "Robert Mayer – Dokumente zur Begriffsbildung des Mechanischen Äquivalents der Wärme". [Robert Mayer – documents on the emergence of concepts concerning the mechanical equivalent of heat] Reprinta historica didactica. Verlag B. Franzbecker, Bad Salzdetfurth (1980) Bd. 1, p. 20–26.

[19] H. Schmolz, H. Weckbach: "Robert Mayer ..." loc.cit p. 66, p. 78.

[20] The *life force* must not be confused with the *vis viva* of the vitalists. In German the kinetic energy was called *lebendige Kraft* at that time, while the *vis viva* was called *Lebenskraft*. In English the distinction is not so clear and sometimes not strictly maintained, although usually the context clarifies the meaning.

Mayer's calculation of the mechanical equivalent of heat

Mayer knew – or thought he knew – that the specific heats of air are $0.267\,\frac{\text{cal}}{\text{gK}}$ and $\frac{0.267}{1.421}\,\frac{\text{cal}}{\text{gK}}$ at constant pressure and volume respectively. To heat 1 cm^3 air at a density of $1.3\ 10^{-3}\text{g/cm}^3$ by 1°C it should therefore take

$$0.347\ 10^{-3}\,\text{cal at fixed pressure, and}$$
$$0.244\ 10^{-3}\,\text{cal at fixed volume.}$$

At constant pressure the volume expands. The difference in heat is $1.03\ 10^{-4}$cal and that difference can lift a 76 cm tube of mercury of mass 1033g which exerts a pressure of 1 atm. At 1°C the lift amounts to $\frac{1}{274}$ cm according to Mariotte's law, which nowadays we call the thermal equation of state of ideal gases, like air. Thus now it is a simple problem of the *rule of three*:

$$1033\text{ g at }1/274\text{cm corresponds to }1.03\ 10^{-4}\text{cal}$$
$$1\text{ g at }\quad H = ?\quad\text{ corresponds to 1cal.}$$

It follows that H = 365 m and so Mayer wrote:

$$1°\text{ heat} = 1\text{ g at 365 m height}$$

Note that Mayer did not measure anything. He took his specific heat from some French experimentalists whom he quotes as Delaroche and Bérard. And the ratio of specific heats he took from Dulong. Both numbers are slightly off and therefore Mayer's mechanical equivalent of heat was low.

Insert 2.1

In words: The fall of a weight from a height of ca. 365 m corresponds to the heating of the same weight of water from 0°C to 1°C. Later, with reference to Joule's better measurements, he changed to 425 m or 1308 Parisian feet. The old value – but not its calculation – is included in Mayer's second paper, see Fig. 2.2, which otherwise is not much clearer than the first one. Anyway that paper established Mayer's priority when Justus von Liebig (1803–1873) published it in his "Annalen der Chemie und Pharmacie". To be sure, Mayer did not give Liebig much of a choice; his accompanying letter would have flattered any hard-nosed editor into acceptance, cf. Fig. 2.3. Those readers who have a command of German may learn from the letter how editors should be approached.

There is a peculiar type of reasoning in the paper. Mayer, rather than just postulate the conversion of motion to heat and make it plausible, attempts to *prove* his discovery from some perceived *theorem of logical cause* or from an assumed axiom *causa aequat effectum.* On another occasion, the conservation of energy – *force* for Mayer – is summarized in the slogan

Ex nihilo nil fit. Nil fit ad nihilum.

ANNALEN
DER
CHEMIE UND PHARMACIE.

XLII. Bandes zweites Heft.

Bemerkungen über die Kräfte der unbelebten Natur;
von J. R. Mayer.

Ausgegeben am 31sten Mai 1842.

Fig. 2.2. Robert Julius Mayer. Cut from the title page of his first published paper

Hochverehrtester Herr Professor!
Beifolgenden kurzen Aufsatz bin ich so frei, Euer Wohlgeboren zur gefälligen Aufnahme in Ihren Annalen für Chemie und Pharmazie vorzulegen. – – –

– – – –. Möchte es Euer Wohlgeboren gefallen, sich vom Durchlesen des kleinen Aufsatzes durch den vielleicht barock scheinenden Anfang nicht abhalten zu lassen und möchten Sie solchen der Erhaltung eines Plätzchens in Ihren weltverbreiteten Annalen für würdig erachten. Sollten Sie indessen von demselben keinen Gebrauch zu machen geneigt sein, so erlaube ich mir die Bitte um gefällige Zurücksendung des Manuskriptchens.

Mit vorzüglichster Hochachtung
verharre ich
Euer Wohlgeboren gehorsamster Diener
Dr. J. R. Mayer

Fig. 2.3. Cut from Mayer's letter accompanying the paper submitted to Liebig

We have to make allowance, however, for Mayer's almost complete isolation. Occasionally he sought scientific advice from physics professors, but then he was fobbed off with the demand to support his theory by experiments and, in one case, he was sent home with the information that the area of science was already so big that an extension was undesirable.[21]

So he was thrown back to his family and a few friends for scientific monologues. They understood nothing and naturally they thought that their husband and friend was more than a little crazy. The pressure on Mayer mounted when his priority claim was ignored by Joule, and Helmholtz, and by a lesser man – a Dr. Otto Seyffer – who ridiculed Mayer's ideas in an article in the daily press.[22] Two of his children died and Mayer came close

---

[21] Reported by Mayer in a letter to his friend W. Griesinger on June 14th 1844. Mayer's correspondence with some of his friends is included in the collection of his works. Reprinta historica didactica. loc cit. Bd. 1, p. 121.

[22] "Augsburger Allgemeine Zeitung" from May 21st, 1849.

to being executed as a spy by some republican radicals who – in the course of the revolution of 1848/49 – briefly won the upper hand in parts of Württemberg. In 1850 all this led to an attempted suicide when Mayer jumped from the third floor of his house into the yard 9 meters below. He survived but was permanently slightly crippled.

Mayer's relatives sought the professional help of an alienist who was a friend of the family. However, the man was also young, and new in his practice, and he needed the money. Therefore he had no intention to let Mayer go anytime soon. He put him behind bars and for good measure kept him in a straightjacket. Eventually, after 13 months of this, Mayer succeeded to escape and he reached home by foot in his nightgown. After that he was indeed a trifle neurotic, patients stayed away from him and the street urchins would taunt him: *There he goes, the dotty Mayer.*

However, my former critical remarks on Mayer's papers must not give the impression that Mayer was anything less than a very original scientist. And despite the evidence of the papers mentioned above, he *could* write well, if he did not force himself to be excessively brief, – and if he did not attempt to use mathematics. The style of his brochure "Die organische Bewegung in ihrem Zusammenhang mit dem Stoffwechsel"[23], published in 1845 by a small Heilbronn printing shop, is still idiosyncratic, but it is clear. Among the subjects which Mayer takes up in that extensive memoir, I mention a few in order to show the scope of his purpose:

- Mayer overcomes Carnot and Clapeyron and paves the way for Clausius when he speaks of the heat engine and says ... *the heat absorbed by the vapour is always bigger than the heat released during condensation. Their difference is the useful work.*
- He explains in detail how he calculated the mechanical equivalent of heat, cf. Insert 2.1. That argument was too brief in his 1842 paper to be understood and appreciated. The calculation is a solid piece of thermodynamics – now very elementary – and it had nothing to do with horses stirring paper pulp in cauldrons, as folklore has it. To be sure, those horses are mentioned in the article, and some rough measurements of the temperature of the pulp, but these were far from good enough to calculate the mechanical equivalent of heat. Incidentally, in this context Mayer mentions Rumford; therefore he knew about Rumford's experience with boring cannon.
- He also reports that a cannon barrel which shoots a ball becomes less hot than if the powder alone is ignited in the barrel. Mayer says that the fact is *common knowledge*. Well, maybe it was at the time. Anyway, the observation makes sense: Part of the chemical energy of the powder is

[23] [Organic motion and metabolism] Verlag der C. Drechslerschen Buchhandlung, Heilbronn (1845).

converted into the kinetic energy of the ball, if there is a ball. Otherwise all goes into heat.

- Mayer extrapolates that observation to the metabolism in animals, and men. The heat liberated by the chemical process of digestion, or of internal combustion of food, can partly be converted into work, he says, whereupon the body becomes colder. In order to support this idea he cites an observation that was published in the "Journal de Chimie médicale, VIII Année, Février", where the author – a man by the name of Douville – measured the temperature of

a negro lazy and inactive in the cabin			37°
ditto	ditto	in the sun	40.20°
ditto	active	in the sun	39.75°.

- Pursuing the idea further, Mayer says that a man sawing wood freezes in the arm which moves the saw. Also a blacksmith who heats a piece of iron to red-heat with three strokes will be cold in the arm that wields the hammer. He says that he has observed that the busy parts of the body sweat less during continual hard work than the inactive ones. For this latter observation he cites biblical proof. Namely when God says to Adam: *In the sweat of your brow you shall eat bread.* Mayer seems to thinks that Adam will henceforth work with his hands and feet, which will therefore sweat less than the head which is involved but little, or not at all.
- In the same memoir Mayer comes out strongly against the *vis viva,* the hypothetical force postulated by physiologists of the time – even Liebig – to explain organic processes, or rather to set them aside as unexplainable.
- The heat of the earth – put in evidence by warm springs and volcanoes – is explained by Mayer as the equivalent of the kinetic energy with which the constituent masses crashed together at the time when the earth was formed. In a rough-and-ready calculation he estimates the original temperature to have been 27600°C, enough for the earth to have been liquid, or actually gaseous.

We could continue the list of Mayer's thoughts on mechanics, astronomy, biology, and physiology by dozens of more item. Maybe they are not all correct, but they are all original. Like the theory of the heat of the earth, or when he thinks that the solar energy stems from the meteors which fall into the sun. Sometimes he capitulates, like when he wonders why planets have orbits with rather small ex-centricities. He suspects that this might be explainable by his ideas on the conversion of motion into heat but cannot do it. Calculations of tidal forces were far beyond his mathematical ability.

Most of the brochure of 1845 is written in a matter-of-fact style, but at the very end Mayer's propensity for hyperbole breaks through again. Thus the work ends with the sentence: *…may the phenomena of life be compared to a wonderful music full of melodious sounds and touching dissonances; only in the concert of all instruments lies harmony and only in harmony lies life.*

For all that, however, Mayer never knew what the nature of heat was. In his brochure "Bemerkungen über das mechanische Äquivalent der Wärme"[24] in 1851 he says that... *the connection between heat and motion is one of quantity rather than quality* and he tends to assume that ... *motion must stop in order to become heat.* Here he was wrong and he could have known it. Indeed, the fledgling mechanical theory of heat existed already and in a short time – in the hands of Maxwell – it should rise to its first peak. By that theory, the kinetic energy of motion of a body was just re-distributed among its atoms when it seemed to disappear; and heat was how that re-distributed motion was felt. Helmholtz, about whom Mayer complains for not having given his work proper credit, explains the relation between heat and atomic motion very well.

Mayer in some way was burned out by that time, he missed the further development of what he had helped to start, although he lived until 1878, one year before Maxwell died. Ironically he did receive some recognition after he had stopped working seriously. John Tyndall (1820–1893), a well-regarded physicist and prolific science author,[25] supported Mayer in his priority quarrel with Joule, and Mayer received the Copley medal from the Royal Society of London. In 1858 Liebig called Mayer *the father of the greatest discovery of the century* and in 1859 Mayer received an honorary doctorate from his old *alma mater* in Tübingen.

The chamber of commerce of Heilbronn elected Mayer to honorary membership, and the king of Württemberg *…whose pleasure it is to reward great achievements*[26] made Mayer a knight of the order of the Württemberg crown. Mayer could now call himself "von Mayer".

Yet, Mayer is largely forgotten, but *not* in his hometown Heilbronn. The people in the town archive look after his memory with loving care.[27] His bronze statue is displayed in a prominent spot of the town, and the monument carries the somewhat pompous quatrain

---

[24] [Remarks on the mechanical equivalent of heat] Verlag von Johann Ulrich Landherr, Heilbronn (1851) Bd. 1, p. 169.

[25] Tyndall is best known for his work on light scattering. It was he who explained the blue colour of the sky, but he also wrote a book on thermodynamics entitled "Heat as a mode of motion" which appeared in 1863.

[26] So Mayer in an autobiographical note. Reprinta historica didactica. loc.cit. Bd. 1, p. 8.

[27] When I visited the archive, I had to park my car precariously. A policeman promptly showed up, but, as soon as he heard that I was interested in Mayer he promised to watch over my car: "Take as long as you like, sir."

Wo Bewegung entsteht, Wärme vergeht
Wo Bewegung verschwindet, Wärme sich findet
Es bleiben erhalten des Weltalls Gewalten
Die Form nur verweht, das Wesen besteht.

James Prescott Joule (1818–1889)

Joule was the son of a rich brewer who was tolerant enough of the scientific interests of his son to furnish him with a home-laboratory. Joule is best known for the discovery of the *Joule heating* of a current that runs through a wire. That heat is proportional to the square of the current. In the course of those studies Joule conceived the idea that there might be a relation between the heating of the current and the mechanical power needed to turn the generator.

And indeed he established that relation and came up with a *mechanical value of heat* which he expresses in the words[28]

> The amount of heat which is capable of raising [the temperature of] one pound of water by 1 degree on the Fahrenheit scale, is equal and may be converted into a mechanical force which can lift 838 pounds to a vertical height of 1 foot.[29]

Joule's memoir is full of tables with carefully recorded observations. He describes his experiments painstakingly, discusses possible sources of experimental error, and attempts to compensate for estimated losses. In that sense his paper has set standards, although to this day thermal and, in particular, caloric measurements are notoriously difficult, time-consuming and inaccurate to boot.

And indeed, in later experiments – reported in a similarly exemplary fashion in the article "On the temperature changes by expansion and compression of air"[30] – Joule obtains the values 820, 814, 795, and 760 instead of the 838 pounds cited in his article of 1843. And there were other values from other experiments so that in 1845 Joule proposed a mean value of 817 pounds[31] as the most likely one. In the letter to the editors of the Philosophical Transactions he says:

---

[28] J.P. Joule: "On the heating effects of magneto-electricity and on the mechanical value of heat." Philosophical Magazine, Series III, Vol. 23 (1843) p. 263ff, 347ff, 435ff.

[29] The paper was read to the Section of Mathematical and Physical Sciences of the British Association, Convention in Cork on August 21st, 1843.

[30] J.P. Joule: Philosophical Magazine, Series III, Vol. 26 (1845), p. 369 ff.

[31] J.P. Joule: "On the existence of an equivalence relation between heat and the ordinary forms of mechanical power". Letter to the editors of the Philosophical Magazine and Journal. Philosophical Magazine. Series III, Vol. 27 (1845), p. 205 ff.

Joule criticizes Carnot's and Clapeyron's analysis of the steam engine, see Chap. 3

He says: *Since I hold the view that only the creator has the power to destruct, I agree with ... Faraday, that any theory that leads to the destruction of force is necessarily false.*[32]

Fig. 2.4. James Prescott Joule. A pious version of the first law

> Each one of your readers who is lucky enough to live in the romantic areas of Wales or Scotland could indubitably confirm my experiments, if he measured the temperature of a waterfall on top and at the bottom. If my results are correct, the fall must create 1° heat for a fall of 817 feet height; and the temperature of the Niagara will therefore be raised 1/5 of a degree by the fall of 160 feet.

Asimov[33] writes that Joule in fact made that experiment at the waterfall himself during his honeymoon when he and his wife visited a scenic waterfall.

In 1850, after many more experiments, Joule came up with 772 which is a really good value, see below.[34]

We have already seen that Joule knew Rumford's work and, in fact, that he tried to calculate the mechanical equivalent of heat from Rumford's observation. This came out too high – 1034 foot-pounds – but it was close enough to Joule's spectrum of values that he could say that Rumford's *result confirms our conclusions satisfactorily*.[35]

In the same postscript Joule says that he observed that water pressed through narrow tubes heats up, and that gave him yet another value, – 770 foot-pounds. And he expresses his believe in the conservation of energy by saying: *I am convinced that the mighty forces of nature are indestructible by virtue of the Creator's*: F I A T!

To this day the conservation of energy is an assumption – well-documented, to be sure, but still an assumption. But like Mayer, Joule feels that he needs to *prove* the law. And since he cannot do that, he comes up with strange formulations: *We may a priori assume that a complete destruction of force is supposedly impossible, since it is obviously absurd,*

---

[32] J.P. Joule: "Temperature changes by expansion and compression of air." Philosophical Magazine Series III, 26 (1845) p. 369 ff.

[33] I. Asimov: "Biographies...." loc.cit.

[34] J.P. Joule (1850) loc.cit.

[35] *Post Scriptum* to Joule's memoir of 1843. loc.cit.

that the properties, with which God has endowed matter, could be destructed.

The attentive reader will have noticed that after Mayer had adjusted his heat-equivalent to Joule's better measurements – as mentioned before – he had

$$1° \text{ heat} = 1 \text{ gram at} \begin{Bmatrix} 425 \text{ m} \\ 1308 \text{ Parisian feet} \end{Bmatrix} \text{height} .$$

Let us see how Mayer came up with those numbers: If 1308 feet is multiplied by 5/9 to convert from °F to °C we obtain 727 feet, – considerably lower than any of Joule's numbers. But then we must realize that an English foot is 30.5 cm while the Parisian one was 32.5 cm. Thus Joule's value, as quoted by Mayer, was indeed 772 English foot-pounds as stated before.

Of course, foot-pounds are out nowadays. The older ones among the readers may remember their university days, when they learned the mechanical equivalent of heat in the form:

$$1 \text{ calorie} = 4.18 \text{ Joule}.$$

Yes, indeed, *Joule* is the modern unit of energy! It is equal to 1 kgm^2/s^2. Joule gets the honour, because he was most accurate for the time and he backed up his figure with a large variety of careful measurements.[36] Actually, the calorie went also out as a unit when the SI units were introduced,[37] and nowadays all energies are measured in Joule, be they mechanical, thermal, chemical, electric, magnetic, or nuclear. This was a great relief indeed for everybody concerned.

A good case can be made that the first law of thermodynamics, the law of conservation of energy, was the greatest discovery of the 19th century. And how was it received? We have already described how Mayer had to grovel in order to have his paper accepted for publication, and Joule fared no better. Asimov writes[38]

> His [Joule's] original statement of his discovery was rejected by several learned journals as well as by the Royal Society and he was forced to present it as a public lecture in Manchester and then get his speech published in full by a reluctant Manchester newspaper editor for whom Joule's brother worked as a music critic.

[36] Of course, 418 m is not Mayer's and Joule's 425 m. The difference lies in the gravitational acceleration 9.81 m/s^2, because Mayer's grams and Joule's pounds were weights, not masses. We have to correct for that.

[37] Système International d'Unites. It was introduced by international agreement in 1960.

[38] I. Asimov: "Biographies" loc.cit.
The lecture was given on April, 28th 1847 in the St. Ann's Church Reading-Room in Manchester. It was published by the Manchester Courier on May 5th and May 12th.

Fortunately for him, the young, up-and-coming scientist William Thomson – later Lord Kelvin (1824–1907) – heard Joule speak and recognized the quality of his research which he continued to advertise successfully. In due course the two men became friends and collaborators.

Joule was eventually able to measure 0.005°F reliably and the two scientists – Joule and Kelvin – used such accurate measurements to show that the temperature drops very slightly when a gas is allowed to expand into vacuum. This is now known as the Joule-Thomson effect – or the Joule-Kelvin effect – and it is due to the fact that the molecules of the gas upon expansion must run uphill in the potential energy landscape that is formed by the molecular attraction.[39] This cooling effect proved to be important for the effort to reach lower and lower temperatures and both James Dewar (1842–1923) and Karl von Linde (1842–1934) made use of it in their efforts to liquefy gases and vapours, see below, Chap. 6.

You cannot be an intelligent man and spend your lifetime measuring temperature and heat without forming an idea what heat *is*. Rumford had already speculated that heat was motion and Joule says:[40] *I hold to the theory which considers heat as a motion of the particles of matter* and he quotes John Locke (1632–1704) who had said it all one and a half century earlier [41]

> Heat is the very brisk agitation of the insensible parts of the object, which produces in us that sensation, from whence we denominate the object hot; so what in our sensation is heat, in the object is nothing but motion.

Largely due to Kelvin's propaganda, Joule's work was widely recognized and appreciated. In 1866 he was awarded the Copley medal of the Royal Society, which Mayer also received, albeit 5 years after Joule. Toward the end of his life Joule's brewery did not go well and he suffered some economic hardship. But he was saved by Queen Victoria who granted him a pension.

Hermann Ludwig Ferdinand (von) Helmholtz (1821–1894)[42]

For centuries people had tried to construct a *perpetuum mobile* by arranging masses – and possibly springs – in the gravitational field, so that they would

[39] This cooling effect is absent in a truely ideal gas, but quite noticeable in a vapour, i.e. a gas close to condensation. That Joule and Kelvin could detect it in air at room temperature does them credit as very careful experimenter. Gay-Lussac had missed the cooling when he made the expansion experiment earlier.

[40] J.P. Joule: "Heating during the electrolysis of water." Memoirs of the literary and Philosophical Society of Manchester. Series II, Vol. 7 (1864) p. 67.

[41] J.P. Joule: (1850) loc.cit.

[42] Helmholtz was ennobled by Kaiser Wilhelm I in 1883. In 1891 he became a *real privy councillor* with the right to be addressed as *Your excellency*. Such were the rewards for successful scientists in 19th century Europe.

turn a wheel (say) and still come back to the original position in order to begin a new cycle. These attempts had always failed and people came to the conclusion that a *perpetuum mobile* was impossible. Therefore as early as 1775 the Paris Academy decided not to review new propositions anymore. The conservation of mechanical energy – kinetic energy, gravitational potential energy, and elastic energy was firmly believed in, no matter how complex the arrangement of masses and springs and wheels was, cf. Fig. 2.5. This could not be proved, of course, since not all possible arrangements could be tried, nor could the equations of motion be solved for complex arrangements.

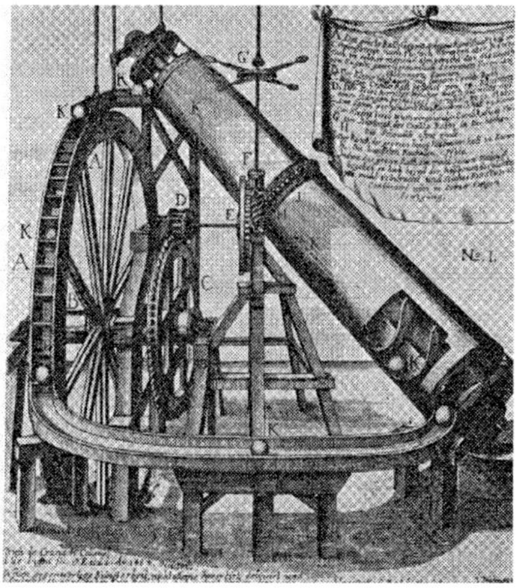

Fig. 2.5. Design of a *perpetuum mobile* by Ulrich von Cranach, 1664

A *perpetuum mobile* was a proposition of mechanics. To be sure, friction and inelastic collisions were recognized as counterproductive, because they absorb work and annihilate kinetic energy, – both produce heat. Helmholtz conceived the idea that

> …what has been called … heat is firstly the … life force [kinetic energy] of the thermal motion [of the atoms] and secondly the elastic forces between the atoms. The first is what was hitherto called free heat and the second is the latent heat.

So far that idea had been expressed before – more or less clearly – but now came Helmholtz's stroke of insight: The bouncing of the atoms and the attractions between them just made a mechanical system more complex than any macroscopic system had ever been.[43] But the impossibility of a

[43] And some of those machines *were* complicated, see Fig. 2.5.

perpetuum mobile should still prevail. Just like energy was conserved in a complex macroscopic arrangement without friction and inelastic collisions, so energy is still conserved – even with friction and inelastic collisions – if the motion of the atoms, and the potential energy of their interaction forces, is taken into account. Friction and inelastic collisions only serve to *redistribute* the energy from its macroscopic embodiment to a microscopic one. And on the microscopic scale there is no friction, nor do inelastic collisions occur between elementary particles.

The idea was set forth by Helmholtz in 1847 in his first work on thermodynamics "Über die Erhaltung der Kraft"[44] which he read to the Physical Society in Berlin. Note that thus all three of the early protagonists of the first law of thermodynamics used the word force rather than energy. Helmholtz's work begins with the sentence: *We start from the assumption that it be impossible – by any combination of natural forces – to create life force* [kinetic energy] *continually from nothing.*

While Helmholtz may have been unaware at first of Mayer's work, he did know Joule's measurements of the mechanical equivalent of heat. He cites them. When his work was reprinted in 1882,[45] Helmholtz added an appendix in which he says that he learned of Joule's work only just before sending his paper to the printer. On Mayer he says in the same appendix that his style *was so metaphysical that his works had to be re-invented after the thing was put in motion elsewhere,* probably meaning by himself, Helmholtz. One thing is true though: Mayer, and to some extent even Joule hemmed and hawed and procrastinated over heat and force; they adduced the *theorem of logical cause* and the *commands of the Creator*. Helmholtz's work on the other hand is crystal clear, at least by comparison.

We have previously reviewed Mayer's and Joule's frustrating attempts to publish their works. Helmholtz fared no better. His paper was dismissed by Poggendorff as *mere philosophy*.[46] Therefore Helmholtz had to publish the work privately as a brochure, see Fig. 2.6.

Helmholtz was not much younger than the other two men, and yet he was a man of the new age. While the others had reached the limit of their capacities – and ambitions – with the discovery of the first law, Helmholtz was keen enough and knew enough mathematics to exploit the new field.

---

[44] [On the conservation of force].

[45] H. Helmholtz: "Über die Erhaltung der Kraft" [On the conservation of force] Wissenschaftliche Abhandlungen, Bd. I (1882).

[46] According to C. Kirsten, K. Zeisler (eds.): "Dokumente der Wissenschaftsgeschichte" [Documents of the history of science] Akademie Verlag, Berlin (1982) p. 6.

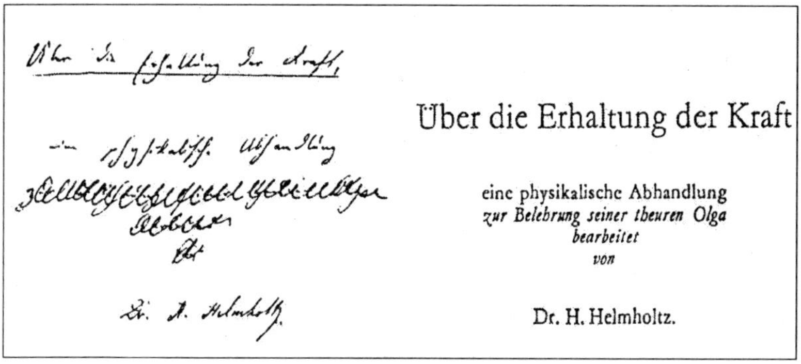

Fig. 2.6. Title page of Helmholtz' brochure. [The dedication to "dear Olga" was scratched out before printing.[47]]

Thus Helmholtz put numbers to Mayer's speculation about the source of energy of solar radiation. First of all he dismissed the idea that the energy comes from the impact of meteors. Rather he assumes that the sun contracts so that its potential energy drops and is converted into heat which is then radiated off. Taking it for granted that the solar energy output is constant throughout the process – and therefore equal to the current value which is $3.6 \cdot 10^{26}$ W – Helmholtz calculates that the sun must have filled the entire orbit of the earth only 25 million years ago, cf. Insert 2.2. The earth would therefore have to be younger than that. Geologists complained; they insisted that the earth had to be much older than a billion years in order to accommodate the perceived geological evolutionary processes, and they were right. It is true that Helmholtz's calculations were faultless, but he could not have known the true source of energy of the sun, which is not gravitational but nuclear.

Helmholtz, on his mother's side a descendant of William Penn, the founder of Pennsylvania, studied medicine and for a while he served as a surgeon in the Prussian army. When he entered academic life it was as a professor of physiology in Königsberg, where he did important work on the functions of the eye and the ear. Without having a formal education in mathematics Helmholtz was an accomplished mathematician, see Fig. 2.7. He worked on Riemannian geometry, and students of fluid mechanics know the Helmholtz vortex theorems which are non-trivial consequences of the momentum balance, – certainly non-trivial for the time. Late in his life he became the first president of the Physikalisch-Technische Reichsanstalt, the German standardizing laboratory.[48]

---

[47] Olga von Velten (1826–1859) became Helmholtz's first wife in 1849.
[48] Now: Physikalisch Technische Bundesanstalt.

Helmholtz was yet another physician turned scientists. He studied the working of the eye and the ear and formulated the "Helmholtz vortex theorems", mathematically non-trivial results for his time.

Lenard[49] says: ... *that Helmholtz, who had no formal mathematical education was able to do this, shows the absolute uselessness of the extensive mathematical instruction in our universities, where the students are tortured with the most outlandish ideas, ... when only a few are capable of getting results with mathematics, and those few do not even need this endless torment.* [50]

Fig. 2.7. Hermann Ludwig Ferdinand von Helmholtz. Also a quote from Lenard, much appreciated by students of thermodynamics

Despite the insight which Helmholtz had into the nature of heat and despite the mathematical acumen which he exhibited in other fields, he did not succeed to write the first law of thermodynamics in a mathematical form, – not at the early stage of his professional career. The last important step was still missing; it concerned the concept of the *internal energy* and its relation to heat and work. That step was left for Clausius to do and it occurred in close connection with the formulation of the *second* law of thermodynamics. The cardinal point of that development was the search for the optimal efficiency of heat engines. We shall consider this in Chap. 3.

Helmholtz's hypothesis on the origin of the solar energy

Although Helmholtz's hypothesis on the gravitational origin of the solar energy is often mentioned when his work is discussed, I have not succeeded to find the argument; it is not included in the 2500 pages of his collected works.[51] Given this – and given the time – one must assume that the calculation was a rough-and-ready estimation rather than a serious contribution to stellar physics. I proceed to present the argument in the form which I believe may be close to what Helmholtz did.

The gravitational potential energy of an outer spherical shell of radius r and mass $\mathrm{d}M_r$ in the field of an inner shell of radius s and mass $\mathrm{d}M_s$ is equal to

$$E_{pot}^{rs} = -G\frac{\mathrm{d}M_r \mathrm{d}M_s}{r}, \quad \text{because} \quad -\frac{\mathrm{d}E_{pot}^{rs}}{\mathrm{d}r} = -G\frac{\mathrm{d}M_r \mathrm{d}M_s}{r^2} = F^{rs}$$

[49] P. Lenard: "Große Naturforscher''. J.F. Lehmann Verlag München (1941).

[50] And yet, in 1921, when M. Planck edited two of Helmholtz's later papers on thermodynamics, he complained about *the shear unbelievable number of calculational errors* in Helmholtz's papers. So, maybe Helmholtz might have profited, after all, from some formal mathematical education.

[51] H. Helmholtz: "Wissenschaftliche Abhandlungen." Vol. I (1882), Vol. II (1883), Vol III (1895).

is the gravitational force on the outer shell. G is the gravitational constant. Therefore the potential energy of the outer shell in the field of all shells with $s < r$ is equal to

$$E_{pot}^{r} = -G\frac{\mathrm{d}M_r}{r}M_r$$

and the potential energy of the whole star is

$$E_{pot} = -G\int_0^R \frac{M_r}{r}\mathrm{d}M_r \underset{\text{by partial integration}}{=} -\frac{1}{2}G\frac{M_R^2}{R} - \frac{1}{2}G\int_0^R \frac{M_r^2}{r^2}\mathrm{d}r.$$

Thus E_{pot} is determined by M_R and R but also by the mass *distribution* M_r within the star. I believe that Helmholtz may have considered ρ as homogeneous, equal to $\frac{M_R}{4/3\pi R^3}$. In that case the calculation is very easy and one obtains $E_{pot} = -\frac{3}{5}G\frac{M_R^2}{R}$. We calculate this value with $G = 6.67 \cdot 10^{-11}\frac{\mathrm{m}^3}{\mathrm{kg\,s}^2}$ for the solar mass $M_R = 2 \cdot 10^{30}$ kg and for the two cases when the sun has its present radius $R = 0.7 \cdot 10^9$m and when it has the radius $R = 150 \cdot 10^9$ m of the earth's orbit. The difference is $\Delta E_{pot} = 22.76 \cdot 10^{40}$ J and, if we suppose that this energy is radiated off at the present rate, see above, we obtain $\Delta t = 20 \cdot 10^6$ years for the time needed for the contraction. That is indeed close to the time given by Helmholtz.

We shall recalculate E_{pot} under a less sweeping assumption in Insert 7.6.

Insert 2.2

Helmholtz remained active until the last years of his life, and he took full advantage of what Clausius was to do. Later on – in Chap. 5 – we shall mention his concept of the free energy – *Helmholtz free energy* in English speaking countries – in connection with chemical reactions.

Electro-magnetic Energy

It was not easy for a person to be a conscientious physicist in the mid-nineteenth century. He had to grapple with the ether or, actually, with up to four types of ether, one each for the transmission of gravitation, magnetism, electricity and light. The ether – or ethers – did not seem to affect the motion of planets,[52] so that matter moved through the ether without any

[52] Actually Isaac Newton (1642–1727) conceived of a viscous interaction between the ether and the moon, and that idea led him to study shear flows in fluids. Thus he discovered *Newton's law of friction* by which the shear stress in the fluid and the shear rate are proportional, with the *viscosity* as the factor of proportionality. Fluids that satisfy this law

interaction, as if it were a vacuum. And yet, the ether could transmit gravitational forces. Its rest frame was supposed to define *absolute space.*

The *luminiferous ether* – also assumed to be at rest in absolute space – carried light and that created its own problem. Indeed, light is a transversal wave and was known to propagate with the speed $c = 3\cdot 10^5 \frac{\text{km}}{\text{s}}$. One had to assume that the ether transmitted vibrations as a wave, like an elastic body. For the speed of propagation to be as big as it was, the theory of elasticity required a nearly rigid body. Therefore physicists had to be thinking of something like a rigid vacuum. Asimov remarks in his customary flamboyant style that *generations of mathematicians ... managed to cover the general inconceivability of a rigid vacuum with a glistening layer of fast-talking plausibility.*[53]

And then there was electricity and magnetism, both exerting forces on charges, currents, and magnets and that seemed to call for two more types of ether. Michael Faraday (1791–1867) and James Clerk Maxwell (1831–1879) were, it seems, not unaffected by such thoughts. Maxwell developed elaborate analogies between electro-magnetic phenomena and vortices in incompressible fluids moving through a *medium.* It is true that Maxwell always emphasized that he was thinking of *analogies* – rather than reality – when he set up his equations in terms of *convergences* in the *medium,* and of *vortices.* However, Maxwell's visualizations were incidental and Heinrich Rudolf Hertz (1857–1894), recognizing the fact, is on record as having said laconically *that the theory of Maxwell is the system of Maxwell equations*, cf. Fig. 2.8. Kelvin was among those who would have preferred something more concrete: a clear relation to a mechanical model.

Maxwell's equations, cf. Fig. 2.8, relate four vector fields[54]

B – magnetic flux density	**E** – electric field
D – dielectric displacement	**H** – magnetic field.

J is the electric current and q is the electric charge density. With all these fields, the Maxwell equations are strongly underdetermined. But then there are two additional relations, the so-called *ether relations,* which close the system, if q and **J** are known. The ether relations connect **D** to **E** and **H** to **B**. They read

$$\boldsymbol{D} - \varepsilon_0 \boldsymbol{E} \quad \text{and} \quad \boldsymbol{H} = \mu_0 \boldsymbol{B} \ ,$$

where $\varepsilon_0 = 8.85\cdot 10^{-12} \frac{\text{As}}{\text{Vcm}}$ and $\mu_0 = 12.5\cdot 10^{-7} \frac{\text{Vs}}{\text{Acm}}$ are constants called the vacuum di-electricity and the vacuum permeability, respectively.

– and there are many of them – are called *Newtonian* to this day. However, Newton could not detect any viscous effect between the ether and the moon.

53 I. Asimov: "The rigid vacuum" in "Asimov on physics" Avon Books, New York (1976).

54 Vectors are denoted by boldface letters, or by their Cartesian components. If the latter notation is used in formulae, summation over repeated indices is implied.

In the vacuum there is neither current nor charge but the fields are there, and they propagate as waves. Indeed, if we apply the curl-operator to the first and third Maxwell equation and make use of the ether relations, we obtain

$$\frac{\partial^2 E_i}{\partial t^2} = \frac{1}{\varepsilon_0 \mu_0} \frac{\partial^2 E_i}{\partial x_j \partial x_j} = 0 \text{ and } \frac{\partial^2 B_i}{\partial t^2} = \frac{1}{\varepsilon_0 \mu_0} \frac{\partial^2 B_i}{\partial x_j \partial x_j} = 0$$

which are the well-known wave equations of mathematical physics. The speed of propagation is $\frac{1}{\sqrt{\varepsilon_0 \mu_0}}$ *which happens to be equal to c, the speed of light.* (!!)

Thus Maxwell was able to relate electro-magnetic wave propagation to light. He says: *The speed of the transversal waves in our hypothetical medium ... is so exactly equal to the speed of light ... that it is difficult to refuse the conclusion that light consists of the wave motion of the medium that is also the agent of electric and magnetic phenomena.*[55]

$$\frac{\partial B_i}{\partial t} + curl_i \boldsymbol{E} = 0 \qquad \frac{\partial B_i}{\partial x_i} = 0$$

$$-\frac{\partial D_i}{\partial t} + curl_i \boldsymbol{H} = J_i \qquad \frac{\partial D_i}{\partial x_i} = q$$

Fig. 2.8. James Clerk Maxwell. Main system of Maxwell equations

As a result, the magnetic and electric ether were cancelled out. What remained was the luminiferous ether – the *rigid vacuum* – and, perhaps, Newton's ether that transmits gravitation. Actually Einstein threw out the luminiferous ether in 1905 as we shall see later, cf. Chap. 7. The gravitational ether is still an embarrassment to physicists today. Nobody believes that it exists, but neither have gravitational waves convincingly been

[55] Retranslated by myself from Giulio Peruzzi: "Maxwell, der Begründer der Elektrodynamik" [Maxwell. The founder of electrodynamics] Spektrum der Wissenschaften, German edition of Scientific American. Biografie 2 (2000).

discovered – to the best of my knowledge – nor the particles that could replace them, the hypothetical gravitons.[56]

This is all quite interesting but it distract us from the main subject in this chapter, which is energy or, here, electro-magnetic energy. The Maxwell equations of Fig. 2.8, combined with the ether relations, imply – as a corollary – four equations which may be interpreted as equations of balance of electro-magnetic momentum and energy, viz.

$$\frac{\partial(\boldsymbol{D}\times\boldsymbol{B})_l}{\partial t}+\frac{\partial((\frac{1}{2}\boldsymbol{E}\cdot\boldsymbol{D}+\frac{1}{2}\boldsymbol{B}\cdot\boldsymbol{H})\delta_{li}-E_iD_l-B_iH_l)}{\partial x_i}=-qE_l-(\boldsymbol{J}\times\boldsymbol{B})_l$$

$$\frac{\partial(\frac{1}{2}\boldsymbol{E}\cdot\boldsymbol{D}+\frac{1}{2}\boldsymbol{B}\cdot\boldsymbol{H})}{\partial t}+\frac{\partial(\boldsymbol{E}\times\boldsymbol{H})_i}{\partial x_i}=-J_iE_i.$$

In this interpretation we have

$(\boldsymbol{D}\times\boldsymbol{H})_l$ – momentum density

$(\frac{1}{2}\boldsymbol{E}\cdot\boldsymbol{D}+\frac{1}{2}\boldsymbol{B}\cdot\boldsymbol{H})\delta_{li}-E_iD_l-B_iH_l$ – pressure tensor

$\frac{1}{2}\boldsymbol{E}\cdot\boldsymbol{D}+\frac{1}{2}\boldsymbol{B}\cdot\boldsymbol{H}$ – energy density

$(\boldsymbol{E}\times\boldsymbol{H})_i$ – energy flux.

The right-hand sides of the equations of balance represent – to within sign – the density of the Lorentz force of an electro-magnetic fields on charges and currents and the power density of the Lorentz force on a current respectively. If the current consists of a single moving charge e, the Lorentz force becomes $e(\boldsymbol{E}+\frac{\mathrm{d}\boldsymbol{x}}{\mathrm{d}t}\times\boldsymbol{B})$ and the power equals $e\frac{\mathrm{d}\boldsymbol{x}}{\mathrm{d}t}\cdot\boldsymbol{E}$.

The trace of the pressure tensor is $3p$, where p is the electro-magnetic pressure. Hence inspection of the balance equations shows that we have

$$\text{electro-magnetic pressure} = {}^1/_3 \text{ electro-magnetic energy density.}$$

This relation was to become important in Boltzmann's investigation of radiation phenomena, cf. Chap. 7.

That the Lorentz force on charged matter and its power should appear in an easily derived corollary – of balance type – of the Maxwell equations places electro-magnetic energy firmly among the multifarious incarnations of energy which altogether are conserved. Maxwell says: *When I speak of the energy of the field, I wish to be understood literally. All energy is identical to mechanical energy, irrespective of whether it appears in the form of motion or as elasticity or any other form.*

---

[56] You can still always make a learned physicist, who is happily expounding the properties of black holes, come to a full stop by asking a simple question. Nothing can escape from a black hole, not even light, which is why it is black. So, you must ask innocently: *But the gravitons do come out, don´t they*?

Maxwell's theory of electro-magnetism was created in three papers[57] between 1856 and 1865 and later summarized and extended in two books,[58] the latter of which appeared posthumously.

The practical impact of Faraday and Maxwell was enormous, although not immediate, and it was twofold: Telecommunication and energy transmission. It is true that electro-magnetic telecommunication *by wire* preceded Maxwell's work. But, of course, *wireless* transmission was firmly based on it after Hertz sent the first *radio*-signal – short for radio-telegraphic signal – from one side of his laboratory to the other one in 1888. Perhaps even more important is the electric generator which was invented by Faraday in 1831 when he rotated a copper disk in a magnetic field, thus inducing a continuous electric current. The reversal of the process could produce – with the appropriate design – rotational motion of a shaft from the current fed into an electric motor.

Generator and electric motor would eventually make it feasible to concentrate steam power generation in some central plant in a city or the countryside, rather than have each consumer set up his own steam engine. But that took time and the electrification of industry and transport – and households – was not complete until well into the 20th century.

Faraday, however, was fully aware of the potential of his invention. There is a story about this, probably apocryphal: In 1844, when Faraday was presented to Queen Victoria, she is supposed to have asked him what one might do with his inventions. *In a hundred years you can tax them* said Faraday.

The scientific impact of Maxwell's equations was equally great, although also delayed. When the equations were closely studied – by H.A. Lorentz and A. Einstein – it turned out that the main set, shown in Fig. 2.8, is invariant under any space-time transformation whatsoever, while the ether relations are invariant only under Lorentz transformations, see below.

The true nature of the Maxwell equations as conservation laws of charge and magnetic flux was identified even later by Gustav Adolf Feodor Wilhelm Mie (1868–1957).[59] Mie put Lorentz's and Einstein's transformation rules into an elegant four-dimensional form. This crowning achievement in electro-magnetism is reviewed by Claus Hugo Hermann

---

[57] J.C. Maxwell: "On Faraday's lines of force." Transactions of the Cambridge Philosophical Society, X (1856).
J.C. Maxwell: "On physical lines of force" Parts I and II, Philosophical Magazine XXI (1861), parts III and IV, Philosophical Magazine (1862).
J.C. Maxwell: "A dynamical theory of the electro-magnetic field" Royal Society Transactions CLV (1864).

[58] J.C. Maxwell: "Treatise on electricity and magnetism" (1873).
J.C. Maxwell: "An elementary treatise on electricity" William Garnett (ed.) (1881).

[59] G. Mie: "Grundlagen einer Theorie der Materie" [Foundations of a theory of matter] Annalen der Physik 37, pp. 511-534; 39, pp. 1-40; 40, pp. 1–66 (1912).

Weyl (1885–1955)[60] and I shall give the briefest possible summary, cf. Insert 2.3. This will help us to appreciate the eventual recognition of energy as mass, or of mass as energy.

Transformation properties of electro-magnetic fields

The most appropriate formulation of electro-magnetism is four-dimensional so that x^A (A = 0,1,2,3) equals (t, x_1, x_2, x_3) where t is time and x_i are Cartesian spatial coordinates of an event. If we introduce the *electro-magnetic field tensor* $\boldsymbol{\varphi}$ and the charge density vector $\boldsymbol{\sigma}$ as

$$\varphi_{AB} = \begin{bmatrix} 0 & -E_1 & -E_2 & -E_3 \\ E_1 & 0 & B_3 & -B_2 \\ E_2 & -B_3 & 0 & B_1 \\ E_3 & B_2 & -B_1 & 0 \end{bmatrix} \qquad \text{and} \qquad \sigma^A = (q, J_1, J_2, J_3)$$

the local form of the conservation laws of magnetic flux and of charge read

$$\underline{\underline{\varepsilon^{ABCD} \frac{\partial \varphi_{CD}}{\partial x^B} = 0}} \qquad \text{and} \qquad \frac{\partial \sigma^A}{\partial x^A} = 0.$$

The latter is formally solved by setting

$$\underline{\underline{\sigma^A = \frac{\partial \eta^{AB}}{\partial x^B} = 0,}} \qquad \text{where} \qquad \eta^{AB} = \begin{bmatrix} 0 & D_1 & D_2 & D_3 \\ -D_1 & 0 & H_3 & -H_2 \\ -D_2 & -H_3 & 0 & H_1 \\ -D_3 & H_2 & -H_1 & 0 \end{bmatrix}$$

is called charge-current potential. For that reason $\boldsymbol{D}$ and $\boldsymbol{H}$ are also known as charge potential and current potential, respectively, as well as by the earlier conventional names dielectric displacement and magnetic field.

Upon inspection the underlined equations are the general Maxwell equations of Fig. 2.8 which are thus recognized as conservation laws of magnetic flux and charge[61] respectively. If φ_{AB} are covariant components and η^{AB} contravariant ones, as indicated by the customary position of the indices, we have for any arbitrary space-time transformation $x'^A = x'^A(x^B)$

$$\varphi'_{CD} = \frac{\partial x^A}{\partial x'^C} \frac{\partial x^B}{\partial x'^D} \varphi_{AB} \qquad \text{and} \qquad \eta'^{CD} = \frac{\partial x'^C}{\partial x^A} \frac{\partial x'^D}{\partial x^B} \eta^{AB},$$

and therefore the general Maxwell equations retain their forms in all frames.

[60] H. Weyl: "Raum-Zeit-Materie" [Space-time-matter] Springer, Heidelberg (1921) English translation: Dover Publications, New York (1950).

[61] For the integral form of these equations of balance the reader might consult I. Müller: "Thermodynamics" Pitman, Boston, London (1985) Chap. 9. Another instructive account of Mie's and Weyl's treatment of electrodynamics and relativity may be found in the memoir by C.A. Truesdell and R. Toupin: "The classical field theories" Handbuch der Physik III/1 Springer. Heidelberg (1960). pp. 660–700 and 736–744.

In particular the transformation rules of $\boldsymbol{E}$ and $\boldsymbol{B}$ read

$$E_i' = -\frac{\partial x^A}{\partial t'}\frac{\partial x^B}{\partial x_i'}\varphi_{AB} \quad \text{and} \quad B_i' = \frac{1}{2}\varepsilon_{ijk}\frac{\partial x^A}{\partial x_j'}\frac{\partial x^B}{\partial x_k'}\varphi_{AB}$$

This defines the components E_i' and B_i' in all frames. Similarly D_i' and H_i' can be calculated from D_i and H_i.

Once the transformation laws of $\boldsymbol{E}, \boldsymbol{B}, \boldsymbol{D}, \boldsymbol{H}$ are known, we may ask for the transformations that leave the ether relations $\boldsymbol{D} = \varepsilon_0 \boldsymbol{E}$ and $\boldsymbol{H} = \mu_0 \boldsymbol{B}$ invariant. It turns out that these are Lorentz transformations, see below.

Insert 2.3

Albert Einstein (1879–1955)

Mayer's haphazard collection of *forces* – fall force, motion, tensile force, heat, magnetism, electricity, and force of chemical separation, three of them *imponderables*, cf. Fig. 2.9 – were now confirmed, actually within Mayer's lifetime as different types of energy: potential, kinetic, elastic, internal, electro-magnetic, and chemical respectively. And energy as a whole was recognized as being conserved, when one type changed into another one. This was a great step of unification, and to a new generation of physicists energy became a familiar concept, like mass, or momentum, which were already well-established conserved quantities of old. In some way all types of energy had to be considered *imponderable*, because a compressed spring (say) did not seem to weigh more than a relaxed one.

But then it turned out – through the work of Einstein – that energy E and mass m were the same; or rather they were two quantities strictly related to each other by the equation

$$\boldsymbol{E = m\,c^2}\,,$$

where c is the speed of light. Thus, if energy is mass, and since mass has weight, now it turned out that *all* energies were *ponderable.*

Indeed, if a body has potential energy or kinetic energy, it is only because its mass is bigger at a height, or when it moves. A compressed spring weighs more than a relaxed one. And, if a body is hot, it is also heavier than if it were cold, because its particles have a bigger speed in the mean. If two atoms are bound together chemically – so that their potential energy is smaller than when they are apart – they have a smaller mass.

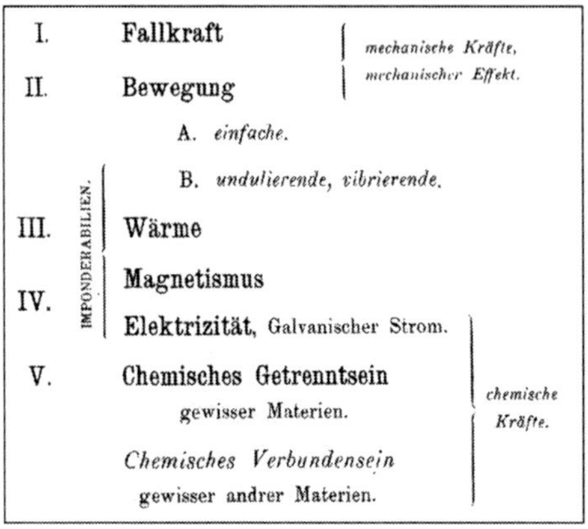

Fig. 2.9. Mayer's collection of *forces*

To be sure, the factor of proportionality c^2 between E and m is so big, and the energy differences are so small, that the mass- and weight-changes in all mentioned cases are too small to be detected. However, this is not so when nuclear forces are involved. Thus the nuclear force between protons and neutrons in a He^4 nucleus – an α-particle – is so strong, and the binding energy is so large, that there is an appreciable *mass defect*: Namely, the masses of two protons and two neutrons are $2{\cdot}1.67239{\cdot}10^{-27}$g and $2{\cdot}1.67470{\cdot}10^{-27}$ g respectively and the mass of the α-particle which they form is $6.64373{\cdot}10^{-27}$ g; consequently there is a mass defect of 0.76% and that is quite noticeable.

The introduction of a "luminiferous ether" will prove to be superfluous inasmuch as the view here to be developed will not require an "absolute stationary space" provided with special properties, nor assign a velocity-vector to a point of the empty space in which electromagnetic processes take place.[62]

Fig. 2.10. Albert Einstein. Dismissal of the ether

[62] A. Einstein: "Zur Elektrodynamik bewegter Körper" [On the electrodynamics of moving bodies] Annalen der Physik 17 (1905) Translation of 1923 in: "The principle of relativity, a collection of original memoirs of the special and general relativity" W. Perrett, G.B. Jeffrey (eds.) Dover Publications. Introductory remarks.

It is true though, that this phenomenon had not yet been noticed in the year 1905, when Einstein presented his paper on what we now call *Special Relativity*.[63] That paper must concern us at this point, because it establishes the relation between energy and mass, which was later used to explain the mass defect, after that phenomenon had been detected. The formula $E = mc^2$ came up in the paper at the very end, almost as an afterthought, and certainly not at all with the fanfare which it deserves for being the most important equation of physics, as which we now recognize it. Actually, the main issue of Einstein's paper was not mass or energy at all, but ether and absolute space. We have to digress in order to explain.

Lorentz Transformation

At that time, the beginning of the 20th century, the universe was supposed to be filled with ether – the luminiferous ether – through which light travelled with the speed c. The ether was supposed to be at rest in *absolute space,* and all bodies moved through the ether without disturbing its state of rest; so also the earth and the sun. The question arose whether the speed of the earth through the ether – the *absolute* speed, as it were – could be measured, and that was the question asked by Albert Abraham Michelson (1852–1931), first alone and then in collaboration with Edward Williams Morley (1838–1923). They sent out a light ray to a mirror at the distance L and measured the time interval before it returned. If the earth, and the light source, and the mirror moved with speed V through the ether, the biggest time interval should have been[64]

$$\Delta t = \frac{L}{c-V} + \frac{L}{c+V} = \frac{2L}{c}\frac{1}{1-\frac{V^2}{c^2}},$$

In the experiment, however, the interval was found to be $\frac{2L}{c}$, irrespective of the direction of the ray, just as if the earth were at rest with the ether which, of course, was unlikely to such a degree that the possibility was not seriously considered.[65] So, the experiment showed that the speed of light is independent of the motion of the source.

---

[63] A. Einstein: "Zur Elektrodynamik ..." loc. cit.

[64] The time interval should have depended on the angle between the light ray and the velocity of the earth. The biggest interval would occur, if that angle were zero.

[65] The actual details of Michelson's measurement are ingenious and cumbersome, because it is not easy to measure Δt. For details the reader may consult Michelson's papers which, incidentally, earned him the Nobel prize of physics in 1907.
A.A. Michelson: "The relative motion of the earth and the luminiferous ether." American Journal of Science 22 (1881), p. 122.

Among the attempts of an explanation there was one that turned out to be heuristically important: George Francis FitzGerald (1851–1901) suggested, in 1890, or so, that distances in the direction of the *ether wind* – the onrushing ether – should be shortened so as to offset the discrepancy between the expected and the measured results of Michelson's experiments. Hendrik Anton Lorentz (1853–1928) made the same assumption in 1895.[66] Lorentz expounds on it by speculating about the influence of the ether on *the action between two molecules or atoms* [so that] *there cannot fail to be a change of dimension as well.*[67]

Einstein does not mention Michelson, but he accepts his experimental result when he speaks about *the unsuccessful attempts to discover any motion of the earth relatively to the "light medium".*[68] And he does not attempt to *explain* Michelson's failure by speculating about the ether; he simply proceeds to identify the transformation of spatial and time coordinates, that is required for two frames K and K′ in uniform relative translation, if the speed of light equals c in both. The problem is somewhat simplified – but not oversimplified – by the assumption that the frames have parallel axes and that their relative motion, with speed V, is along the x-axis. For that case Einstein obtains

$$x_1' = \frac{x_1 - Vt}{\sqrt{1-\frac{V^2}{c^2}}}, \quad x_2' = x_2, \quad x_3' = x_3, \quad t' = \frac{t - \frac{V}{c^2}x_1}{\sqrt{1-\frac{V^2}{c^2}}}, \quad \text{or inversely}$$

$$x_1 = \frac{x_1' + Vt'}{\sqrt{1-\frac{V^2}{c^2}}}, \quad x_2 = x_2', \quad x_3 = x_3', \quad t = \frac{t' + \frac{V}{c^2}x_1'}{\sqrt{1-\frac{V^2}{c^2}}}.$$

This is the *Lorentz transformation*, so called, because Lorentz[69] derived it from the requirement that the Maxwell equations of electro-magnetism

A.A. Michelson, E.W. Morley: "Influence of motion of the medium on the velocity of light" American Journal of Science 31 (1886), p. 377.

[66] H.A. Lorentz: "Versuch einer Theorie der elektrischen und optischen Erscheinungen in bewegten Körpern" [Attempt of a theory of electrical and optical phenomena in moving bodies.] Leiden 1895 §§89–92. Translation of 1923 in: "The principle of relativity, ..." under the title "Michelson's interference experiment" Dover Publications. loc.cit.
Lorentz acknowledges FitzGerald's priority grudgingly by saying: *As FitzGerald kindly tells me, he has for a long time dealt with his hypothesis in his lectures.* The then hypothetical phenomenon became known as the FitzGerald contraction, but is more often called the Lorentz contraction.

[67] I believe that Lorentz fools himself here. Indeed in Michelson's experiments the rod carrying the light source and the mirror were of brass and stone in different experiments; it seems quite inconceivable that the ether would have affected both materials in the same manner.

[68] A. Einstein: "Zur Elektrodynamik..." loc.cit.

[69] H.A. Lorentz: "Electro-magnetic phenomena in a system moving with any velocity less than light." English version of Proceedings of the Academy of Sciences of Amsterdam, 6 (1904). Reprinted in: "The principle of relativity, ..." Dover Publications. loc. cit.

should have the same form in all uniformly translated frames. Einstein does not mention Lorentz except in a later reprinting of his papers, where he says in a footnote: *The memoir by Lorentz was not at this time known to the author* [i.e. Einstein].[70]

For $\frac{V}{c} << 1$ the Lorentz transformation becomes the *Galilei transformation* of classical mechanics. But generally, for higher velocities, it differs from the Galilei transformation subtly, and in a manner difficult to grasp intuitively. Let us consider this:

A sphere of radius R at rest in frame K′ with the centre in the origin has the surface ${x'_1}^2 + {x'_2}^2 + {x'_3}^2 = R^2$. According to the Lorentz transformation that sphere, seen from the frame K, has the surface of an ellipsoid with a contracted axis in the direction of the motion, viz.

$$\frac{{x_1}^2}{1-\frac{V^2}{c^2}} + {x_2}^2 + {x_3}^2 = R^2 .$$

Conversely a sphere at rest in K is given by ${x_1}^2 + {x_2}^2 + {x_3}^2 - R^2$, but viewed from frame K′ it appears as the ellipsoid

$$\frac{{x'_1}^2}{1-\frac{V^2}{c^2}} + {x'_2}^2 + {x'_3}^2 = R^2 .$$

Thus, according to Einstein, no frame of absolute rest exists; it is the *relative motion of the frames* that is responsible for the contractions. No ether is mentioned and no suggestive explanation is offered. This is cold comfort for people who understand and argue intuitively. Einstein presents reason, pure and undiluted, a mathematical deduction from a convincing observation, that is all, – no speculation.

What happens with time intervals is even more counter-intuitive: Let there be two events at some fixed point with x'_1 which are apart in time by $\Delta t'$ in frame K′. By the Lorentz transformation the interval is equal to $\Delta t = \frac{1}{\sqrt{1-\frac{V^2}{c^2}}}\Delta t' > \Delta t'$.

Thus the observer in K will see the time interval lengthened, a phenomenon that is known as *time dilatation*. The phenomenon is often discussed in scientific feuilletons as giving rise to the *twin paradox*: Twin 1 remains at home – at a fixed place x'_1 – while twin 2 goes on a long trip with high speed along the x_1-axis and then returns, again with high speed. His heart beat is lengthened by the time dilatation and therefore his metabolism is slowed down, so that after his return he is still a young man, while his brother, twin 1, has aged. That observation is amazing, and strange, but not paradoxical yet. The paradox appears when

---

[70] I have never been able to see Einstein's papers in the Annalen der Physik, because whenever I looked for them – in the libraries of several countries – they were stolen; cut out, or torn out, the ultimate accolade! But then, the papers have been reprinted many times and some re-printings carry footnotes by Einstein, so also the Dover publication cited above. That is a good thing, because some of the footnotes are quite illuminating.

we realize that both twins are in *relative motion*. Thus twin 2 remains firmly at some point x_1 and considers his brother as travelling – relative to him. The interval between heart beats of twin 2 is therefore $\Delta t = \sqrt{1-\frac{v^2}{c^2}}\Delta t' < \Delta t'$ in his frame K, so that *he* has aged, while twin 1 is still young after the return. That is a genuine paradox, if there ever was one.[71]

Eleven years later, in 1916, Einstein would declare himself not entirely satisfied with the arid reasoning exhibited in his work on special relativity. At the beginning of his memoir on general relativity he says:[72] *In classical mechanics, and no less in special relativity, there is an inherent epistemological defect which was, perhaps for the first time, clearly pointed out by Ernst Mach.* It is not enough to state that uniformly moving frames – inertial frames – are special; we should like to know what makes them so, irrespective of whether they are related by Galilei- or Lorentz-transformations. Einstein explains that he sees *distant masses and the motion of frames with respect to those as the seat of the causes* for the phenomena occurring in frames. Thus non-inertial frames feel gravitational forces from the distant masses, while inertial frames feel no effect at all, – and that defines them.

$E = m\,c^2$

Maxwell's ether relations are invariant under Lorentz transformations,[73] while the general set of Maxwell equations in Fig. 2.8 is generally invariant, against all analytic transformations, see above. Einstein felt that there was a problem, because Newton's equation – the basis of mechanics – are Galilei-invariant. He says somewhat awkwardly: [74] ... *the laws of electrodynamics ... should be valid for all frames of reference for which the equations of mechanics hold good. We will raise this conjecture the purport of which will hereafter be called "Principle of Relativity" to the status of a postulate.* Since electrodynamics was trustworthy – not least because of Michelson's

---

[71] I have been told that the twins will turn out to be equally old after their reunion when the inevitable periods of acceleration at the beginning, middle and end of the trip are taken into account. And, of course, that acceleration is only suffered by the twin who *really* travels. Accelerations are the subject of the *general theory of relativity*, and we shall not go into this any further.

[72] A. Einstein: "Die Grundlage der allgemeinen Relativitätstheorie" Annalen der Physik 49 (1916). English translation: "The foundation of the general theory of relativity" in: "The principle of relativity, ..." Dover Publications. loc. cit.

[73] The invariance of the speed of light in Lorentz frames is, of course, a corollary of the invariance of the ether relations.

[74] A. Einstein: "Zur Elektrodynamik ..." loc.cit.

experiment – mechanics had to be modified so as to become Lorentz invariant. The question was: How?

Mechanics and electrodynamics are largely separate, of course, but they do have points of contact, like when a moving charge e is accelerated by an electro-magnetic force F_i in an electric field E_i and a magnetic flux density B_i. This force is called the Lorentz force and we have

$$F_i = e\left(E_i + \varepsilon_{ijk}\frac{dx_j}{dt}B_k\right) \quad \text{or} \quad F_i' = e\left(E_i' + \varepsilon_{ijk}\frac{dx_j'}{dt'}B_k'\right)$$

in frame K and K′ respectively. Thus Newton's equations in K and K′ should read

$$m\frac{d^2x_1}{dt^2} = F_i \quad \text{or} \quad m'\frac{d^2x_1'}{dt'^2} = F_i',$$

and one should follow from the other one by a Lorentz transformation. It turned out that this requirement could not be satisfied, not even with different masses m and m' as indicated in the equations. If, for simplicity, the charge is at rest in K′ – so that its velocity in K equals $(\frac{dx_1}{dt},0,0)$ – it is possible to show, cf. Insert 2.4, that the Lorentz transformation from K′ to K gives

$$\frac{m'}{\sqrt{1-\frac{1}{c^2}(\frac{dx_1}{dt})^2}^{\,3}}\frac{d^2x_1}{dt^2} = F_1 \quad \text{and} \quad \frac{m'}{\sqrt{1-\frac{1}{c^2}(\frac{dx_1}{dt})^2}}\frac{d^2x_{2,3}}{dt^2} = F_{2,3}\,.$$

That result led Einstein to postulate a *longitudinal* mass for the x_1-direction and a *transverse* mass for the other two directions.[75]

The distinction between two masses – a transversal and a longitudinal one – can be avoided. Indeed, both equations – the one for $x_1(t)$ and those for $x_2(t)$, $x_3(t)$ – may be combined in one as

$$\frac{d}{dt}\left(\frac{m'}{\sqrt{1-\frac{1}{c^2}(\frac{dx_1}{dt})^2}}\frac{dx_i}{dt}\right) = F_i$$

so that there is only *one* velocity-dependent mass m which is related to the *rest mass* m' by $m = \frac{m'}{\sqrt{1-\frac{1}{c^2}(\frac{dx_1}{dt})^2}}$. That formal simplification of the new equation of motion – which amounts to a momentum balance – was

[75] The notions of transverse and longitudinal mass had already been introduced by Lorentz in his paper: "Electro-magnetic phenomena …" (1904) loc.cit. which Einstein later said he had been unaware of, see above.

suggested by Planck. Says Einstein in a later footnote:[76] *The definition of force here given* [in his 1905 paper] *is not advantageous, as was first shown by M. Planck. It is more to the point to define force in such a way that the laws of momentum and energy assume the simplest form.*

Transverse and longitudinal masses

The invariance of the Maxwell equations implies, of course, the invariance of the speed of light as a corollary, but it implies more: Namely the transformation laws for the electric field components, cf. Insert 2.3

$$E_1' = E_1, \qquad E_2' = \frac{E_2 - VB_3}{\sqrt{1 - \frac{V^2}{c^2}}}, \qquad E_3' = \frac{E_3 + VB_2}{\sqrt{1 - \frac{V^2}{c^2}}}.$$

On the other hand, if a mass is momentarily at rest in K′, – that is the simple case under consideration – its accelerations in K′ and K are dictated by the Lorentz transformation and it is a simple matter to calculate the relation. It reads

$$\frac{d^2x_1'}{dt'^2} = \frac{1}{\sqrt{1 - \frac{1}{c^2}\left(\frac{dx_1}{dt}\right)^2}^{\,3}} \frac{d^2x_1}{dt^2}, \quad \frac{d^2x_2'}{dt'^2} = \frac{1}{1 - \frac{1}{c^2}\left(\frac{dx_1}{dt}\right)^2} \frac{d^2x_2}{dt^2}, \quad \frac{d^2x_3'}{dt'^2} = \frac{1}{1 - \frac{1}{c^2}\left(\frac{dx_1}{dt}\right)^2} \frac{d^2x_3}{dt^2}.$$

Insertion into Newton's law

$$m' \frac{d^2x_i'}{dt'^2} = eE_i' \quad \text{provides} \quad \begin{aligned} &\frac{m'}{\sqrt{1 - \frac{1}{c^2}\left(\frac{dx_1}{dt}\right)^2}^{\,3}} \frac{d^2x_1}{dt^2} = F_1 \\ &\frac{m'}{\sqrt{1 - \frac{1}{c^2}\left(\frac{dx_1}{dt}\right)^2}} \frac{d^2x_2}{dt^2} = F_2 \\ &\frac{m'}{\sqrt{1 - \frac{1}{c^2}\left(\frac{dx_1}{dt}\right)^2}} \frac{d^2x_3}{dt^2} = F_3\ , \end{aligned}$$

so that the inertial mass is different in the direction of the relative motion of the frames and perpendicular to that direction. Einstein speaks of *transverse* and *longitudinal* masses. One can avoid this unfamiliar concept when one rephrases Newton's law from

"mass·acceleration = force" to "rate of change of momentum = force."

Insert 2.4

[76] In A. Einstein: "The principle of relativity, ..." Dover Publications. loc.cit.

So what about the law of energy? Multiplication of the 1-component of the momentum balance by $\frac{dx_1}{dt}$ provides an expression for the power of the force on the moving mass, viz.

$$\frac{dmc^2}{dt} = F_1 \frac{dx_1}{dt},$$

and, since the power is known to produce a rate of change of energy in mechanics, we must interpret mc^2 as energy

$$E = mc^2 = \frac{m'c^2}{\sqrt{1-\frac{1}{c^2}(\frac{dx_1}{dt})^2}} \approx m'c^2 + \frac{m'}{2}\left(\frac{dx_1}{dt}\right)^2.$$

Of course, the first term of the approximate formula is huge compared to the second one, but it is also constant, so that we obtain the familiar energy balance of classical mechanics: The rate of change of the kinetic energy $\frac{m'}{2}(\frac{dx_1}{dt})^2$ equals the power of the force.

Special relativity – the theory of frames of reference related by the Lorentz transformation – says nothing about mass and *potential energy* except by implication: Indeed, if a body has a big mass because it moves fast, that movement may be due to a fall from a great height. And if mass, or energy is conserved, the body must have had the big mass before it fell, simply by resting in a high place.[77] Considerations like these have led to an extrapolation of the formula $E = mc^2$ to all types of energy other than kinetic energy; for example to the binding energy in nuclei which manifests itself in the mass defect.

Annus Mirabilis

The year 2005 – when I write this – has been declared the *Einstein year* by physicists all over the world in order to celebrate the centenary of the *annus mirabilis* when Einstein published three salient papers, of which we have just discussed *one*. The other two concern thermodynamics as well, and they will be discussed below, cf. Chaps. 7 and 9.

It is quite unusual that the anniversary of a scientific achievement like this should be celebrated in this manner. Occasions of such type are more common for the feats of politicians, or generals or, perhaps, football players and sports coaches. But now it is upon us, the *annus mirabilis*. In Germany, where Einstein was born and where he spent some of his productive years, – not 1905 though! – the centennial is taken seriously to the extent that most public buildings in Berlin carry words of wisdom from Einstein. On the

---

[77] Einstein's general relativity – the theory of accelerated frames and gravitation – makes such arguments explicit.

chancellery it says in huge bright red letters: *The state must serve the people, not people the state.* And buses, trains, trams and moving vans carry the slogan: *If you wish to have a happy life, set yourself a destination.*

It is true that some of these maxims are somewhat trite, but Einstein *was* capable of pregnant wise-cracks, like when he discarded the probabilistic aspects of quantum mechanics by saying *God does not throw dice.* Or when he expressed doubt about Heisenberg's uncertainty principle: *Our Lord may be subtle, but He is not malicious.* Incidentally, on both occasions Einstein was wrong, at least according to current wisdom. He was indubitably right, however, when he advised physicists that their *theories should be as simple as possible, but not simpler.*

Despite the present-day fanfare, the fact is that Einstein could not get a professorship until four years after the *annus mirabilis* – and *not* because he was not trying! It was eight years before a special position was created for him at the Kaiser Wilhelm Institute in Berlin – at the instigation of Max Planck. In 1916 Einstein published his paper[78] on General Relativity – as opposed to Special Relativity – and that is perhaps his greatest achievement. Einstein became world-famous in 1919 when his prediction, made in 1911,[79] about light rays being deflected by gravitational fields was confirmed by an observation during a solar eclipse.

Einstein anticipated the loss of his position and the impending banishment from Germany by not returning from a trip to the United States when Hitler came to power in 1933. From then on he lived and taught in Princeton until his death.

The above-mentioned mass-defect occurs not only in the fusion of light elements but also in the fission of heavy ones like uranium. And in 1939 Otto Hahn (1879–1968) and Lise Meitner (1878–1968) reported that they had achieved fission. The collateral conditions were such that a chain reaction of fission could conceivably occur, and that provided the feasibility for nuclear explosions.

The possibility of a chain reaction had been conceived by Leo Szilard (1898–1964), an admirer of the science fiction stories by H.G. Wells (1866–1946),[80] in one of which the term *atomic bomb* is first used.[81] Szilard, himself an able physicist, knew of Hahn's and Meitner's work, and he feared that Germany might develop and use a fission bomb in the impending second world war. He convinced Einstein – then an absolute legend as a scientist and a public figure – to sign a letter to President

---

[78] A. Einstein: "Die Grundlagen der allgemeinen Relativitätstheorie" [On the foundation of the general theory of relativity] Annalen der Physik 49, (1916).

[79] A. Einstein: "Über den Einfluß der Schwerkraft auf die Ausbreitung des Lichtes" [On the influence of gravitation on the properties of light]. Annalen der Physik 35, (1911).

[80] Herbert George Wells (1866–1946) was a scientific visionary and social prophet, best known for his classic short story: "The time machine" first published in 1895.

[81] According to I. Asimov: "The finger of God." In: "The sun shines bright." Avon Books (1981).

Roosevelt, in which an American crash program for the development of the bomb was recommended. The letter succeeded and on December 6th, 1941 – the eve of the Japanese attack on Pearl Harbour – President Roosevelt signed *Project Manhattan* into existence.

As is was, German scientists never worked more than half-heartedly on a fission bomb, and the Manhattan project was successfully concluded too late, shortly after Germany's capitulation. But there was still Japan, and two bombs were available, incongruously called *thin man* and *fat boy*. So they were dropped on Hiroshima and Nagasaki on August 6th and August 9th, 1945 when 300.000 civilians died.

Unlike other scientists who lent their support to scientific warfare, Szilard, and Einstein, and the physicists of the Manhattan project – among them Compton, Fermi, and Bohr – are largely excused, or even praised for their commitment. One might say that the theory of relativity asserts itself here in one of its more popular versions: *Everything is relative*, or else: It is imperative to be on the winning side.

Maybe, however, it is fair to say that a fair number of the scientists, who had promoted the bomb project, had second thoughts afterwards, and campaigned for the decommissioning of the atomic arsenal. Among them were Einstein, Fermi, and Bohr. The politicians brushed their initiative aside and, when Bohr would not give up, Winston Churchill (1874–1956) threatened to put him in jail.[82]

After the second world war nuclear *fission* was employed as an energy source in power plants, and now a growing proportion of the human demand for energy is covered in this way.[83] The mass defect inherent in *fusion* of light elements has been utilized in the hydrogen bomb – so far not used in war. Despite energetic – and vastly expensive – research in the field, *controlled* fusion for the conversion of nuclear energy into *useful power* could not so far be realized. The problem is that enormous temperatures must be reached before the charged nuclei can overcome their repulsive electric forces so as to be able to fuse.[84] The centre of the sun is the only place in our planetary system where such temperatures are available and, indeed, nuclear fusion is the process that supplies the energy of the sun, cf. Chap. 7.

In the 1990's two physicists from Provo, Utah, USA claimed to have achieved fusion – on their laboratory table and at room temperature – by somehow overcoming the repulsion *catalytically*, as it were, inside metals.

---

82 According to I. Asimov: "Biographies ..." loc. cit. p. 614.

83 Except in those unfortunate countries with a virulent green, or environmentalist party, which, more often than not, is also anti-nuclear.

84 Actually, the difficulty is not so much to *reach* the high temperatures, but it is difficult to *contain* the hot gas. All conventional container walls would melt and, in fact vaporize.

Although extremely unlikely, this is conceivable in a general way.[85] In the event, however, it turned out to be an error, or a fake. Anyway, the *cold-fusion*-experiment could not be repeated. Actually, however, before the bubble burst, several laboratories worldwide jumped on the bandwagon and reported having seen cold fusion as well. Among ordinary non-nuclear physicists there was some furtive malicious gloating over the simplicity of the process, because for years they had seen the funding of their own projects refused, while a near infinite amount of money was poured into ineffective fusion research, – hot fusion naturally. Their world turned grey again after the truth emerged. But controlled fusion seems still a long way off.

[85] It is true that chemical reactions can sometimes be catalysed by contact with a metal, but the energy barriers to be overcome in such cases are much, much smaller than the nuclear ones.

3 Entropy

It may seem strange that the entropy – which is one of the most subtle concepts of theoretical physics, or natural philosophy – first emerged in the context of an engineering proposition. Namely the question of how to improve the efficiency of heat engines. We shall see how that came about.

Actually the entropy has never shed its hybrid position between physics and engineering: The students of mechanical engineering keep a (temperature-entropy)-diagram among their files, which they are taught to use for the lay-out of power plants and jet nozzles. The chemical engineers are familiar with the *entropy of mixing* which they use to construct phase diagrams, and all physicists know that nature strikes a compromise between entropy and energy when it drives the sap into the tree-tops by osmosis.

Heat Engines

It was Denis Papin (1647–1712) – a student of Christaan Huygens (1629–1695) – who first condensed water and lifted a weight by doing so.[1] Papin owned a long brass tube of diameter 5cm. Some water at the bottom was evaporated, and thus lifted a piston, which was then fixed by a bolt. Afterwards the tube was taken from the fire, the vapor condensed and a Torricelli vacuum formed inside, i.e. a low pressure equal to the vapor pressure appropriate to the extant temperature. When the bolt was removed, the air pressure drove the piston downward and was thus able to lift a weight of sixty pounds. This in a nutshell is the manner in which the *motive power of steam* works: by creating a vacuum through condensation.

Denis Papin knew the properties of saturated vapor well, so that he also knew that water under pressures beyond 1atm boils at a higher temperature than 100°C. He made use of this phenomenon in a pressure cooker: In a closed vessel some tough meat is heated in water. The accumulating steam raises the pressure and thus the boiling point of water, so that the meat finds itself immersed in water as hot as 150°C (say). Thus it becomes sufficiently cooked in a short time. Papin was invited

---

[1] We shall not enter into speculations about whether and how Hero of Alexandria – in the first century A.D. – employed steam power in the automatic working of doors and statues, which priests used to impose on gullible worshippers, cf. I. Asimov: "Biographies..." loc.cit. p. 38.

to demonstrate his *digester* for the Royal Society of London and he cooked an impressive meal for King Charles II.[2]

However, Papin's brass tube was not a steam engine yet; it did only one stroke at a time. Proper steam engines were developed later when a pressing need arose in England in the early 18th century. England was suffering a kind of energy crisis: *The country was deforested and what trees remained were needed for the navy and could not be used for fuel.*[3] At the same time the output from the coal mines was in decline, and threatened to cease altogether, because of difficulties with drainage at the depth where the pits had arrived. That situation provided a strong incentive for inventors, and so the steam engine came just in time. It was developed by the engineer Thomas Savery (1650–1715) and by Thomas Newcomen (1663–1729), a clever and skilful blacksmith. The machine was at first exclusively used to pump water from mines, so that coal could be brought up from a greater depth, previously inaccessible. Therefore it may not have mattered so much, that a good part of the coal was used to heat the boiler of the engine. Indeed Newcomen's engine was quite wasteful of fuel.[4]

In due time, however, the steam engine was employed by the iron industry to power bellows, and hammers for crushing the ore. Thus coal became a commodity to be paid for by the owners of the iron works, and therefore the efficiency of the engine had to be improved.

The Newcomen machine worked by injection of cold water into the cylinder, cf. Fig. 3.1. Thus the steam was condensed and a good vacuum was developed, which pulled down the piston in a powerful stroke. Afterwards new steam from the boiler pushed the piston back up, before water was injected again, etc.

James Watt (1736–1819) recognized the reason for the wastefulness of the process: A good part of the precious new hot steam condensed while reheating cylinder and piston, which had just been cooled by the injected water. Watt improved the machine by inventing a separate cooler, or condenser, into which the steam was pushed before condensation. The condensed water was then pumped back into the boiler. Watt also introduced other improvements, like

- keeping the cylinder wall warm by heating it with the incoming steam,
- introducing an ingenious system of valves so that the piston could work in both the down-stroke and the up-stroke,

[2] According to I. Asimov: "Biographies..." loc.cit. p. 204.

[3] According to I. Asimov: ibidem p. 145.

[4] Yet the machines were successful. *By 1775 sixty of them had been erected in Cornwall alone and there were about one hundred in the Tyne basin*. According to R.J. Law: "The Steam Engine". A Science Museum booklet. Her Majesty's Stationary Office, London (1965) p. 10.

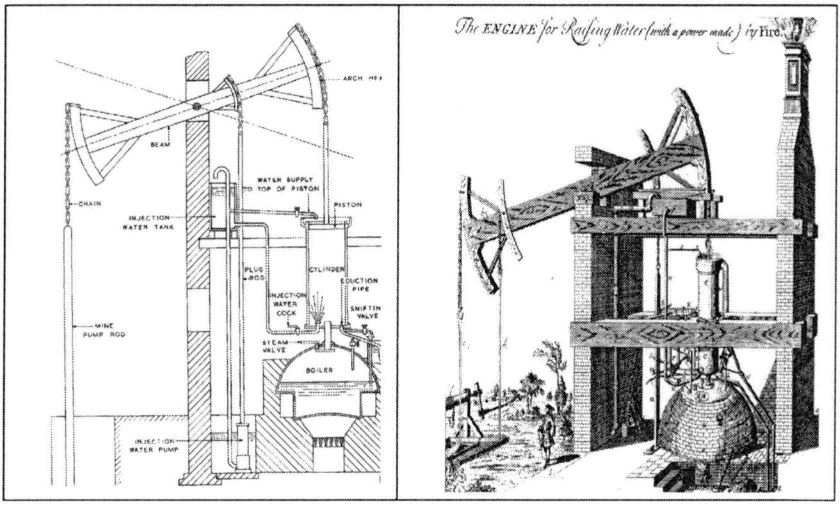

Fig. 3.1. The Newcomen engine

- closing the steam valve before the end of the stroke; it is true that this provided less work per cycle, but it was an efficient measure nevertheless, because still less steam was consumed.

Above all, however, Watt has made the steam engine into more than a pump. He converted the up- and down-movement of the piston into the rotation of a wheel in his famous *rotative engine* with a sun-and-planet transmission gear. This extended the efficacy of the engine greatly, because it could now be used to drive lathes, drills, spinning wheels and looms, – then ships and locomotives. Thus Watt's machine became the motor of the *industrial revolution.*

James Watt was born in Glasgow. He received an abbreviated education as an instrument maker in London, whereupon he became a laboratory assistant at the University of Glasgow. He repaired and improved a model of the Newcomen machine which had broken down, and was thus able to attract the attention of Joseph Black, the discoverer of the latent heat, see above. Black became Watt's first mentor and financier, and he introduced him to an industrialist, Dr. John Roebuck, with whom Watt went into a $\frac{1}{3}, \frac{2}{3}$ partnership, – one third for Watt. Later the $\frac{2}{3}$ share was taken over by Matthew Boulton, and the two partners started a successful business selling steam engines. Law writes [5] ... *the customer paid for all the materials and found the labour for erection. The firm sent drawings and an erector. They also supplied important parts like the valves and the valve gear... As*

[5] R.J. Law: "The Steam Engine" loc.cit. p. 13.

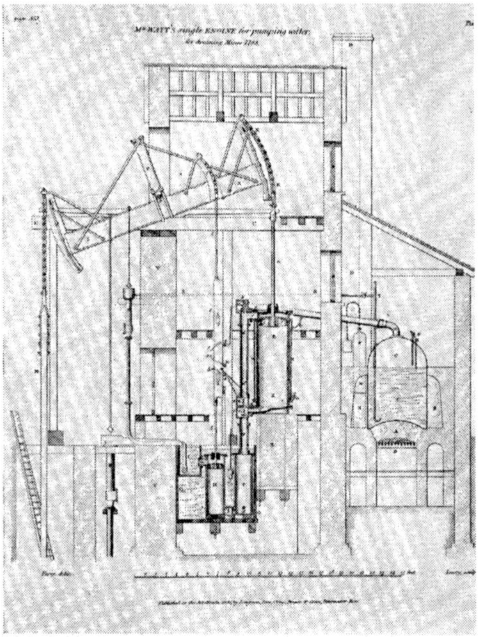

Watt's single acting pumping engine 1788.

Fig. 3.2. Watt's steam engine

payment, they claimed one third of the saving in coal over the old [Newcomen] *engines.*

This was enough to make Watt a rich man, because, indeed, Watt's engine was three to four times more efficient than Newcomen's.[6] Watt retired in 1800. He was famous by then, and much honoured for his life's work. Thus he was elected to membership of the Royal Society of London and he received an honorary doctorate from the University of Glasgow where he had previously served in the lowly position of a laboratory assistant.

Liquid water and steam are particularly well-suited for the conversion of heat into work, because the heat absorbed and emitted – by boiler and cooler respectively – is exchanged isobarically. And a large portion of those isobars are also isotherms, because they lie in the two-phase region of wet steam, where boiling liquid and saturated vapour coexist. This makes the process somewhat similar to a Carnot process, which has maximum efficiency, see below.

[6] Actually, the efficiencies were all quite low: In Newcomen's case about 2% and 5–7% in Watt's case. A modern power station reaches between 45% and 50%. The engineers have done a good job indeed over the past 200 years.

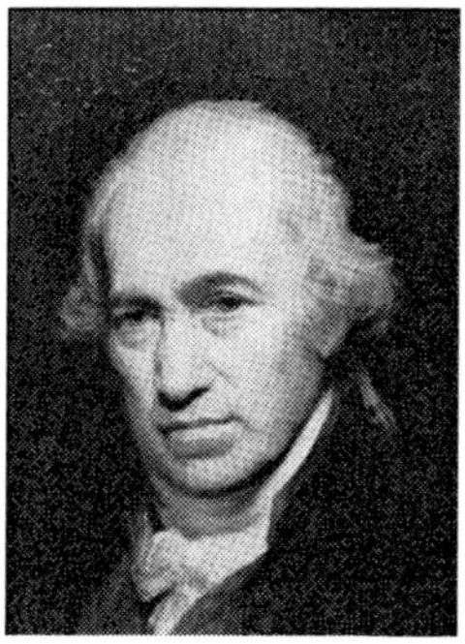

In 1783 he tested a strong horse and decided that it could raise a 150-pound weight nearly four feet in a second. He therefore defined a ''horsepower'' as 550 foot-pounds per second. This unit of power is still used, particularly for automobiles. However, the unit of power in the metric system is called 1 Watt, in honour of the Scottish engineer. One horsepower equals 746 Watt.

Fig. 3.3. James Watt. A quote from Asimov [7]

Heat can also be converted into work by an air engine or, more generally, a gas engine. Figure 3.4 shows the prototypical *Joule process* schematically, where an adiabatic compressor furnishes hot air which is then further heated by isobarically absorbing the heat Q_+. Afterwards the gas cools by adiabatic expansion in the working cylinder which pushes it into a heat exchanger, where it gives off the heat Q_- isobarically. In its alternation between adiabatic and isobaric steps the process is much like the process in the steam engine. However, in the Joule process the isobars are in no way similar to isotherms, since no phase transition occurs.

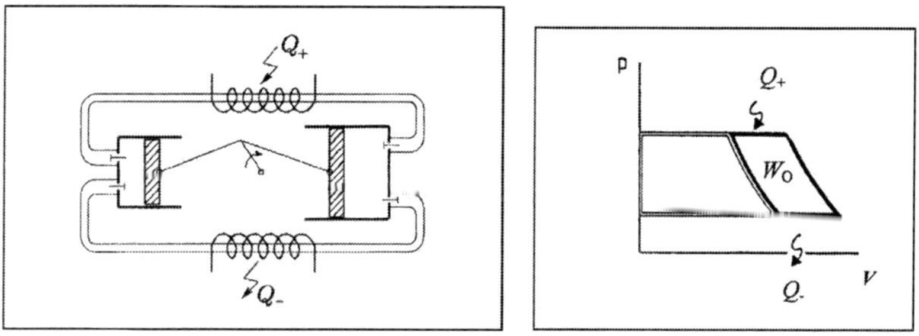

Fig. 3.4. The Joule process and a (pressure,volume)-diagram of the Joule process in an ideal gas

None of the engineers who invented or improved the steam engine – or the air engine – was in any way distracted by any soul-searching about the nature of heat, or whether or not there was a caloric. They proved that heat could produce work *by doing it*, – and doing it better and better as time went on.

The efficiency of the engines climbed up slowly but surely through many ingenious improvement and in the 1820s it had arrived at 18%. At that time

[7] I. Asimov: ''Biographies...'' loc.cit. p. 187.

Sadi Carnot, a physicist educated at the École Polytechnique in Paris, posed himself the question, how far this improvement could possibly go and he attempted to find an answer.

Nicolas Léonard Sadi Carnot (1796–1832)

Sadi Carnot was named after the 13th century Persian poet Saadi Musharif ed Din who was *en vogue* in the France of the directorate. His father Lazare Carnot was one of the directors, and later he became one of Napoléon's loyal and efficient generals. The father was also an accomplished mathematician who published a book on mechanical machines in 1803: ''Fundamental principles on equilibrium and movement.'' In that book Lazare Carnot strongly supported the view that a *perpetuum mobile* was impossible.

By hereditary taint, perhaps, the son picked up the question whether the *possible improvements* [of heat engines] *might have an assignable limit.*

And, in 1824, Sadi Carnot published a book in which he addressed the problem: ''Réflexions sur la puissance motrice du feu et sur les machines propres à développer cette puissance''[8] Everything seemed conceivable at the time:

- The process in which heating and cooling occurred at constant pressures might be improved by letting the heat exchange occur at constant volumes or constant temperatures, and
- perhaps working agents like sulphur or mercury might have an advantage over water.

Carnot came to correct conclusions concerning both propositions. About the first one he says:

> The best manner to employ a heat engine, whose working agent assumes temperatures between T_{Low} and T_{High} in the process, is *the* engine – which we now call a Carnot engine – which exchanges heat *only* at those temperatures.
>
> Because, so Carnot, [that process is] *...le plus avantageux possible, car il ne s'est fait aucun rétablissment inutile d'équilibre dans la calorique.*[9]

---

[8] S. Carnot: [Reflections on the motive power of fire and on machines fitted to develop that power] à Paris chez Bachelier, Libraire. Quai des Augustin, No. 55 (1824). English translation by R.H. Thurston: ''Reflections on the motive power of fire by Sadi Carnot and other papers on the second low of thermodynamics by É. Clapeyron and R. Clausius.'' E. Mendoza (ed.) Dover Publ. New York (1960). pp. 1–59.

[9] S. Carnot: ''Réflexions...'' loc.cit. p. 35.

The argument goes like this: Carnot plausibly postulates that a machine is optimal when the temperature of the working agent is always homogeneous and, if it changes in time, that change must be connected with a change in volume.[10] Other changes in temperature are useless, and even detrimental. It is clear that the steam engine does not satisfy that optimality condition, since the cold feed-water from the condenser enters the hot boiler, so that a *rétablissement inutile* must occur. Actually Carnot shows a lot of insight and ingenuity here, because in a lengthy footnote he proposes to preheat the feed-water by condensing a part of the vapour after partial expansion and at a temperature intermediate between boiler and the principal condenser.[11] This kind of feed-water preheating – actually in several steps – is done routinely in modern power stations; it is known as *Carnotization* of the steam engine process. To be sure, in order to be practical, the procedure requires expansion in a turbine, not in a steam cylinder, but the principle was recognized by Carnot.

Concerning Carnot's second proposition, – the one on the potential advantage of using an agent other than water – he comes to the conclusion that

When a Carnot engine is used, all agents provide the same work.

> In Carnot's words: *La puissance motrice de la chaleur est indépendente des agens mis en oeuvre pour la réaliser ; sa quantité est fixée uniquement par les temperatures entre lesquels se fait en dernier résultat le transport du calorique.*[12]

This statement is proved by letting two Carnot engines – with different agents, but the same heat exchanges, and in the same temperature range – work against each other, one as a heat engine and one as a refrigerator, or heat pump. If one engine requires more work than the other one produces, we should be able to *create motive power without consumption either of caloric or of any other agent whatever. Such a creation is entirely contrary to ideas now accepted, to the laws of mechanics and of sound physics. It is inadmissible. It would be perpetual motion.*[13]

It was his insight into the working of heat engines that permitted Carnot to come to these conclusions. For the above arguments it was quite irrelevant, whether he knew what heat was, – and he didn't! Indeed, Carnot believed in the caloric theory of heat and he thought that the caloric entering the boiler came out of the cooler *unchanged in amount*. Therefore it was natural for Carnot to draw an analogy between the motive power of heat and that of a waterfall, – *une chute d'eau*, see Fig. 3.5.

---

[10] S. Carnot: ibidem p. 23.
[11] S. Carnot: ibidem p. 26.
[12] S. Carnot: ibidem p. 38.
[13] S. Carnot: ibidem p. 21.

… on peut comparer avec assez de justesse la puissance motrice de la chaleur à celle d´une chute d´eau : toutes deux ont un maximum que l´on ne peut pas dépasser, quelle que soit d´une part la machine employée à recevoir l´action de l´eau, et quelle que soit de l´autre la substance employée à recevoir l´action de la chaleur.
La puissance motrice d´une chute d´eau dépend de sa hauteur et de la quantité du liquide; la puissance motrice de la chaleur dépend aussi de la quantité de calorique employé, et de ce … que nous appellerons en effet la hauteur de sa chute, c´est-à-dire de la différence de température…

Fig. 3.5. Sadi Carnot. His reflections about the fall of heat[14]

This misconception, and the false information, which Carnot had about the specific heat of gases, and the latent heat of water vapour, invalidates much of the second half of his paper.[15] He tied himself into knots over the specific heats of gases, which he thinks he can prove to be logarithmic functions of the density when in reality they are constants, independent of both density and temperature.

However, Carnot did ask the right questions. Thus he was interested to know how the location of the temperature *range* of the Carnot engine affected the efficiency. He states that *a given fall of the caloric* [a given temperature difference] *produces more motive power at inferior than at superior temperatures.*[16] This is true, but unfortunately Carnot invalidates the statement in his marginal analysis,[17] where he proves that – for temperature-independent specific heats – the efficiency is *independent* of the temperature range. The whole argument is a mess.

The best concrete result, which Carnot reached, concerned a Carnot engine working in the infinitesimal temperature range $\mathrm{d}t$ at t. In his notation the efficiency e is given by $e = F'(t)\mathrm{d}t$, where $F'(t)$ is a universal function, sometimes called the *Carnot function.* Carnot could not determine that function. Thus, although he proved that the efficiency of a Carnot engine is maximal, he did not know the value of the maximum, – not even for an infinitesimal cycle. The Carnot function, however, partly because of its universal character, provided a strong stimulus for further research on the

[14] S. Carnot: ''Réflexions…'' loc.cit. p. 28.
[15] Carnot refers repeatedly to the experimental results of MM. Delaroche and Bérard, who thought that they had measured the specific heat of air to be dependent on pressure. We recall that Mayer was led to a wrong value of the mechanical equivalent of heat by measurements of the same two men, see Chap. 2.
[16] S. Carnot: ''Réflexions…'' loc.cit. p. 72.
[17] S. Carnot: ibidem pp. 73–78.

subject. Both Clapeyron and Kelvin recognized the need to know the values of that function, but were frustrated in their attempts to either measure or calculate it, cf. Inserts 3.1, 3.2. The problem was left open for Clausius to solve – twenty five years after Carnot.

It seems likely that Carnot, before publication of his work, did have second thoughts about the validity of his "Reflections", particularly about the caloric theory. E. Mendoza who has had access to Carnot's manuscript, quotes Carnot's *original* summary, where Carnot concludes: *The fundamental law that we have proposed ... seems to us to have been placed beyond doubt.*[18] In the *published* version this triumphant sentence is replaced by the more thoughtful one: *The fundamental law that we have proposed seems to us to require ... new verification. It is based upon the theory of heat as it is understood today ...* [whose] *foundation does not appear to be of unquestionable solidity.*

Carnot died in 1832 at the age of 36 years in a cholera epidemic. He left behind unpublished notes, in which he shows himself sceptical of the caloric theory,[19] and where he speculates • on the conversion of heat into work, • on the conservation of *motion*, and • on the impossibility to produce work by cooling a heat bath without transmitting heat to a reservoir of lower temperature. Had he lived longer, it seems likely that he might have anticipated Clausius's work by nearly 30 years.

As it was, however, his book which he tried to sell for 3 francs, found no readers and Carnot would have been entirely forgotten, perhaps, were it not for Clapeyron, like Carnot a former student of the École Polytechnique.

Benoît Pierre Émile Clapeyron (1799–1864)

Mendoza[20] writes that the list of men associated with the early period of the École Polytechnique in Paris – founded in 1794 as a school for army engineers – reads like the author index of a book on mathematical physics. *Among the first instructors were Lagrange, Fourier, Laplace, Berthollet, Ampère, Malus, and Dulong; among former students who stayed on as instructors were Cauchy, Arago, Désormes, Coriolis, Poisson, Gay-Lussac, Petit, and Lamé; other students included Fresnel, Biot, Sadi Carnot, and Clapeyron.* Maybe Clapeyron is not the most eminent one among these men, but his contribution was competent and he is remembered for it.

[18] E. Mendoza: Footnote to Carnot's "Reflections": Dover (1960) loc.cit p. 46.

[19] E. Mendoza (ed.): Appendix to Carnot's "Reflections": Selection from the posthumous manuscripts of Carnot: Dover (1960) loc.cit. p. 60.

[20] E. Mendoza : Introduction to Carnot's "Reflections" Dover loc.cit. p. ix.

Clapeyron's work[21] is a big step forward from Carnot in clarity, but it marks time with respect to the caloric theory. An interesting feature of the work is the introduction of the graphical representation of reversible thermodynamic processes in a (pressure,volume)-diagram, such that the work of the process equals the area below its graph. This is a method of visualization which is still used today, see Fig. 3.4 above. Apart from that, Clapeyron's *analysis* is also perfect. I believe that his paper could have become a classic, if only the physics had not been below par.

However, even so, for some arguments involving heat, it does not matter whether heat is caloric or *motion*. Thus Clapeyron was able to establish a valid relationship between the slope of the vapour pressure curve $p(t)$ and the latent heat, or heat of evaporation $R(t)$, see Insert 3.1. That relation contains the Carnot function $F'(t)$ and it makes it possible to find the values of that function, if only $R(t)$ and $p(t)$ are measured. Extensive measurements of that type were published by Regnault[22] in 1847, but that did not help Clapeyron in 1834, of course. His results remain indeterminate, because as he says *...unfortunately there are no experiments which allow us to determine the values of that function* [the Carnot function] *at all values of the temperature.*

The Clausius-Clapeyron equation

Clapeyron considered a Carnot process of wet steam. That process consists of horizontal isobars and steep adiabates. The isobars are also isotherms, since the vapour pressure depends only on temperature: $p = p(t)$. If the process is infinitesimal – with the temperature difference $\mathrm{d}t$, and the evaporation of the mass fraction $\mathrm{d}x$ of liquid water on the isothermal branch – we have

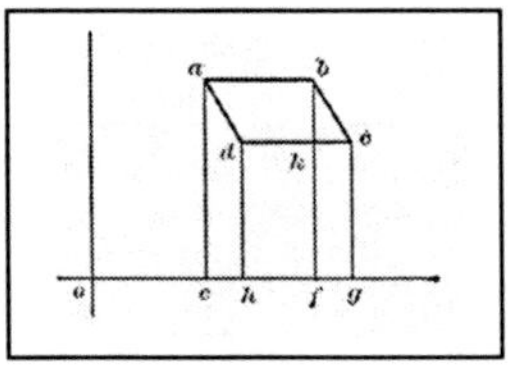

$R\,\mathrm{d}x$ – for the heat absorbed, and

$\frac{\mathrm{d}p}{\mathrm{d}t}\mathrm{d}t\,[(V'' - V')\mathrm{d}x\,]$ – for the work done.

Fig. 3.6. (p,V)-diagram of infinitesimal Carnot process in wet steam

R is the latent heat of evaporation and $\mathrm{d}p$ is the height of the small cycle, cf. Fig. 3.6.

[21] E. Clapeyron: "Mémoire sur la puissance motrice de la chaleur" Journal de l'École Polytechnique. Vol XIV (1834) pp. 153–190. Translations: (English) "Memoir on the motive power of heat." Scientific Memoirs Vol. 1 (1837) pp. 347–376. (German) "Über die bewegende Kraft der Wärme." Annalen der Physik und Chemie Vol 135 (1843).

[22] H.V. Regnault: "Relations des expériences ...pour déterminer les principales lois et les données numériques qui entrent dans le calcul des machines à vapeur." Mémoires de l'Académie des Sciences de l'Institut de France , Paris, Vol. 21 (1847) pp. 1–748.

V' and V'' are the volumes of the boiling water and of the saturated vapour of which the wet steam is composed.

The ratio of the two quantities is the efficiency e. And by Carnot's results it is equal to $e = F'(t)\mathrm{d}t$, see above. Therefore we have

$$\frac{\mathrm{d}p}{\mathrm{d}t} = F'(t)\frac{R}{V''-V'} \ .$$

Thus Carnot's universal function could be calculated from measurements of R, of $\frac{\mathrm{d}p}{\mathrm{d}t}$, – the slope of the vapour pressure curve – and of the vapour volume V'', all at temperature t. Kelvin attempted such calculations, see below.

Clausius found later – in 1850, cf. Insert 3.3 – that $F'(t)$ equals $\frac{1}{T}$, where T is the absolute temperature, and therefore the relation

$$\frac{\mathrm{d}p}{\mathrm{d}T} = \frac{R}{T(V''-V')}$$

is called the Clausius-Clapeyron relation. Nowadays it is used to calculate the latent heat R of a new refrigerant (say) from the vapour pressure curve; the latter is easier to measure than R.

Insert 3.1

William Thomson (1824–1907), Lord Kelvin since 1892

Kelvin has accompanied the development of thermodynamics for more than half a century, starting from his graduation in 1845. He went to Paris after graduation to work and study under Regnault, the careful and influential experimenter, whom we have already mentioned. Later Kelvin encouraged and supported Joule, and together the two men discovered the Joule Thomson effect in real gases, see Chaps. 2 and 6. Kelvin suggested the absolute temperature scale that bears his name, and he was a forerunner of the second law with the idea that there is a continuous degradation, or dissipation of energy into heat. However, Kelvin missed out himself on the *paradigmatic changes*[23] in thermodynamics. To be sure, when they occurred, he was often the first, or one of the first, to interpret and rephrase them, and apply them. Therefore a history of thermodynamics is incomplete without a prominent place for Kelvin. Perhaps his greatest achievement is that he suggested the possibility of *convective equilibrium*, see Chap. 7,

[23] This term has been made popular by Thomas S. Kuhn in his book: "The structure of scientific revolutions." The University of Chicago Press, Chicago and London. Third edition (1996).

which goes a long way to determine the structure of stars and the conditions in the lower atmosphere of the earth. Another original result of his is the Thomson formula for super-saturation in the processes of boiling and condensation on account of surface energy. However, here I choose to highlight Kelvin's capacity for original thought by a proposition he made for an absolute temperature scale, – an alternative to the Kelvin scale which we all know; see Insert 3.2. The proposition is intimately linked to the Carnot function $F'(t)$ which Kelvin attempted to calculate from Regnault's data. The new scale would have been logarithmic, and *absolute zero* would have been pushed to $-\infty$, a fact that gives the proposition its charm.

Kelvin's alternative absolute temperature scale

We recall the Carnot function $F'(t)$, a universal function of the temperature t, which neither Carnot nor Clapeyron had been able to determine. After Regnault's data were published, Kelvin used them to calculate $F'(t)$ for 230 values of t between 0°C and 230°C.[24] He proposed to rescale the temperature, and to introduce $\tau(t)$ such that the Carnot efficiency $F'(t)\mathrm{d}t$ for a small *fall* dt *of caloric* would be equal to $c\mathrm{d}\tau$, where c is a constant, independent of t or τ. Kelvin found that feature appealing. He says: *This* [scale] *may justly be termed an absolute scale.* By integration $\tau(t)$ results as

$$\tau(t) = \tau(0) + \frac{1}{c}\int_0^t F'(x)dx \cdot$$

Had Kelvin been able to fit an analytic function to Regnault's data, and to his calculations of $F'(t)$, he would have found a hyperbola

$$F'(t) = \frac{1}{273\,°C + t}$$

and his new scale would have been logarithmic:

$$\tau(t) = \tau(0) + \frac{1}{c}\ln\frac{273\,°C + t}{273\,°C} \quad .$$

$\tau(0)$ and c need to be determined by assigning τ-values to two fix-points, e.g. melting ice and boiling water.

However, not even the 230 values, which Kelvin possessed, were good enough to suggest the hyperbola in a convincing manner.

Therefore Kelvin had to wait for Clausius to determine $F'(t)$ in 1850, cf. Insert 3.3. When Kelvin's papers were reprinted in 1882, he added a note in which indeed he proposes the logarithmic temperature scale.

---

[24] W. Thomson: "On the absolute thermometric scale founded on Carnot's theory of the motive power of heat, and calculated from Regnault's observations." Philosophical Magazine, Vol. 33 (1848) pp. 313–317.

Compared to this daring proposition Kelvin's previous introduction of the absolute scale $T(t)=(273+\frac{t}{°C})°K$ seems straightforward, and rather plain. As it was, however, the logarithmic scale was never seriously considered, not even by Kelvin.

One might think that nobody really wanted the temperature scale on a thermometer to look like a slide rule. Yet, in the meteorological range between −30°C and +50°C the function $\tau(t)$ is nearly linear. And also, for $t \to -273$°C the rescaled temperature τ tends to $-\infty$, which is not a bad value for the absolute minimum of temperature. One could almost wish that Kelvin's proposition had been accepted. That would make it easier to explain to students why the minimum temperature cannot be reached.

Insert 3.2

Rudolf Julius Emmanuel Clausius (1822–1888)

By 1850 the efforts of Rumford, Mayer, Joule and Helmholtz had finally succeeded to create an overwhelming feeling that something was wrong with the idea that heat passes from boiler to cooler *unchanged in amount*: Some of the heat, in the passage, ought to be converted to work. But how to implement that new knowledge? Kelvin despaired:[25] *If we abandon* [Carnot's] *principle we meet with innumerable other difficulties ... and an entire reconstruction of the theory of heat* [is needed].

Clausius was less pessimistic: [26] *I believe we should not be daunted by these difficulties. ...* [and] *then, too, I do not think the difficulties are so serious as Thomson* [Kelvin] *does.* And indeed, it took Clausius surprisingly slight touches in surprisingly few spots of Carnot's and Clapeyron's works to come up with an expression for the Carnot function $F'(t)$ which determines the efficiency e of a Carnot cycle between t and $t+dt$. We recall that Carnot had proved $e{-}F'(t)dt$. And Clausius was the first person to argue convincingly that $F'(t)=\frac{1}{273°C+t}=\frac{1}{T}$ holds, cf. Insert 3.3.

[25] W. Thomson: "An account of Carnot's theory of the motive power of heat." Transactions of the Royal Society of Edinburgh 16 (1849). pp. 5412–574.

[26] R. Clausius: "Über die bewegende Kraft der Wärme und die Gesetze, welche sich daraus für die Wärme selbst ableiten lassen." Annalen der Physik und Chemie 155 (1850). pp. 368–397. Translation by W.F. Magie: "On the motive power of heat, and on the laws which can be deduced from it for the theory of heat." Dover (1960). Loc.cit. pp. 109–152.

Clausius's derivation of the internal energy and the calculation of the Carnot function

When a body absorbs the heat dQ it changes the temperature by dt and the volume by dV, as dictated by the heat capacity C_v and the latent heat λ [27] so that we have

$$dQ = C_v(t,V)\,dt + \lambda(t,V)\,dV.$$

Truesdell, who had the knack of a pregnant expression, calls this equation the *doctrine of the latent and specific heat*.[28] Applied to an infinitesimal Carnot process $abcd$ this reads, cf. Fig. 3.7:

$$ab \cong (dV, dt=0) \quad dQ_{ab} = \boxed{C_v(t,V)\,dt} + \lambda(t,V)\,dV$$
$$bc \cong (\delta'V, dt) \quad \boxed{dQ_{bc}} = -C_v(t,V+dV)\,dt + \lambda(t,V+dV)\,\delta'V$$
$$cd \cong (d'V, dt=0) \quad dQ_{cd} = \boxed{C_v(t-dt, V+\delta V)\,dt} - \lambda(t-dt, V+\delta V)\,d'V$$
$$da \cong (\delta V, dt) \quad \boxed{dQ_{da}} = C_v(t,V)\,dt - \lambda(t,V)\,\delta V$$

All framed quantities are zero, since the process is composed of isotherms and adiabates. Thus with a little calculation – expanding the coefficients – Clausius arrived at formulae for

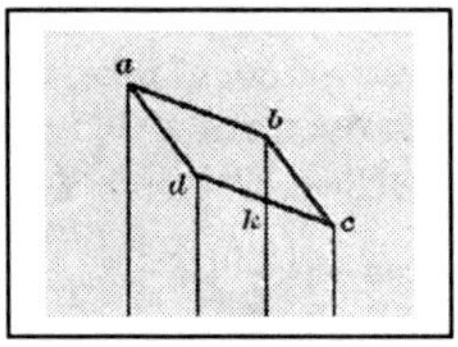

heat exchanged: $dQ_{ab} + dQ_{cd} = \left(\dfrac{\partial \lambda}{\partial t} - \dfrac{\partial C_V}{\partial V}\right) dt\,dV$

heat absorbed: $dQ_{ab} = \lambda\, dV$

work done: $dp\,dV = \dfrac{\partial p}{\partial t}\,dV\,dt.$

Fig. 3.7. (p,V)-diagram of an infinitesimally small Carnot cycle in a gas.

The work was calculated as the area of the parallelogram.

By the first law the heat exchanged equals the work done: Hence

$$\frac{\partial \lambda}{\partial t} - \frac{\partial C_V}{\partial V} = \frac{\partial p}{\partial t} \quad \text{or} \quad \frac{\partial(\lambda - p)}{\partial t} - \frac{\partial C_V}{\partial V} = 0$$

which may be considered as the integrability condition of the differential form

$$dU = C_v\,dt + (\lambda - p)\,dV \quad \text{or} \quad dU = dQ - p\,dV.$$

Thus Clausius arrived at the notion of the state function *internal energy* U, generally a function of t and V. Clausius assumed – correctly – that *in an ideal gas* U depends only on t. Therefore $\lambda = p$ holds and the efficiency e of the Carnot process is

---

[27] In modern thermodynamics the term *latent heat* is reserved as a generic expression for the heat of a phase transition – like heat of melting, or heat of evaporation –, but this was not so in the 19th century.

[28] C. Truesdell: "The tragicomical History of Thermodynamics 1822–1854". Springer Verlag New York (1980) [The specific heat is the heat capacity per mass.].

$$e = \frac{\text{work done}}{\text{heat absorbed}} = \frac{\frac{m}{V}\frac{k}{\mu}}{\lambda}dt = \frac{1}{273\ {}^\circ\mathrm{C} + t}dt$$

and the universal Carnot function $F'(t)$ is now calculated once and for all:

$$F'(t) = \frac{1}{273^\circ\mathrm{C} + t} = \frac{1}{T}.$$

[It is true that Clausius in 1850 calculated the *work done* only for an ideal gas. The above generalization to an arbitrary fluid came in 1854.[29]]

Insert 3.3

Notation and mode of reasoning of Clausius is nearly identical to that of Clapeyron with the one difference, – an essential difference indeed – that the total heat exchange of an infinitesimal Carnot cycle is not zero; rather it is equal to the work. Thus the heat Q is not a *state function* anymore, i.e. a function of t and V (say). To be sure, there *is* a state function, but it is not Q. Clausius denotes it by U, cf. Insert 3.3, and he calls U the *sum of the free heat and of the heat consumed in doing internal work,* meaning the sum of the kinetic energies of all molecules and of the potential energy of the intermolecular forces.[30] Nowadays we say that U is the *internal energy* in order to distinguish it from the kinetic energy of the flow of a fluid and from the potential energy of the fluid in a gravitational field.

A change of U is either due to heat exchanged or work done, or both:

$$\mathrm{d}U = \mathrm{d}Q - p\mathrm{d}V.$$

With this relation the first law of thermodynamics finally left the compass of verbiage – like *heat is motion* or *heat is equivalent to work*, or *impossibility of the perpetuum mobile*, etc. – and was cast into a

[29] R. Clausius: "Über eine veränderte Form des zweiten Hauptsatzes der mechanischen Wärmetheorie". Annalen der Physik und Chemie 169 (1854). English translation: "On a modified form of the second fundamental theorem in the mechanical theory of heat." Philosophical Magazine (4) 12, (1856).

[30] It was Kelvin who, in 1851, has proposed the name energy for U: W. Thomson: "On the dynamical theory of heat, with numerical results deduced from Mr. Joule's equivalent of a thermal unit, and M. Regnault's observations on steam." Transactions of the Royal Society of Edinburgh 20 (1851). p. 475.
Clausius concurred: ... *in the sequel I shall call U the energy*. It is quite surprising that Clausius let himself be preceded by Kelvin in this matter, because Clausius himself was an inveterate name-fixer. He invented the *virial* for something or other in his theory of real gases, see Chap. 6, and he proposed the *ergal* as a word for the potential energy, which seemed too long for his taste. And, of course, he invented the word *entropy,* see below.

mathematical equation, albeit for the special case of reversible processes and for a closed system, i.e. a body of fixed mass.

Clausius reasonably – and correctly – assumes that U is independent of V in an ideal gas and a linear function of t, so that the specific heats are constant. Because, he says: ...*we are naturally led to take the view that the mutual attraction of the particles... no longer acts in gases,* so that U does not *feel* how far apart the particles are, or how big the volume is. For an ideal gas we may write[31]

$$U(T,V) = U(T_R) + m\ z\tfrac{k}{\mu}(T - T_R),$$

where T_R is a reference temperature, usually chosen as 298K. The factor z has the value $^3/_2$, $^5/_2$, and 3 for one-, two-, or more-atomic gases respectively.

Actually Clausius could have proved his *view* – at least as far as it relates to the V-independence of U – from Gay-Lussac's experiment, mentioned in Chap. 2, on the adiabatic expansion of an ideal gas into an empty volume, where U must be unchanged after the process, and the temperature is *observed* to be unchanged, although the density does change, of course. As it is, Clausius mentions the (p, V, t)-relation of Mariotte and Gay-Lussac on every second page, but he seems to be unaware of Gay-Lussac's expansion experiment, or he does not recognize its significance.

In his paper of 1850, which we are discussing, Clausius deals with ideal gases and saturated vapour. Having determined the universal Carnot function, he is able to write the Clausius-Clapeyron equation, cf. Insert 3.1. Also he can obtain the adiabatic (p,V,t)-relation in an ideal gas, whose prototype is pV^γ = const, – well-known to all students of thermodynamics – where $\gamma = C_p/C_v$ is the ratio of specific heats. Later, in 1854,[32] Clausius applies this knowledge to calculate the efficiency e of a Carnot cycle of an ideal gas in any range of temperature, no matter how big; certainly not infinitesimal. He obtains, cf. Insert 3.4

$$e = 1 - \frac{T_{Low}}{T_{High}},$$

so that even the maximal efficiency is smaller than one, unless $T_{Low} = 0$ holds of course, which, however, is clearly impractical.

[31] This is a modern version which, once again, is somewhat anachronistic. Clausius was concerned with air and he used the poor value of the specific heat – given by Delaroche and Bérard – which had already haunted the works of Carnot and Mayer.
To do full justice to the specific heats, even of ideal gases, one could write a book all by itself. But that would be a different book from the present one.

[32] R. Clausius: (1854) loc.cit.

Efficiency of a Carnot cycle of a monatomic ideal gas

We refer to Fig. 3.8 which shows a graphical representation of a Carnot cycle between temperatures T_{High} and $T_{Low.}$ For a monatomic ideal gas we have for the work and the heat exchanged on the four branches

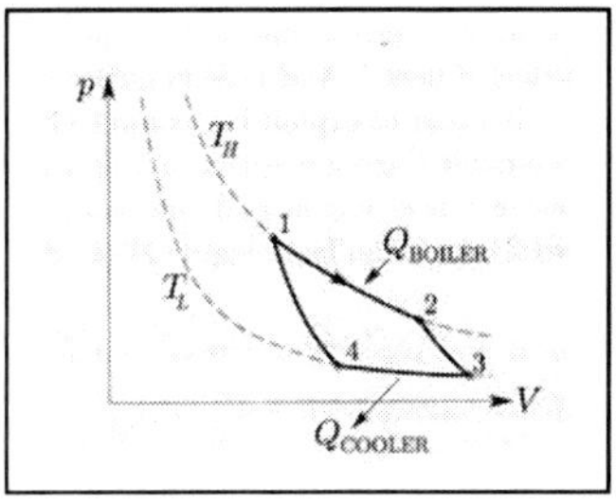

Fig. 3.8 Graph of a Carnot process

$$W_{12} = -m\frac{k}{\mu}T_H \ln\frac{V_2}{V_1}, \qquad Q_{12} = m\frac{k}{\mu}T_H \ln\frac{V_2}{V_1}$$

$$W_{23} = m\frac{3}{2}\frac{k}{\mu}(T_L - T_H), \qquad Q_{23} = 0$$

$$W_{34} = -m\frac{k}{\mu}T_L \ln\frac{V_4}{V_3}, \qquad Q_{12} = m\frac{k}{\mu}T_L \ln\frac{V_4}{V_3}$$

$$W_{41} = m\frac{3}{2}\frac{k}{\mu}(T_H - T_L), \qquad Q_{41} = 0$$

Therefore the efficiency comes out as

$$e = \frac{m\frac{k}{\mu}T_H \ln\frac{V_2}{V_1} + m\frac{k}{\mu}T_L \ln\frac{V_4}{V_3}}{m\frac{k}{\mu}T_H \ln\frac{V_2}{V_1}} = 1 - \frac{T_L}{T_H} .$$

The last equation results from the observation that $\frac{V_2}{V_1} = \frac{V_3}{V_4}$ holds.

Insert 3.4

With all this – by Clausius's work of 1850 thermodynamics acquired a distinctly modern appearance. His assumptions were quickly confirmed by experimenters,[33] or by reference to older experiments, which Clausius had either not known, or not used. Nowadays a large part of a modern course on thermodynamics is based on that paper by Clausius: the part that deals with ideal gases, and a large portion of the part on wet steam.

For Clausius, however, that was only the beginning. He proceeded with two more papers[34,35] in which he took five important steps forward:

---

[33] W. Thomson, J.P. Joule: "On the thermal effects of fluids in motion." Philosophical Transactions of the Royal Society of London 143 (1853).

[34] R. Clausius: (1854) loc.cit.

[35] R. Clausius: "Über verschiedene für die Anwendungen bequeme Formen der Hauptgleichungen der mechanischen Wärmetheorie". Poggendorff's Annalen der Physik 125 (1865). English translation by R.B. Lindsay: "On different forms of the fundamental equations of the mechanical theory of heat and their convenience for application". In: "The Second Law of Thermodynamics." J. Kestin (ed.), Stroudsburgh (Pa), Dowden Hutchinson and Ross (1976).

• away from infinitesimal Carnot cycles • away from ideal gases • away from Carnot cycles altogether, • away from cycles of whatever type, and • away from reversible processes.[36] In the end he came up with the concept of entropy and the properties of entropy, and that is his greatest achievement. We shall presently review his progress.

Among the people, whom we are discussing in this book, Clausius was the first one who lived and worked entirely in the place that was to become the natural habitat of the scientist: The autonomous university with tenured professors,[37] often as public or civil servants. With Clausius the time of *doctor-brewer-soldier-spy* had come to an end, at least in thermodynamics. General and compulsory education had begun and universities sprang up to satisfy the need for higher education and they had to be staffed. Thus one killed two birds with one stone: When a professor was no good as a scientist, he could at least teach and thus earn part of his keep. On the other hand, if he was good, the teaching duties left him enough time to do research.[38] Clausius belonged to the latter category. He was a professor in Zürich and Bonn, and his achievements are considerable: He helped to create the kinetic theory of ideal and real gases and, of course, he was the discoverer of entropy and the second law. His work on the kinetic theory was largely eclipsed by the progress made in that field by Maxwell in England and Boltzmann in Vienna. And in his work on thermodynamics he had to fight off numerous objections and claims of priority by other people, who had thought, or said, or written something similar at about the same time. By and large Clausius was successful in those disputes. Brush calls Clausius *one of the outstanding physicists of the nineteenth century.*[39]

---

[36] Reversible processes are those – in the present context of single fluids – in which temperature and pressure are always homogeneous, i.e. spatially constant, throughout the process, and therefore equal to temperature and pressure at the boundary. If that process runs backwards in time, the heat absorbed is reversed (sic) into heat emitted, or vice versa. A hallmark of the reversible process is the expression $-p\mathrm{d}V$ for the work $\mathrm{d}W$. That expression for $\mathrm{d}W$ *is* not valid for an irreversible process, which may exhibit turbulence, shear stresses and temperature gradients inside the cylinder of an engine (say) during expansion or compression. Irreversibility usually results from rapid heating and working.

[37] Tenure was intended to protect freedom of thought as much as to guarantee financial security.

[38] The system worked fairly well for one hundred years before it was undermined by job-seekers or frustrated managers, who failed in their industrial career. They are without scientific ability or interest, and spend their time attending committee meetings, reformulating curricula, and tending their gardens.

[39] Stephen G. Brush: "Kinetic Theory" Vol I. Pergamon Press, Oxford (1965).

Second Law of Thermodynamics

Clausius keeps his criticism of Carnot mild when he says that ... *Carnot has formed a peculiar opinion* [of the transformation of heat in a cycle]. He sets out to correct that opinion, starting from an axiom which has become known as the *second law of thermodynamics*:

Heat cannot pass by itself from a colder to a warmer body.

This statement, suggestive though it is, has often been criticized as vague. And indeed, Clausius himself did not feel entirely satisfied with it. Or else he would not have tried to make the sentence more rigorous in a page-long comment, which, however, only succeeds in removing whatever suggestiveness the original statement may have had.[40] We need not go deeper into this because, after all, in the end there will be an unequivocal *mathematical* statement of the second law.

The technique of exploitation of the axiom makes use of Carnot's idea of letting two reversible Carnot machines compete, – one a heat engine and the other one a heat pump, or refrigerator, cf. Fig. 3.9; the pump becomes an engine when it is reversed and *vice versa*; and the heats exchanged are changing sign upon reversal. Both machines work in the temperature range between T_{Low} and T_{High} and one produces the work which the other one consumes, cf. Fig. 3.9. Thus Clausius concludes that both machines must exchange the same amounts of heat at both temperatures, lest heat flow from cold to hot, which is forbidden by the axiom. So the efficiencies of both machines are equal, – if they work as heat engines. And, since nothing is said about the working agents in them, the efficiency must be universal. So far this is all much like Carnot's argument.

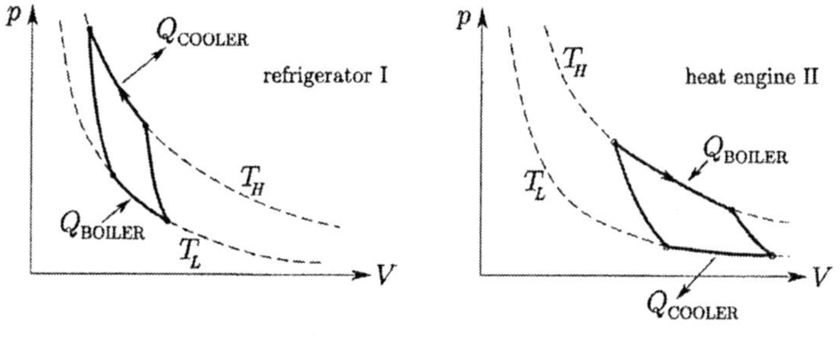

Fig. 3.9. Clausius's competing reversible Carnot engines

[40] E.g. see R. Clausius: "Die mechanische Wärmetheorie" [The mechanical theory of heat] (3.ed.) Vieweg Verlag, Braunschweig (1887) p. 34.

But then, unlike Carnot, Clausius knew that the work W_0 of the heat engine is the difference between Q_{boiler} and $|Q_{cooler}|$ so that the efficiency of any engine, – not necessarily a reversible Carnot engine – is given by

$$e = \frac{W_0}{Q_{boiler}} = 1 - \frac{|Q_{cooler}|}{Q_{boiler}}.$$

Q_{cooler} could conceivably be zero; at least, if it were, that would not contradict the first law, which only forbids W_0 to be *bigger* than Q_{boiler}. However, if the engine is a reversible Carnot engine with its universal efficiency, that efficiency is equal to that of an ideal gas – see above – so that we must have

$$\frac{Q_{boiler}}{T_{High}} = \frac{|Q_{cooler}|}{T_{Low}}.$$

It is clear from this equation that it is not the *heat* that passes through a Carnot engine unchanged in amount; rather it is Q/T, the *entropy.*

Clausius sees two types of *transformations* going on in the heat engine: The conversion of heat into work, and the passage of heat of high temperature to that of low temperature. Therefore in 1865[41] he proposes to call $\frac{Q}{T}$ the *entropy, … after the Greek word τροπή = transformation, or change* and he denotes it by *S*. He says that he has intentionally chosen the word to be similar to *energy*, because he feels that the two quantities … *are closely related in their physical meaning.* Well, maybe they appeared so to Clausius. However, it seems very much the question, in what way two quantities with different dimensions can be *close.*

The last equation shows that $|Q_{cooler}|$ cannot be zero, except for the impractical case $T_{Low} = 0$. Thus even for the optimal engine – the Carnot engine – there must be a cooler. Far from getting *more* work than the heat supplied to the boiler, we now see that we cannot even get that much: The boiler heat cannot all be converted into work. Therefore we cannot gain work by just cooling a single heat reservoir, like the sea. Students of thermodynamics like to express the situation by saying, rather flippantly:

1st law: You cannot win.
2nd law: You cannot even break even.

All of this still refers to cycles, or actually Carnot cycles. In Insert 3.5 we show in the shortest possible manner, how Clausius extrapolated these results to arbitrary cycles, and how he was able to consolidate the notion of entropy as a state function $S(T,V)$, whose significance is not restricted to cycles. The final result is the mathematical expression of the second law

[41] R. Clausius: (1865) loc.cit.

and it is an *inequality*: For a process from (T_B, V_B) to (T_E, V_E) the entropy growth cannot be smaller than the sum of heats exchanged divided by the temperature, where they are exchanged:

$$S(T_E, V_E) - S(T_B, V_B) \geq \int_B^E \frac{\mathrm{d}Q}{T} \quad \text{[equality holds for reversible processes].}$$

Clausius's derivation of the second law

Since $Q_{cooler} < 0$, the relation

$$\frac{Q_{boiler}}{T_{High}} = \frac{|Q_{cooler}|}{T_{Low}} \quad \text{may be written as} \quad \frac{Q_{boiler}}{T_{High}} + \frac{Q_{cooler}}{T_{Low}} = 0 \; .$$

In order to extrapolate this relation away from Carnot cycles to arbitrary cycles, Clausius decomposed such an arbitrary cycle into Carnot cycles with infinitesimal isothermal steps, cf. Fig.3.10. On those steps the heat dQ is exchanged such that dS=dQ/T is passing from the warm side to the cold one. Summation – or integration – thus leads to the equation

$$\oint \mathrm{d}S = \oint \frac{\mathrm{d}Q}{T} = 0$$

Hence follows for an open reversible process – not a cycle – between the points B and E

$$S(T_E, V_E) - S(T_B, V_B) = \int_B^E \frac{dQ}{T} \, ,$$

where $S(T_E, V_E) - S(T_B, V_B)$ is independent of the path from B to E, so that the entropy function $S(T, V)$ is a *state function.* After the internal energy $U(T, V)$ this is the second state function discovered by Clausius.

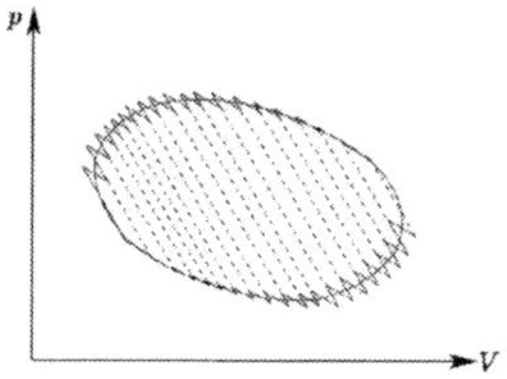

Fig. 3.10. Smooth cycle decomposed into narrow Carnot cycles

It remains to learn how this relation is affected by irreversibility. For that purpose Clausius reverted to the two competing Carnot engines, – one driving the other one. But now, one of them, the heat engine, was supposed to work irreversibly. In that case the process in the heat engine cannot be represented by a graph in a (p,V)-diagram, and therefore we show it *schematically* in Fig. 3.11. It turns out that the system of two engines contradicts Clausius's axiom, if the heat pump absorbs more heat at the low temperature than the heat engine delivers there. And now the reverse case cannot be excluded, because the engine changes its heat exchanges when it is made to work as a pump. Therefore for the irreversible heat engine we have

$$\frac{Q_{boiler}}{T_{High}} + \frac{Q_{cooler}}{T_{Low}} < 0 ,$$

It follows that the efficiency of the irreversible engine is lower than that of the reversible engine, and *a fortiori* – by the same sequence of arguments as before – that in an arbitrary irreversible process between points B and E we have

$$S(T_E, V_E) - S(T_B, V_B) > \int_B^E \frac{dQ}{T} .$$

The two relations for the change of entropy – one for the reversible and the other for the irreversible process – may be combined in a single ≥alternative, as we have done in the main text..

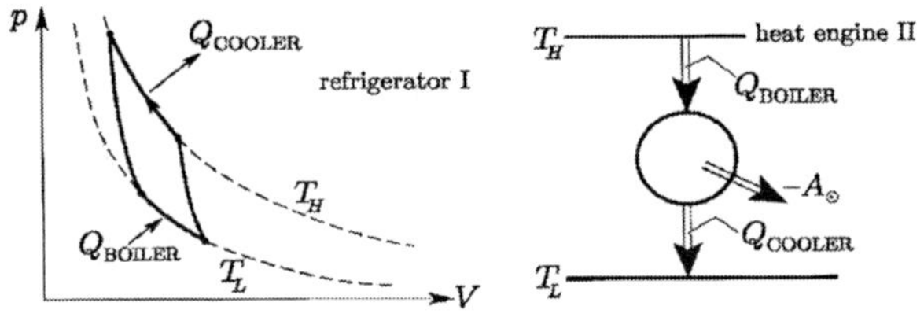

Fig. 3.11. Two competing Carnot engines with an irreversible heat engine

Insert 3.5

Exploitation of the Second Law

An important corollary of the second law concerns a reversible process between B and E, when those two point are infinitesimally close. In that case we have

$$\mathrm{d}S = \frac{\mathrm{d}Q}{T}$$

and when we eliminate $\mathrm{d}Q$ between that relation and the first law in the form $\mathrm{d}Q = (\mathrm{d}U + p\mathrm{d}V)$, we obtain

$$\mathrm{d}S = \frac{1}{T}(\mathrm{d}U + p\mathrm{d}V).$$

This equation is called the *Gibbs equation*.[42] Its importance can hardly be overestimated; it saves time and money and it is literally worth billions to the chemical industry, because it reduces drastically the number of measurements, which must be made in order to determine the internal energy $U = U(T,V)$ as a function of T and V.

Let us consider this:

Both the *thermal equation of statep*=$p(T,V)$ and the *caloric equation of state* $U = U(T,V)$ are needed explicitly for the calculation of nearly all thermodynamic processes, and they must be measured. Now, it is easy – at least in principle – to determine the thermal equation, because p, T, and V are all measurable quantities and they need only be put down in tables, or diagrams, or – in modern times – on CD's. But that is not so with the caloric equation, because U is not measurable. $U(T,V)$ must be calculated from caloric measurements of the heat capacities $C_V(T,V)$ and $C_p(T,V)$. Such measurements are difficult and time-consuming, – hence expensive – and they are unreliable to boot. And this is where the Gibbs equation helps. It helps to reduce – drastically – the number of caloric measurements needed, cf. Insert 3.6 and Insert 3.7.

Calculating $U(T,V)$ from measurements of heat capacities

The heat capacities C_V and C_p are defined by the equation $dQ = C dT$. Thus they determine the temperature change of a mass for a given application of heat dQ at either constant V or p. In this way C_V and C_p can be measured. By $dQ = dU + p\,dV$, and since we do know that U is a function of T and V – we just do not know the form of that function – we may write

$$C_V = \left(\frac{\partial U}{\partial T}\right)_V \quad \text{and} \quad C_p = \left(\frac{\partial U}{\partial T}\right)_V + \left(\left(\frac{\partial U}{\partial V}\right)_T + p\right)\left(\frac{\partial V}{\partial T}\right)_V \quad \text{or}$$

$$\left(\frac{\partial U}{\partial T}\right)_V = C_V \quad \text{and} \quad \left(\frac{\partial U}{\partial V}\right)_T = \frac{C_p - C_V}{\left(\frac{\partial V}{\partial T}\right)_p} - p.$$

Having measured $C_V(T,V)$ and $C_p(T,V)$ and $p(T,V)$ we may thus calculate $U(T,V)$ by integration to within an additive constant.

The integrability condition implied by the Gibbs equation provides

$$\frac{\partial U}{\partial V} = -p + T\frac{\partial p}{\partial T}.$$

Hence follows that the V-dependence of U, hence C_p, need not be measured: It may be calculated from the thermal equation of state. Moreover, differentiation with respect to T provides the equation

[42] Actually the equation was first written and exploited by Clausius, but Gibbs extended it to mixtures, see Chap. 5; the extension became known as *Gibbs's fundamental equation* and, as time went by, that name was also used for the special case of a single body.

$$\left(\frac{\partial C_V}{\partial V}\right)_T = T\left(\frac{\partial^2 p}{\partial T^2}\right)_V ,$$

so that the V-dependence of C_V is also determined by $p(T,V)$. Therefore the only caloric measurements needed are those of C_V as a function of T for *one* volume, V_0 (say). The number of caloric measurements is therefore considerably reduced, and that is a direct result of the Gibbs equation and the second law.

Insert 3.6

Once we know the thermal and caloric equations of state we may calculate the entropy $S(T,V)$, or $S(T,p)$ – by integration of the Gibbs equation – to within an additive constant. Thus for an ideal gas of mass m we obtain

$$S(T,p) = S(T_R, p_R) + m\left((z+1)\frac{k}{\mu}\ln\frac{T}{T_R} - \frac{k}{\mu}\ln\frac{p}{p_R}\right) .$$

Therefore the entropy of an ideal gas grows with $\ln T$ and $\ln V$: The isothermal expansion of a gas increases its entropy.

Clausius-Clapeyron equation revisited

If the Gibbs equation is applied to the reversible evaporation of a liquid under constant pressure – and temperature – it may be written in the form

$$(U\text{-}TS\text{+}pV)'' = (U\text{-}TS\text{+}pV)' ,$$

where, once again, ′ and ″characterize liquid and vapour. Thus the combination U-TS+pV, called *free enthalpy* or *Gibbs free energy,* is continuous across the interface between liquid and vapour, along with T and p. Therefore the vapour pressure must be a function of temperature only. We have $p=p(T)$ and the derivative of that function is given by the Clausius-Clapeyron equation, cf. Insert 3.1. When we realize that the heat of evaporation equals $R=T(S''-S')$, we may write the Clausius-Clapeyron equation in the form

$$\frac{U''-U'}{V''-V'} = -p + T\frac{\mathrm{d}p}{\mathrm{d}T} ,$$

which is clearly – for steam – the analogue to the integrability condition of Insert. 3.6. The relation permits us to dispense with measurements of the latent heat of steam and to replace them with the much easier (p,T)-measurements.

Insert 3.7

There is a school of thermodynamicists – the axiomatists – who thrive on formal arguments, and who would never let considerations of measurability enter their thoughts.[43] One can hear members of that school say, that the temperature T is

[43] Such an attitude is not uncommon in other branches of physics as well. Thus in mechanics there is a school of thought that considers Newton's law $F = m\,a$ as the definition of the force rather than a physical law between measurable quantities.

defined as $\left(\frac{\partial U}{\partial S}\right)_V$. That interpretation of the Gibbs equation ignores the fact that we should never know anything about either U or S, let alone $U=U(S,V)$, unless we had determined them first by measurements of p, V, T, and $C_V(T, V_0)$ in the manner described above.

Actually, the measurability of T is a consequence of its continuity at a diathermic wall, i.e. a wall permeable for heat. That continuity is the *real* defining property of temperature, and it gives temperature its central role in thermodynamics.

The chief witness of the formal interpretation of temperature is Gibbs, unfortunately, the illustrious pioneer of thermodynamics of mixtures. He, however, for all his acumen, was an inveterate theoretician, and I believe that he never made a single thermodynamic measurement in his whole life. We shall come back to this discussion in the context of chemical potentials, cf. Chap. 5, which have a lot in common with temperature.

Continuing our discussion of the consequences of the second law, we now come to another important corollary, namely that the *entropy in an adiabatic process*, – where $\mathrm{d}Q = 0$ holds –, *cannot decrease.* It grows until it reaches a maximum. We know from experience that, when we leave an adiabatic system alone, it tends to a state of homogeneity – the *equilibrium,* – in which all driving forces for heat conduction and expansion have run down.[44] That is the state of maximum entropy.

And so Clausius could summarize his work in the triumphant slogan:[45]

Die Energie der Welt ist constant.
Die Entropie der Welt strebt einem Maximum zu.

Die Welt [the universe] was chosen in this statement as being the ultimate thermodynamic system, which presumably is not subject to heating and working, so that $\mathrm{d}U = 0$ holds, as well as $\mathrm{d}S > 0$.

So the world has a purpose, or a destination, the *heat death*, see Fig. 3.12, not an attractive end!

It is often said that the world goes in a circle ...such that the same states are always reproduced. Therefore the world could exist forever. The second law contradicts this idea most resolutely ... The entropy tends to a maximum. The more closely that maximum is approached, the less cause for change exists. And when the maximum is reached, no further changes can occur; the world is then in a dead stagnant state.

Fig. 3.12. Rudolf Clausius and his contemplation of the heat death

---

[44] See Chap. 5 for a formal proof and for an explanation of what exactly homogeneity means.
[45] R. Clausius: (1865) loc.cit. p. 400.

Terroristic Nimbus of Entropy and Second Law

Concerning the heat death modern science does not seem to have made up its mind entirely. Asimov[46] writes:

> Though the laws of thermodynamics stand as firmly as ever, cosmologists ...[show] a certain willingness to suspend judgement on the matter of heat death.

At his time, however, Clausius's predictions were much discussed. The teleological character of the entropy aroused quite some interest, not only among physicists, but also among philosophers, historians, sociologists and economists. The gamut of reactions ranged from uneasiness about the bleak prospect to pessimism confirmed. Let us hear about three of the more colourful opinions:

The physicist Josef Loschmidt (1821–1895)[47] deplored

> ... the terroristic nimbus of the second law ..., which lets it appear as a destructive principle of all life in the universe.[48]

Oswald Spengler (1880–1936), the historian and philosopher of history devotes a paragraph of his book "The Decline of the West"[49] to entropy. He thinks that ... *the entropy firmly belongs to the multifarious symbols of decline,* and in the growth of entropy toward the heat death he sees the scientific equivalent of the *twilight of the gods* of Germanic mythology:

> The end of the world as the completion of an inevitable evolution – that is the twilight of the gods. Thus the doctrine of entropy is the last, irreligious version of the myth.

And the historian Henry Adams (1838–1918) – an apostle of human degeneracy, and the author of a meta-thermodynamics of history – commented on entropy for the benefit of the *ordinary, non-educated historian.* He says:

> .... this merely means that the ash-heap becomes ever bigger.

---

[46] I. Asimov: "Biographies" loc.cit. p. 364.

[47] J. Loschmidt: "Über den Zustand des Wärmegleichgewichts eines Systems von Körpern mit Rücksicht auf die Schwerkraft." [On the state of the equilibrium of heat of a system of bodies in regard to gravitation.] Sitzungsberichte der Akademie der Wissenschaften in Wien, Abteilung 2, 73: pp. 128–142, 366–372 (1876), 75: pp. 287–298, (1877), 76: pp. 205–209, (1878).

[48] If the author of this book had had his way in the discussion with the publisher, this citation of Loschmidt would have been either the title or the subtitle of the book. But, alas, we all have to yield to the idiosyncrasies of our real-time terrorists, – and to the show of paranoia by our opinionators.

[49] O. Spengler: "Der Untergang des Abendlandes: Kapitel VI. Faustische und Apollinische Naturerkenntnis. § 14: Die Entropie und der Mythos der Götterdämmerung." Beck'sche Verlagsbuchhandlung. München (1919) pp. 601–607.

Well, maybe it does. But then, Adams was an inveterate pessimist, to the extent even that he looked upon optimism as a sure symptom of idiocy.[50]

The entropy and its properties have not ceased to stimulate original thought throughout science to this day:

- biologists calculate the entropy increase in the diversification of species,
- economists use entropy for estimating the distribution of goods,[51]
- ecologists talk about the dissipation of resources in terms of entropy,
- sociologists ascribe an entropy of mixing to the integration of ethnic groups and a heat of mixing to their tendency to segregate.[52]

It is true that there is the danger of a lack of intellectual thoroughness in such extrapolations. Each one ought to be examined properly for mere shallow analogies.

Modern Version of Zeroth, First and Second Laws

Even though the historical development of thermodynamics makes interesting reading, it does not provide a full understanding of some of the subtleties in the field. Thus the early researchers invariably do not make it clear that the heat dQ and the work dW are applied to the *surface* of the body. Nor do they state clearly that the T and the p occurring in their equations, or inequalities, are the homogeneous temperature and the homogeneous pressure on the surface which may or may not be equal to those in the interior of the body; they *are* equal in equilibrium or in reversible processes, i.e. slow processes, but not otherwise.

The kinetic energy of the flow field inside the body is never mentioned by either Carnot or Clausius although, of course, its conversion into heat was paramount in the minds of Mayer, Joule and Helmholtz.

All this had to be cleaned up and incorporated into a systematic theory. That was a somewhat thankless task, taken on by scientists like Duhem, and

---

[50] According to S.G. Brush: "The Temperature of History. Phases of Science and Culture in the Nineteenth century." Burt Franklin & Co. New York (1978).

[51] N. Georgescu-Roegen: "The Entropy Law and the Economic Process." Harvard University Press, Cambridge, Mass (1971).

[52] I. Müller, W. Weiss: "Entropy and Energy – A Universal Competition, Chap. 20: Socio-thermodynamics." Springer, Heidelberg, (2005).
A simplified version of socio-thermodynamics is presented at the end of Chap. 5.

Jaumann[53] and Lohr.[54,55] These people recognized the first and second laws for what they are: Balance equations, or conservation laws on a par – formally – with the balance equations of mass and momentum.

Generically an equation of balance for some quantity $\Psi = \int_V \rho\psi \mathrm{d}V$ in a volume V, whose surface ∂V – with the outer normal n_i – moves with the velocity u_i, has the form

$$\frac{\mathrm{d}}{\mathrm{d}t}\int_V \rho\psi \mathrm{d}V = -\int_{\partial V}(\rho\psi(v_i - u_i) + \Phi_i)n_i \mathrm{d}A + \int_V \sigma \mathrm{d}V .$$

ρ is the mass density and ψ is the specific value of Ψ, such that $\rho\psi$ is the density of Ψ.[56] The velocity of the body, a fluid (say), is v_i. In the surface integral, $\rho\psi(v_i - u_i)n_i$ is the convective flux of Ψ through the surface element dA and $\Phi_i n_i$ is the non-convective flux. σ is the source density of Ψ; it vanishes for conservation laws.

For mass, momentum, energy, and entropy the generic quantities in the equation of balance have values that may be read off from Table 3.1.

t_{li} is called the stress tensor, whose leading term is the pressure $-p\delta_{li}$; that is the *only* term in t_{li}, if viscous stresses are ignored. E_{kin} is the kinetic energy of the flow field and q_i is the heat flux. Mass, momentum and energy are conserved, so that their source-densities vanish.[57] Note that the internal energy is not conserved, because it may be converted into kinetic energy. The entropy source is assumed non-negative which represents the growth property of entropy.

---

[53] G. Jaumann: ''Geschlossenes System physikalischer und chemischer Differentialgesetze" [Closed system of physical and chemical differential laws] Sitzungsbericht Akademie der Wissenschaften Wien, 12 (IIa) (1911).

[54] E. Lohr: ''Entropie und geschlossenes Gleichungssystem" [Entropy and closed system] Denkschrift der Akademie der Wissenschaften, 93 (1926).

[55] While Lohr is largely forgotten, Gustav Jaumann (1863–1924) lives on in the memory of mechanicians as the author of the *Jaumann derivative*, a ''co-rotational" time derivative, i.e. the rate of change of some quantity – like density or velocity – as seen by an observer locally moving and rotating with the body; that derivative plays an important role in rheology and in theories of plasticity. Jaumann was a student of Ernst Mach and carried Mach's prejudice against atoms far into the 20th century, thus making himself an outsider of any serious scientific circle. He died in a mountaineering accident.

[56] It has become customary in thermodynamics to denote global quantities – those referring to the whole body – by capital letters, and specific quantities – referred to the mass – by the corresponding minuscules.

[57] We ignore gravitation and radiation. See, however, Chap. 7, where radiation is treated. Gravitation changes thermodynamic in some subtle and, indeed, interesting ways, since the pressure field cannot be homogeneous in equilibrium, – neither on ∂V, nor in V. However, here is not the place to treat gravitational effects, because we do not wish to encumber our arguments. Let is suffice to say that in gases and vapours the gravitational effects are usually so small as to be negligible.

Table 3.1. Canonical notation for specific values of mass, momentum, energy and entropy and their fluxes and sources

Ψ	ψ	Φ_i	Σ
mass m	1	0	0
momentum P_l	v_l	$-t_{li}$	0
energy $U+E_{kin}$	$u+{}^1/_2\, v^2$	$-t_{li}v_l+q_i$	0
internal energy U	u	q_i	$t_{li}\dfrac{\partial v_l}{\partial x_i}$
entropy S	s	$\dfrac{q_i}{T}$ [58]	$\sigma \geq 0$

In order to clarify the special status of Clausius's first law, the equation for dU, we first observe that viscous forces did not enter Clausius's mind in connection with the first law. Also he considered *closed systems*, whose surfaces move with the velocity of the body on the surface so that no convective flux appears. Therefore Clausius would have written the equation of balance of energy in the form

$$\frac{\mathrm{d}(U+E_{kin})}{\mathrm{d}t} = \dot{Q} + \dot{W}, \text{ where}$$

$$\dot{Q} = -\int_{\partial V} q_i n_i \mathrm{d}A \text{ is the heating, and}$$

$$\dot{W} = -\int_{\partial V} p\, v_i n_i \mathrm{d}A \text{ is the working of pressure.}$$

The balance of internal energy should then have the form

$$\frac{\mathrm{d}U}{\mathrm{d}t} = \dot{Q} + \dot{W}_{\text{int}} \text{ where } \dot{W}_{\text{int}} = -\int_V p \frac{\partial v_l}{\partial x_l}\mathrm{d}V \text{ is the internal working.}$$

If we assume that the pressure is homogeneous on ∂V, the first equation becomes[59]

[58] This form of the entropy flux is nearly universally accepted, although the kinetic theory of gases furnishes a different form; the difference is small and we ignore it for the time being. See, however, Chap. 4.

[59] Note that $\int_{\partial V} v_l n_l \mathrm{d}A = \int_V \frac{\partial v_l}{\partial x_l}\mathrm{d}V = \frac{\mathrm{d}V}{\mathrm{d}t}$.

$$\frac{\mathrm{d}(U + E_{kin})}{\mathrm{d}t} = \dot{Q} - p\frac{\mathrm{d}V}{\mathrm{d}t};$$

and if we assume that the pressure is homogeneous throughout V, the second equation becomes

$$\frac{\mathrm{d}U}{\mathrm{d}t} = \dot{Q} - p\frac{\mathrm{d}V}{\mathrm{d}t}.$$

By comparison it follows that, for a homogeneous pressure p in V, there is no change of kinetic energy of the flow field which, of course, is reasonable. Indeed, according to the momentum balance, there is no acceleration in this case. Thus now, under all these restrictive assumptions – and with $\dot{Q}\mathrm{d}t = \mathrm{d}Q$ – we have obtained the Clausius form of the first law.

All these assumptions were *tacitly* made by Clausius, and his forerunners, and the majority of his followers to this very day. Indeed, among students thermodynamics has acquired the reputation of a difficult subject just *because* of the many tacit assumptions. The difficulty is not inherent in the field, however; it is due to sloppy teaching.

According to Table 3.1, the entropy balance contains a non-negative source density and a non-convective flux which is assumed to be given by q_i/T, so that we may write

$$\frac{\mathrm{d}S}{\mathrm{d}t} + \int_{\partial V} \frac{q_i n_i}{T}\mathrm{d}A \geq 0.$$

This inequality is known as the *Clausius-Duhem inequality*. If T is homogeneous on ∂V, we may write

$$\frac{\mathrm{d}S}{\mathrm{d}t} - \frac{\dot{Q}}{T} \geq 0, \qquad \text{where} \qquad \dot{Q} = -\int_{\partial V} q_i n_i \mathrm{d}A$$

and that is – again with $\dot{Q}\,\mathrm{d}t = \mathrm{d}Q$ – the form obtained by Clausius. He considered *only* this case. If T is not homogeneous on ∂V, the natural extension of his inequality was conceived by Pierre Maurice Marie Duhem (1861–1916).

Duhem was professor of theoretical physics in Bordeaux. He worked successfully in thermodynamics at the time when Gibbs was still unknown in Europe. However, he is also known as a philosopher of science, who expressed the view that the laws of physics are but symbolic constructions, neither true nor wrong representations of reality. He advocated metaphysical hypotheses for a provisional understanding of nature. Somehow Duhem's ideas found their way from Bordeaux to Vienna, where they were welcomed by Ernst Mach who thought that science should concentrate exclusively on finding relations between observed phenomena, see Chap. 4. Duhem's thoughts helped to underpin this kind of positivistic thinking in what became known as the *Vienna circle*, a niche for philosophers belly-aching about truth in the laws of natural science. A latter-day representative of the school was Karl Raimund Popper (1902–1994) – Sir Karl since 1964 – in whose writings the dilemma is largely reduced to the question of how, or whether, and why we know that the sun will rise tomorrow, *after approximately 90,000 pulse beats,* – or will it everywhere and always? Popper wrote a book about this important problem.[60]

The energy balance implies that the normal component of the heat flux q_i is continuous at a diathermic wall, i.e. a wall permeable to heat. The Clausius-Duhem inequality on the other hand implies that the normal component of $\frac{q_i}{T}$ is also continuous, provided that no entropy is produced in the wall. Therefore T must also be continuous. In this manner the zeroth law, cf. Chap. 1, may be said to represent a corollary of the Clausius-Duhem inequality. Its continuity is the defining property of temperature, and by virtue of the continuity, the temperature is measurable by contact thermometers. That is the reason why temperature plays a privileged role among thermodynamic variables. We shall review this role of temperature in Chap. 8, cf. Insert 8.3.

What is Entropy?

A physicist likes to be able to grasp his concepts plausibly and on an intuitive level. In that respect, however, the entropy – for all its proven and recognized importance is a disappointment. The formula $\mathrm{d}S = \frac{\mathrm{d}Q}{T}$ does not lend itself to a suggestive interpretation.

What is needed for the modern student of physics, is an interpretation in terms of atoms and molecules. Like with temperature: It is all very well to explain that temperature is defined by its continuity at a diathermic wall, but the "ahaa"-experience comes only after it is clear that temperature measures the mean kinetic energy of the molecules, – and then it comes immediately.

Such a molecular interpretation of entropy was missing in the work of Clausius. It arrived, however, with Boltzmann, although one must admit, that the interpretation of entropy was considerable more subtle than that of temperature. Let us consider this in the next chapter.

---

[60] K.R. Popper: "Objective knowledge – an evolutionary approach." Clarendon Press, Oxford (1972).

4 Entropy as $S = k \ln W$

Greek and Roman philosophers had conceived of atoms, and they developed the idea in more detail than we are usually led to believe. In the thinking of Leukippus and Demokritus in the 5th and 4th century B.C., the atoms of air move in all directions, and only occasionally they change their paths when they hit each other. To the ancients this fairly modern view implied a kind of determinism, which was incompatible with the idea of God, or gods, playing out their pranks, benevolent or otherwise. Therefore in later times, in the hands of Epicurus (341–270 B.C.) and Lucretius (95–55 B.C.), the atomistic philosophy of the "Natura Rerum" – this is the title of Lucretius's long poem – adopted an anti-religious and even atheistic flavour, which rendered it politically and socially unacceptable. Therefore atomism faded away, and in the end it came to represent no more than a footnote in ancient philosophy.

In the Age of Reason, by the work of Pierre Simon Marquis de Laplace (1749–1827), determinism came back with a vengeance in the form of Laplace's demon: ... *an intelligent creature capable of knowing all forces ... and all places of all things in the world, and equipped with the intelligence to analyse those data. Thus he can evaluate the motion of the greatest celestial bodies as well as of the tiniest atom; nothing is hidden for this demon: future and past lie open to his eyes.*

And just like in antiquity, this kind of determinism was considered as running counter to religion. Laplace was a minister under Napoléon, to whom he presented a part of his voluminous "Traité de Mécanique Céleste." Napoléon is supposed to have remarked that he saw no mention of God in the book. *I had no need of that hypothesis* said Laplace. When Lagrange[1] – a colleague and frequent co-worker of Laplace – heard about this exchange, he exclaimed: *Ah, but it is a beautiful hypothesis just the same. It explains so many things.*

Those enlightened men of post-revolutionary France clearly had their fun at the expense of religion.

---

[1] Joseph Louis Comte de Lagrange (1736–1813) was an eminent mechanician and mathematician. Napoléon made him a count to reward him for his achievements.

Renaissance of the Atom in Chemistry

And yet, when the concept of atoms was firmly established – at least in chemistry – that was the achievement of a devout Quaker, John Dalton (1766–1844). Dalton proposed that, in a chemical reaction, atoms combine to form molecules of a compound without losing their identity. Using the evidence collected by others, notably by Joseph Louis Proust (1754–1826), Dalton was able to determine the relative atomic and molecular masses of many elements and compounds. Once conceived, the ideas is extremely simple to explain and exploit: Carbon monoxide is made from carbon and oxygen in the definite proportion of 3 to 4 by mass, or weight. Thus, if we believe that carbon monoxide molecules are made up of one atom each of carbon and oxygen, the oxygen atom must be 1.33 times as massive as the carbon atom, which is correct.

Occasionally this type of reasoning can go wrong, however, as it did for Dalton with hydrogen and oxygen that form water in the proportion of 1 to 8. Thus Dalton concluded that the oxygen atom is 8 times as massive as the hydrogen atom. The proper number is 16, as we all know, because water has *two* hydrogen atoms for *one* oxygen atom. We shall soon see how that error was ironed out by Gay-Lussac and Avogadro.

In 1808 Dalton published his results in a book "New System of Chemical Philosophy", in which he gave relative atomic and molecular masses, most of them correct.

It became common practice to denote by M_r the ratio of the mass μ of any atom or molecule to the mass μ_0 of a hydrogen atom.[2] And M_rg is defined as the mass of what is called a "mol". If a mol contains L molecules, so that its mass is $L\mu$, we have

$$M_r\text{g} = L\mu \qquad \text{and hence} \qquad L = \frac{1\text{g}}{\mu_0}.$$

Therefore a mol of any element or compound has the same number of atoms or molecules.

The absolute mass – in kg (say) – of the atoms could not be had in that way, and it took another half century before that was found. We shall come to this shortly.

Dalton's atoms were rather immediately accepted by chemists. However, some hard-nosed physicists waged a losing battle against the *atomic hypothesis* that lasted all through the 19th century.

[2] Later the reference mass was based on the oxygen atom and still later – now – on the carbon atom. The reasons do not concern us, and the changes of M_r are minute.

Dalton was colour-blind and he studied that condition, which is sometimes still called Daltonism.

When he was presented to King William IV, Dalton's Quaker ethics did not allow him to put on the required colourful court dress. His friends had to convince him that the dress was grey before the ceremony could go ahead.

Fig. 4.1. John Dalton

For ideal gases there is a kind of corollary to Dalton's *law of definite proportions*, and that helped to correct Dalton's errors, e.g. the one on the composition of water. The chemist Gay-Lussac – pioneer of the thermal equation of state of ideal gases – was dealing with reactions, whose reactants are all gases; he found, that simple and definite proportions also hold for volumes. Thus one litre of hydrogen combines with one litre of chlorine, both at the same pressure and temperature, and give hydrogen chloride. Or two litres of hydrogen and one litre of oxygen combine to water, or three litres of hydrogen and one litre of nitrogen provide ammonia. These observations could most easily be understood by assuming that equal volumes contain equal numbers of atoms or molecules. Therefore the water molecule should contain two hydrogen atoms, – not one (!) – and ammonia should contain three. That conclusion was drawn by the chemist Jöns Jakob Berzelius (1779–1848) from Stockholm, and by Amadeo Avogadro (1776–1850), Conte di Quaregna. Avogadro was the physicist who is responsible for the mantra still taught to schoolchildren:

Equal volumes of different gases at the same pressure and temperature contain equally many particles.

Also Avogadro was first to use the words atom and molecule in the sense we are used to. Berzelius, on the other hand, is the chemist who introduced the now familiar nomenclature, like H_2O for water, or NH_3 for ammonia.

So, chemistry – such as it was at those days – had been put into perfect shape in a short time by the use of the concept of atoms. But, alas, such is human nature that Dalton, who had started it all, was pretty much the only chemist who could not bring himself to accept Gay-Lussac's and Avogadro's and Berzelius's reasoning and nomenclature. He stuck to his view that water contained only one hydrogen atom, and to a cumbersome notation.

Elementary Kinetic Theory of Gases

In physics it was Daniel Bernoulli (1700–1782), who first put to use the ancient atomistic idea of the randomly flying molecules of a gas. He explained the pressure of a gas on the wall of the container by the change of momentum of the molecules during their incessant bombardment of the wall. Bernoulli also related the temperature to the square of the (mean) speed of the molecules, and he was thus able to interpret the thermal equation of state of ideal gases, – the law found by Boyle, Mariotte, Amontons, Charles, and Gay-Lussac.

Daniel Bernoulli came from a family of illustrious mathematicians. His father Johann (1667–1748) started variational calculus, and his uncle Jakob (1654–1705) progressed significantly in the calculus of probability; he discovered the *law of large numbers*, and is the author of the *Bernoulli distribution*, whose limit for large numbers is the Gauss distribution or – in a gas – the Maxwell distribution, which is so important in the kinetic theory of gases. Also Jakob solved the non-linear ordinary differential equation, which carries his name and which we shall encounter in Chap. 8 in the context of acceleration waves. Daniel's best-remembered theorem is the *Bernoulli equation*, which states that the pressure of an incompressible fluid drops when the speed increases. The theorem is part of Daniel Bernoulli's book on hydrodynamics – published in 1738 – in which the kinetic theory of gases represents Sect. 10.[3] That section was largely ignored by scientists, and it sank into oblivion for more than a century.

Two other pioneers of the kinetic theory of gases fared no better. They were John Herapath (1790–1868), an engineer and amateur scientist and John James Waterston (1811–1883), a military instructor in the services of the East India Company in Bombay. The former did a little less than what Bernoulli had done and the latter did a little more. Both sent their works to the Royal Society of London for publication in the Philosophical Transactions and both found themselves rejected. Waterston received a less than complimentary evaluation to the effect that his work was *nothing but nonsense*.[4,5]

---

[3] D. Bernoulli: "Hydrodynamica, sive de vivibus et motibus fluidorum commentarii. Sectio decima: De affectionibus atque motibus fluidorum elasticorum, praecique autem aëris". English translation of Sect. 10: "On the properties and motions of elastic fluids, particularly air". In S. Brush: "Kinetic theory" Vol I, Pergamon Press, Oxford (1965).

[4] According to D. Lindley: "Boltzmann's atom". The Free Press, New York (2001), p. 1.

[5] S.G. Brush reviews, at some length, the efforts of both scientist to publish their works in his comprehensive memoir: "The kind of motion we call heat, a history of the kinetic theory of gases in the 19th century" Vol I pp. 107–159. North Holland Publishing Company, Amsterdam (1976).
Brush reports that Lord Rayleigh found Waterston's paper in the archives of the Royal Society, and published it in 1893, 48 years after it was submitted. Rayleigh added an introduction in which he gave good advise for junior scientists by saying that

However, in the 1850's the kinetic theory gained some ground. Clausius wrote his influential paper – later much praised by Maxwell [6] – "On the kind of motion we call heat" [7] which provided a clearly written and quite convincing kinetic interpretation • of temperature, • of the thermal equation of state of a gas, • of adiabatic heating upon compression, • of the liquid and solid state of matter in terms of molecular motion, and • of condensation and evaporation. Stephen Brush has chosen Clausius's title as the motto and main title for his comprehensive two-volume history of the kinetic theory of gases.[8] Actually Clausius had been anticipated by August Karl Krönig (1822—1879), [9] – at least in the derivation of the equation of state and in the interpretation of temperature. Krönig had considered a strongly simplified caricature of a gas in which the particles all move with the same speed and in only six directions – orthogonal to the six walls of a rectangular box – see Insert 4.1. Even earlier, Joule had described a similarly simple model in 1851.[10] It seems that Joule was the first to show that the mean speed $\overline{c}$ of the molecules of a gas at temperature T is such that $\frac{\mu}{2}\overline{c}^2 = \frac{3}{2}kT$ holds. At room temperature $\overline{c}$ is thus of the order of magnitude of several hundred meters per second. That result met with considerable scepticism which was aired most poignantly by Christoph Hendrik Diederik Buys-Ballot (1817-1890), a meteorologist with a butler. He argued that

---

> *.... highly speculative investigations, especially by an unknown author, are best brought before the world through some other channel than a scientific society.*

Rayleigh praises the *marvellous courage of the author,* i.e. Waterston, and provides some additional council.

> *A young author, who believes himself capable of great things, would usually do well to secure the favourable recognition of the scientific world by work whose scope is limited, and whose value is easily judged, before embarking on higher flights.*

I do not know whether Lord Raleigh was serious, when he wrote these sentences, or whether, perhaps, he was being sarcastic.

[6] J.C. Maxwell: "On the dynamical theory of gases". Philosophical Transactions of the Royal Society of London 157 (1867).

[7] R. Clausius: "Über die Art der Bewegung, welche wir Wärme nennen," Annalen der Physik 100, (1857) pp. 353–380.

[8] S.G. Brush: "The kind of motion we call heat." loc.cit.

[9] A.K. Krönig: "Grundzüge einer Theorie der Gase". [Basic theory of gases] Annalen der Physik 99 (1856), p. 315.

[10] J.P. Joule: "Remarks on the heat and the constitution of elastic fluids". Memoirs of the Manchester Literary and Philosophical Society. November 1851. Reprinted in Philosophical Magazine. Series IV, Vol. XIV (1857), p. 211.

> ... if he were sitting at one end of a long dining room and a butler brought in dinner at the other end, it would be some moments before he could smell what he was about to eat. If atoms were flying at hundreds of meters per second ... he should smell the dinner as soon as he saw it.

Clausius was able to answer that objection by arguing that the molecules of the fragrant dinner fumes – like all gas molecules – do not fly on straight lines for any great length of time. They collide with other molecules and are turned sideways and backwards, and those deflections lead to a zig-zag path of an atom, which is much longer between two given points than the distance of the points. Therefore, the atom needs more time between those points along its path than it would, if it did not collide. In making this argument Clausius developed the concept of the mean free path l of an atom or a molecule:[11] If the particle is imagined to be a spherical *billiard ball* of radius r, it sweeps out a cylinder of volume $l\pi r^2$ between two collisions – in the mean – so that such a cylinder should contain just one other particle. Therefore, we must have $l\pi r^2 n \approx 1$, where $n=N/V$ is the number density of particles. Hence follows the mean free path l and Clausius suspected its value to be a small fraction of a millimetre for a gas under normal conditions of p=1atm and T=298 K (say). But he could not be sure, since r and n were unknown. Therefore his argument remained somewhat inconclusive.

Mean speed of molecules

We consider an ideal gas at rest in a rectangular box of volume $V = LD^2$ with the number density N/V of atoms and assume that the inter-atomic forces are negligible. The pressure p of the gas on the walls results from the bombardment with gas atoms. $-pD^2$ is the force of the right wall on the gas. By Newton's law that force is equal to the rate of change of momentum of the atoms that hit the area D^2. For simplicity we assume that all atoms have the same speed $\bar{c}$ and that one sixth of them are flying perpendicular to each of the six walls. The change of momentum of a single atom of mass μ in an elastic collision with the right wall of area D^2 is equal to $-2\mu\bar{c}$ and the collision rate on that wall is $\bar{c}D^2 \frac{1}{6}\frac{N}{V}$. Therefore the rate of change of momentum is

[11] R. Clausius: "Über die mittlere Länge der Wege, welche bei Molekularbewegung gasförmiger Körper von den einzelnen Molekülen zurückgelegt werden, nebst einigen anderen Bemerkungen über die mechanische Wärmetheorie". Annalen der Physik (2) 105 (1858) pp. 239–258. English translation: "On the mean free path of the molecular motion in gaseous bodies; also other remarks on the mechanical theory of heat." In S. Brush: "Kinetic theory." Vol I, Pergamon Press, Oxford (1965).

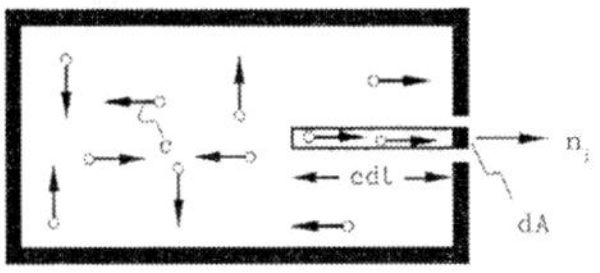

$$-2\mu\bar{c}^2 \tfrac{1}{6}\tfrac{N}{V} D^2 = -\tfrac{m}{V}\tfrac{1}{3}\bar{c}^2 D^2 ,$$

This must be set equal to $-pD^2$ and it follows that

$$pV = \tfrac{1}{3}\bar{c}^2 .$$

Fig. 4.2. On the pressure of a gas

Comparison with the thermal equation of state of an ideal gas provides

$$\tfrac{1}{3}\bar{c}^2 = \tfrac{k}{\mu}T$$

Therefore the mean kinetic energy of the atoms equals $\frac{3}{2}kT$.

Consider air [12] in $V = 1\text{m}^3$ and with the mass $m = 1.2\text{kg}$, the mass of air in 1m^3 at p = 1atm and T = 298 K: We obtain $\bar{c}$ = 503m/s and that may be considered the mean speed of the air molecules.

Insert 4.1

So, now it became imperative to determine how many particles there are in a given volume of a gas in some agreed-upon reference state of pressure and temperature. To be sure, one mol contains $L = 1\text{g}/\mu_0$ particles but how big is μ_0, the mass of a hydrogen atom?

Avogadro's observation meant that the thermal equation of state of a gas contains a universal constant. We may put that equation into the form $pV = NkT$, where – according to Avogadro – k is a universal constant, now called the Boltzmann constant.[13] The value of k was unknown, since N was unknown and not measurable. Of course, one might try to be clever and replace the particle number N by the mass $m = N\mu$ of the gas. The mass *can* be measured by weighing, but that does not help, because in terms of mass the formula reads $pV = m\, k/\mu T$ and thus, while k/μ may now be calculated, k and μ individually are still unknown.

Actually, that was the dilemma of the atomic hypothesis which delayed its acceptance among physicists: The whole idea had too much of a hypothetical flavour; neither the masses of the atoms were known, nor their radii r, nor was it known how many there were in a given volume. It is true that the speeds of the particles were known, see above, but the distances and the distances between collisions were again unknown.

---

[12] Air is made up of molecules, not atoms. Moreover, in air there are at least two types of molecules, nitrogen and oxygen. But this does not matter for the present consideration of orders of magnitude.

[13] Boltzmann had just been born when Avogadro died. At that time the universal constant was called the *ideal gas constant*, denoted by R which is k/μ_0. I avoid that constant and trust that the reader will tolerate the anachronism, perhaps.

One could do something about the radius, because it seemed to be a good assumption, that in the liquid phase, where the density ρ_{liq} was known, the particles lay close together so that $\rho_{\text{liq}} r^3 \approx \mu$ had to hold, at least by order of magnitude. But, once again, that determines r only, if μ is known and vice versa. Thus the game seemed to be destined to go on and on in an vicious spiral: Each new thought added a new quantity which could not be determined unless one of the previous quantities was known.

However, now the end of the spiral was near, because Maxwell did calculate the viscosity η of a gas. He obtained $\eta = \frac{1}{3}\mu \frac{N}{V} l\bar{c}$, cf. Insert 4.3, and η *could be measured.* Thus now we have 5 equations for the 5 unknowns *k, N, μ, r, l* viz.

$$\frac{pV}{NT} = k, \quad m = N\mu, \quad \rho_{\text{liq}} r^3 = \mu, \quad \frac{N}{V} l\pi r^2 = 1, \quad \eta = \frac{1}{3}\mu\frac{N}{V} l\bar{c}.$$

It was Josef Loschmidt[14] (1821–1895) who recognized that Maxwell's formula for the viscosity could be used to close the argument, and he calculated the missing values. I have repeated Loschmidt's calculation for 1 litre of air at p = 1atm, T = 298K, m = $1.2 \cdot 10^{-3}$kg, $\bar{c}$ = 503m/s, cf. Insert 4.1, and with $\rho_{\text{liq}} \approx 10^3$kg/m^3 and η = $1.8 \cdot 10^{-5}$Ns/m^2 – all measurable, or calculable, or reasonably estimable values – and have obtained

$$k = 1.1 \cdot 10^{-23}\text{J/K}, \quad N = 3.2 \cdot 10^{22}, \quad \mu = 37.8 \cdot 10^{-27}\text{kg}, \quad r = 3.4 \cdot 10^{-10}\text{m}, \quad l = 0.9 \cdot 10^{-7}\text{m}.$$

Since air is a mixture of particles with an average relative molecular mass M_r = 29, the mass of the hydrogen atom comes out as $\mu_0 = 1.3 \cdot 10^{-27}$kg and therefore the number of particles in a mol is $L = 7.7 \cdot 10^{23}$. That number is officially called the Avogadro number, although in Austria and Germany, where Loschmidt lived, it is also known as the Loschmidt number. None of these values is really good by modern standards, due to the coarseness of the assumptions and of the input values. Even so, the orders of magnitude are fine,[15] and that was all physicists could do in the mid-nineteenth century.

It was Kelvin who emphasized the enormous size of the number most poignantly, when he suggested to dilute a glass full of marked water molecules with all the water of the seven seas. Afterwards each glass of sea water would still contain about 100 marked molecules!

---

[14] J. Loschmidt: "Zur Grösse der Luftmoleciile." [On the size of air molecules] Zeitschrift für mathematische Physik 10, (1865), p. 511.

[15] The modern value of the Avogadro constant is L = $6.0221367 \cdot 10^{23}$, so that we have $\mu_0 = 1.660540 \cdot 10^{-27}$kg. The proper value of the Boltzmann constant is $k = 1.38044 \cdot 10^{-23}$J/K.

James Clerk Maxwell (1831–1879)

None of the scientists before Maxwell had recognized in his calculations that the atoms of a gas move with different speeds. This was not because they thought that the speeds were all equal. Rather they did not know how to account for different speeds mathematically. This changed when Maxwell took up the question.

In a recent biography[16] we read that *there are no anecdotes to tell about Maxwell because he led a quiet life, devoted to his family and science.* And he died prematurely of cancer at the age of only 48 years. The interest in his person is based on the admiration of his scientific work. Indeed, Maxwell was a genius – both as a mathematician and as a physicist – who is best known for his formulation of the equations governing electro-magnetism, a theory of vast scientific and technical importance, without which modern life would be inconceivable, see Chap. 2 above. Boltzmann, Maxwell's congenial contemporary, and occasional correspondent was so much moved to enthusiasm over the Maxwell equations – that is what they are called to this day – that he exclaimed: *War es ein Gott, der diese Zeichen schrieb*?[17] Maxwell had put the keystone on Faraday's collection of the phenomena of electro-magnetism and suggested that light were an electro-magnetic phenomenon, – truly a revolutionary discovery. We have discussed this in Chap. 2.

However, here it is not Maxwell's electro-magnetism that is of interest. Rather it is his equally profound – albeit, perhaps, less momentous – contribution to the kinetic theory of gases.[18] In an early work Maxwell – stimulated by the offer of an award in an open competition – had studied the rings of Saturn.[19] He proved that they could not consist of flat, hollow disks. Such rigid disks would be broken up by tidal forces. Therefore, what appeared to be solid rings, had to consist of numerous small solid rocks and icy lumps that travel around Saturn on elliptical orbits like so many satellites, which have different orbital speeds. Occasionally the lumps might collide and thus be kicked inwards or outwards, thereby carrying their orbital momentum into the faster or slower adjacent ellipses. Maxwell became well-known by winning the competition. Also the work found him

---

[16] Giulio Peruzzi: "Maxwell, der Begründer der Elektrodynamik" [Maxwell the founder of electrodynamics] Spektrum der Wissenschaften, German edition of Scientific American. Biografie 2 (2000).

[17] *Was it a God who wrote these marks*? This is a quotation from Goethe's Faust.

[18] Maxwell wrote several papers on the kinetic theory. Here we are concerned with the first one: J.C. Maxwell: "Illustrations of the dynamical theory of gases." Philosophical Magazine 19 and 20, both (1860).

[19] J.C. Maxwell: "On theories of the constitution of Saturn's rings." Proceedings of the Royal Society of Edinburgh IV (1859).
J.C. Maxwell: "On the Stability of the Motion of Saturn's Rings." An Essay which obtained the Adams Prize for the year 1856, in the University of Cambridge.

well-acquainted with the properties of swarms of particles when he turned his attention to gases.

Maxwell introduced the function $\varphi(c_i)\mathrm{d}c_i$ (i = 1,2,3) for the fraction of atoms in a gas that have velocity components in the i-direction between c_i and $c_i + \mathrm{d}c_i$, and he proved, cf. Insert 4.2, that the form of the function $\varphi(c_i)$ in equilibrium is given by a Gaussian, whose peak lies at zero velocity and whose height and width is determined by temperature

$$\varphi_{equ}(c_i)\frac{1}{\sqrt{2\pi\frac{k}{\mu}T}}\exp\left(-\frac{\mu c_i^2}{2kT}\right), \qquad (i = 1,2,3).$$

The fraction of atoms with velocities between (c_1, c_2, c_3) and ($c_1 + \mathrm{d}c_1$, $c_2 + \mathrm{d}c_2$, $c_3 + \mathrm{d}c_3$) is then a function of the speed c

$$f_{equ}(c_1,c_2,c_3)\mathrm{d}c_1\mathrm{d}c_2\mathrm{d}c_3 = \frac{1}{\sqrt{2\pi\frac{k}{\mu}T}^{\,3}}\exp\left(-\frac{\mu c^2}{2kT}\right)\mathrm{d}c_1\mathrm{d}c_2\mathrm{d}c_3 .$$

Accordingly the fraction of atoms $F_{equ}(c)\mathrm{d}c$ with speeds between c and $c + \mathrm{d}c$ is given by

$$F_{equ}(c)dc = \frac{4\pi c^2}{\sqrt{2\pi\frac{k}{\mu}T}^{\,3}}\exp\left(-\frac{\mu c^2}{2kT}\right)dc .$$

Thus most atoms have a small *velocities* and only few are moving fast. But small *speeds* are also rare, since only few velocities represent small speeds. The mean speed follows as $\overline{c} = \sqrt{3\frac{k}{\mu}T}$.[20]

All three of these equilibrium distribution functions are often called Maxwell distributions, or simply Maxwellians.

The Maxwell distribution

Consider a gas of N atoms in the volume V which is at rest as a whole and possesses the internal energy $U = N\,{}^3/_2\,kT$, because the atoms have velocities (c_1,c_2,c_3). The gas is in equilibrium, and therefore homogeneous with an isotropic distribution of atomic velocities.

Let $\varphi_{equ}(c_i)\mathrm{d}c_i$ be the fraction of atoms with the velocity component i between c_i and $c_i + \mathrm{d}c_i$, such that $\varphi_{equ}(c_1)\varphi_{equ}(c_2)\varphi_{equ}(c_3)$ determines the fraction of atoms with the

[20] Actually this is the root mean square velocity. There are slight differences between $\overline{c}$, and the mean speed, and the most probable speed which we ignore.

velocity (c_1, c_2, c_3). Because of isotropy that product can only depend on the speed $c = \sqrt{c_1^2 + c_2^2 + c_3^2}$. We thus have

$$n(c) = \varphi_{equ}(c_1)\, \varphi_{equ}(c_2)\, \varphi_{equ}(c_3).$$

Logarithmic differentiation with respect to c_i provides

$$\frac{1}{c}\frac{\mathrm{d}\ln n}{\mathrm{d}c} = \frac{1}{c_i}\frac{\mathrm{d}\ln \varphi_{equ}}{\mathrm{d}c_i},$$

such that both sides must be constants. Hence follows by integration

$$\varphi_{equ}(c_i) = A\exp(-Bc_i^2)\,.$$

The two constants A and B may be calculated from

$$1 = \int_{-\infty}^{\infty} A\exp(-Bc_i^2)\mathrm{d}c_i \quad \text{and} \quad \tfrac{1}{2}kT = \int_{-\infty}^{\infty} \tfrac{\mu}{2}c_i^2 A\exp(-Bc_i^2)\mathrm{d}c_i$$

so that we obtain

$$\varphi_{equ}(c_i)\frac{1}{\sqrt{2\pi\frac{k}{\mu}T}}\exp\left(-\frac{\mu c_i^2}{2kT}\right).$$

Insert 4.2

Brush[21] says that Maxwell's derivation of the equilibrium distribution – replayed in Insert 4.2 – *mystified his contemporaries* because of its novelty and originality and he suggests. ... *that the proof may have been simply copied from a book on statistics by Quételet* [22] *or from a review* [of that book] *by Herschel in the Edinburgh Review*.[23] In that review John Herschel, – the son of the eminent astronomer Friedrich Wilhelm Herschel (1738–1822) – calculates the probability of the deviation of a ball dropped from a height in order to hit a mark; his analysis is very similar to Maxwell's.

Then, still in the same paper, Maxwell proceeded to propose an ingenious interpretation for the friction in a gas, cf. Insert 4.3.[24] That interpretation could have been motivated by the investigation of Saturn's rings, although they are not mentioned in the paper. Newton had assumed that the force needed to maintain a velocity gradient in a fluid or a gas – or in the ether for that matter – is proportional to the value of the velocity

---

[21] S.G. Brush: (1976) loc.cit. p. 342.

[22] A. Quételet: "La théorie des probabilités appliquées aux sciences morales et politiques".

[23] J. Herschel: Edinburgh Review 92 (1850).

[24] Insert 4.3 presents a caricature of Maxwell's argument, which I have found useful when explaining the mechanism of gaseous friction to students.

gradient. The factor of proportionality is the viscosity η and it is a common experience that the viscosity, or *shear resistance* of water (say), or honey decreases with increasing temperature. The same behaviour was expected for gases.

Viscous friction in a gas, a caricature

The mechanics of viscous friction can be appreciated from the consideration of two trains of equal masses M with velocities V_1 and V_2 passing each other on parallel adjacent tracks. People change the momentum of the trains by stepping from one to the other at the equal mass rate μ in both directions. Upon arrival in the new train, a person must support himself against either the forward or the backward wall in order to stay on his feet. Thereby he accelerates or brakes the new train, and thus the two trains eventually equalize their velocities. The equations of motion for the trains read

$$M\frac{dV_1}{dt} = \mu(V_2 - V_1) \qquad M\frac{dV_2}{dt} = \mu(V_1 - V_2) \qquad \text{hence} \qquad M\frac{d(V_1 - V_2)}{dt} = -2\mu(V_1 - V_2).$$

It follows that the velocity difference of the trains decreases exponentially due to a "shear force" proportional to the actual value of the difference. The jumping rate μ is the factor of proportionality; if it increases, the braking is more efficient.

Basically the same argument was used by Maxwell to calculate the shear force τ between two gas layers moving with a y-dependent flow velocity $V(y)$ in the x-direction. The result reads

$$\tau = \frac{1}{3}\rho l \bar{c}\frac{dV}{dy},$$

where ρ is the mass density and l the mean free path, $\bar{c}$ is the mean speed of the atoms which jump between the layers, much like the passengers of the train model do.

Insert 4.3

We have anticipated Maxwell's result for the viscosity $\eta = \frac{1}{3}\mu\frac{N}{V}l\bar{c}$ above, when we reported Loschmidt's calculation of the size of the air molecules. $\bar{c}$ here is the mean speed of the atoms – at least in order of magnitude – and l is the mean free path which we have previously introduced by the equation $N/V l\pi r^2 = 1$. Thus the viscosity is independent of the density of the gas and it grows with temperature, since the mean speed is proportional to $\sqrt{T}$, see above. Maxwell says *…this consequence of a mathematical theory is quite surprising* and he was doubtful, because *… the* *only experiment which I know does not seem to confirm the result.* In the event, however, the theory was right and the experiment was wrong. Maxwell had been too timid and Boltzmann says when he reports the event:[25]

> ... observations revealed only the lack of confidence of Maxwell in the power of his own weapons.

It is true that the viscosity of liquids drops with growing temperature, but for gases this relationship is reversed. When this was confirmed by new experiments – by Maxwell himself – the fact provided a boost of confidence in the kinetic theory of gases. To be sure, the square root growth of η on temperature is an artifact of the simple model. In 1867 Maxwell revisited the argument in a more systematic manner, when he derived equations of transfer with collision terms,[26] see below. For that purpose he had to study the dynamics of a binary collision between two atoms interacting at a short distance r with a repulsive force of the type $1/r^s$[27]. It turned out that the temperature dependence of the viscosity is then given by T^n with $n = \frac{2}{s-1} + \frac{1}{2}$. An infinite value of s corresponds to the billiard ball model, while $s=5$ corresponds to the so-called Maxwellian molecules. Maxwell concluded from his own experiments that η was proportional to T, which must have been wishful thinking, because forces of the type $1/r^5$ make for a particularly simple form of the collision term in the equations of transfer, see below.[28]

There is an element of probability in the kinetic theory, or the mechanical theory of heat, which was previously absent from mechanics: Indeed, when we say that $N\varphi(c_i)\mathrm{d}c_i$ is the number of atoms with the i-component c_i of velocity, we do not expect that statement to be strictly true. Since the atoms are perpetually changing their velocities in frequent collisions, the number of atoms with c_i is fluctuating and $N\varphi(c_i)\mathrm{d}c_i$ is merely the mean value or expectation value of that number. Accordingly $\varphi(c_i)\mathrm{d}c_i$ is the *probability* for a single atom to have the velocity component c_i. Assuming that the velocity components of an atom are independent we have that

$$\varphi(c_1)\,\varphi(c_2)\,\varphi(c_3)\mathrm{d}c_1\mathrm{d}c_2\mathrm{d}c_3$$

is the probability of an atom to have the velocity (c_1,c_2,c_3). So physicists had to learn the rules of probability calculus. For some of them there were

---

[25] L. Boltzmann: "Der zweite Hauptsatz der mechanischen Wärmetheorie" [The second law of the mechanical theory of heat]. Lecture given at the Kaiserliche Akademie der Wissenschaften on May 29, 1886.

[26] J.C. Maxwell: loc.cit. (1867).

[27] Actually in reality the force has repulsive and attractive branches, see below. However, for a rarefied gas the simple power potential is often good enough.

[28] Modern measurements give n a value of about 0.8; for argon n is equal to 0.816.

peculiar scruples. So also for Maxwell, a deeply religious man with the somewhat bigoted ethics that often accompanies piety. In a letter he wrote:

> … [probability calculus], of which we usually assume that it refers only to gambling, dicing, and betting, and should therefore be wholly immoral, is the only *mathematics for practical people* which we should be.

The Boltzmann Factor. Equipartition

True to that recommendation Maxwell employed probabilistic arguments when he returned to the kinetic theory in 1867. Indeed, probabilistic reasoning led him to an alternative derivation of the equilibrium distribution – different from the derivation indicated in Insert 4.2 above. The new argument concerns elastic collisions of two atoms with energies

$$\frac{\mu}{2}c^2, \frac{\mu}{2}c^{1^2} \text{ which after the collision have the energies } \frac{\mu}{2}c'^2, \frac{\mu}{2}c'^{1^2}.$$

Boltzmann was not satisfied. He acknowledges Maxwell's arguments and calls them *difficult to understand because of excessive brevity.* Therefore he repeats them in his own way, and extends them. Let us consider his reasoning:[29] Boltzmann concentrates on energy in general – rather than only translational kinetic energy – by considering $G(E)\mathrm{d}E$, the fraction of atoms between E and E+dE. The transition probability P that two atoms – with E and E^1 – collide and afterwards move off with E', E'^1 is obviously[30] proportional to $G(E)\,G(E^1)$. Therefore we have

$$P_{E.E^1 \to E'.E'^1} = c\,G(E)G(E^1)\,.$$

The probability for the inverse transition is[31]

$$P_{E',E'^1 \to E,E^1} = cG(E')G(E'^1)\,.$$

In equilibrium both transition probabilities must be equal so that $\ln G(E)$ is a *summational collision invariant*. Indeed, in equilibrium we have

$$G(E)G(E^1) = G(E')G(E'^1) \quad \text{hence} \quad \ln G(E) + \ln G(E^1) = \ln G(E') + \ln G(E'^1)\,.$$

[29] L. Boltzmann: "Studien über das Gleichgewicht der lebendigen Kraft zwischen bewegten materiellen Punkten." [Studies on the equilibrium of kinetic energy between moving material points] Wiener Berichte 58 (1868) pp. 517–560.

[30] Actually, what is obvious to one person is not always obvious to others. And so there is a never-ending but fruitless discussion about the validity of this multiplicative ansatz.

[31] The most difficult thing to prove in the argument is that the factors of proportionality – here denoted by c – are equal in both formulae. We skip that.

Since E itself is also such an invariant – because of energy conservation during the collision – it follows that $\ln G_{equ}(E)$ must be a linear function of E, i.e.

$$G_{equ}(E) = a\exp(-bE) = \frac{1}{kT}\exp\left(-\frac{E}{kT}\right).$$

The constants a and b follow from the requirement

$$\int_0^\infty G_{equ}(E)\mathrm{d}E = 1 \quad \text{and} \quad \int_0^\infty EG_{equ}(E)\mathrm{d}E = kT\,.$$

Boltzmann noticed – and could prove – that the argument is largely independent of the nature of the energy E. Thus E may simply be equal to $\frac{\mu}{2}c^2$ – as it was for Maxwell – but then it may also contain the three additive contributions of the rotational energy of a molecule and the contributions of the kinetic and elastic energy of a vibrating molecule. According to Boltzmann all these energies contribute the equal amount $^1/_2kT$ – on average – to the energy U of a body. This became known as the *equipartition theorem.*

The problem was only that the theory did not jibe with experiments. To be sure, the specific heat $c_v = \frac{\partial U}{\partial T}$ of a monatomic gas was $^3/_2kT$ but for a two-atomic gas experiments showed it to be equal to $^5/_2kT$ when it should have been $3kT$. Boltzmann decided that the rotation about the connecting axis of the atoms should be unaffected by collisions, thus begging the question, as it were, since he did not know why that should be so. And vibration did not seem to contribute at all. The problem remained unsolved until quantum mechanics solved it, cf. Chap. 7.

If Boltzmann was not satisfied with Maxwell's treatment, Maxwell was not entirely happy with Boltzmann's improvement. Here we have an example for a fruitful competition between two eminent scientists.

Maxwell acknowledges Boltzmann's *ingenious treatment* [which] *is, as far as I can see, satisfactory*:[32] But he says: ... *a problem of such primary importance in molecular science should be scrutinized and examined on every side...This is more especially necessary when the assumptions relate to the degree of irregularity to be expected in the motion of a system whose motion is not completely known.* And indeed, Maxwell's treatment does offer two interesting new aspects:

---

[32] J.C. Maxwell: "On Boltzmann's theorem on the average distribution of energy in a system of material points." Cambridge Philosophical Society's Transactions XII (1879).

He extends Boltzmann's argument to particles in an external field, the force field of gravitation (say), and thus could come up with the equilibrium distribution of molecules of the earth's atmosphere which reads

$$f_{equ} = \frac{1}{\sqrt{2\pi \frac{k}{\mu} T}^3} \exp\left(-\frac{\mu c^2}{2kT} - \frac{\mu g z}{kT} \right).$$

The second exponential factor is also known as the *barometric formula,* it determines the fall of density with height in an isothermal atmosphere. In the same paper Maxwell provided a new aspect of a statistical treatment, which foreshadows Gibbs's canonical ensemble, see below.

So between them, Boltzmann and Maxwell derived what is now known as the

$$\text{Boltzmann factor}: \exp\left(-\tfrac{E}{kT}\right).$$

It represents the ratio of probabilities for states that differ in energy by E – in equilibrium, of course.

For practical purposes in physics, chemistry, and materials science the Boltzmann factor is perhaps Boltzmann's most important contribution; it is more readily applicable than his statistical interpretation of entropy, although the latter is infinitely more profound philosophically. We proceed to consider this now.

Ludwig Eduard Boltzmann (1844–1906)

For those who had reservations about probability in physics, bad times were looming, and they arrived with Boltzmann's most important work.[33]

Maxwell and Boltzmann worked on the kinetic theory of gases at about the same time in a slightly different manner and they achieved largely the same results, – all except one! That one result, which escaped Maxwell, concerned entropy and its statistical or probabilistic interpretation. It provides a deep insight into the strategy of nature and explains irreversibility. That interpretation of entropy is Boltzmann's greatest achievement, and it places him among the foremost scientists of all times.

[33] L. Boltzmann: "Weitere Studien über das Wärmegleichgewicht unter Gasmolekülen". [Further studies about the heat equilibrium among gas molecules] Sitzungsberichte der Akademie der Wissenschaften Wien (II) 66 (1872) pp. 275–370.

Boltzmann about Maxwell:
immer höher wogt das Chaos der Formeln.[34]

Maxwell about Boltzmann:
... I am much inclined to put the whole business in about six lines

Fig. 4.3. James Clerk Maxwell

Maxwell had derived equations of transfer for moments of the distribution function in 1867,[35] and Boltzmann in 1872 formulated the transport equation for the distribution function itself, which carries his name. What emerged was the Maxwell-Boltzmann transport theory, so called by Brush.[36] Neither Maxwell's nor Boltzmann's memoirs are marvels of clarity and systematic thought and presentation, and both privately criticized each other for that, cf. Fig. 4.3. Therefore we proceed to present the equations and results in an modern form. The knowledge of hindsight permits us to be brief, but still it is inevitable that we write lengthy formulae in the main text, which is otherwise avoided. Basic is the distribution function $f(\boldsymbol{x},\boldsymbol{c},t)$ which denotes the number density of atoms at the point $\boldsymbol{x}$ and time t which have velocity $\boldsymbol{c}$. The *Boltzmann equation* is an integro-differential equation for that function

$$\frac{\partial f}{\partial t}+c_i\frac{\partial f}{\partial x_i}=\int (f'f'^1-ff^1)\sigma g\sin\theta \mathrm{d}\theta \mathrm{d}\varphi \mathrm{d}c^1 .$$

The right hand side is due to collisions of atoms with velocities $\boldsymbol{c}$ and $\boldsymbol{c}^1$ which, after the collision, have velocities $\boldsymbol{c}'$ and $\boldsymbol{c}'^1$. The angle φ identifies the plane of the binary interaction, while θ is related to the angle of deflection of the path of an atom in the collision. θ ranges between 0 and $\pi/2$. σ is the cross section for a (θ,φ)-collision and g is the relative speed of the colliding atoms. The f' s in the collision integral are the values of the distribution function for the velocities $\boldsymbol{c}'$, $\boldsymbol{c}'^1$ and $\boldsymbol{c}$, $\boldsymbol{c}^1$ respectively as

---

[34] *...ever higher surges the chaos of formulae.*
[35] J.C. Maxwell: (1867) loc.cit.
[36] S.G. Brush: (1976) loc.cit. p. 422 ff.

indicated. The form of the collision term represents the *Stosszahlansatz*[37] which was mentioned before; it is particularly simple for Maxwellian molecules, because in their case σg is a function of θ only, rather than a function of θ and g. The combination $ff'^1 - ff^1$ in the integrand reflects the difference of the probabilities for collisions

$$\boldsymbol{c}'\boldsymbol{c}'^1 \rightarrow \boldsymbol{c}\boldsymbol{c}^1 \text{ and } \boldsymbol{c}\boldsymbol{c}^1 \rightarrow \boldsymbol{c}'\boldsymbol{c}'^1.$$

This must have been easy for Boltzmann, since logically it is adapted from the argument which he had used before for the derivation of the Boltzmann factor, see above.

Generic *equations of transfer* follow from the Boltzmann equation by multiplication by a function $\psi(\boldsymbol{x},\boldsymbol{c},t)$ and integration over $\boldsymbol{c}$. We obtain

$$\frac{\partial \int \psi f \mathrm{d}\boldsymbol{c}}{\partial t} + \frac{\partial \int \psi f c_k \mathrm{d}\boldsymbol{c}}{\partial x_k} - \int \left(\frac{\partial \psi}{\partial t} + c_k \frac{\partial \psi}{\partial x_k} \right) f \mathrm{d}\boldsymbol{c}$$

$$= \frac{1}{4} \int (\psi + \psi^1 - \psi' - \psi'^1)(f' f'^1 - f f^1) \sigma g \sin\theta \mathrm{d}\theta \mathrm{d}\varphi \mathrm{d}\boldsymbol{c}^1 \mathrm{d}\boldsymbol{c} .$$

This equation has the form of a balance law for the generic quantity Ψ with

density $\int \psi f \mathrm{d}\boldsymbol{c}$,

flux $\int c_k \psi f \mathrm{d}\boldsymbol{c}$,

intrinsic source $\int \left(\frac{\partial \psi}{\partial t} + c_k \frac{\partial \psi}{\partial x_k} \right) f \mathrm{d}\boldsymbol{c}$, and

collision source $\frac{1}{4} \int (\psi + \psi^1 - \psi' - \psi'^1)(f' f'^1 - f f^1) \sigma g \sin\theta \mathrm{d}\theta \mathrm{d}\varphi \mathrm{d}\boldsymbol{c}^1 \mathrm{d}\boldsymbol{c}$

ψ^1, ψ', and ψ'^1 stand for $\psi(\boldsymbol{x},\boldsymbol{c}^1,t)$, $\psi(\boldsymbol{x},\boldsymbol{c}',t)$, and $\psi(\boldsymbol{x},c'^1,t)$.

Stress and heat flux in the kinetic theory

In terms of the distribution function the densities of mass, momentum, and energy can obviously be written as

[37] That cumbersome word – even for German ears – describes a formula for the number of collisions which lead to a particular scattering angle by the binary interaction of atoms. The expression is not due to Maxwell, of course, nor to Boltzmann. As far as I can find out it was first used by P. and T. Ehrenfest in "Conceptual Foundations of the Statistical Approach in Mechanics." Reprinted: Cornell University Press, Ithaca (1959).
The word seems to be untranslatable, and so it has been joined to the small lexicon of German words in the English language like Kindergarten, Zeitgeist, Realpolitik and, indeed, Ansatz.

$$\rho = \int \mu f \mathrm{d}c,\quad \rho v_i = \int \mu c_i f \mathrm{d}c,\quad \rho\left(u+\frac{1}{2}v^2\right) = \int \frac{\mu}{2} c^2 f \mathrm{d}c.$$

u is the specific *internal* energy formed with $C_i = c_i - v_i$

$$\rho u = \int \frac{\mu}{2} C^2 f \mathrm{d}c\,.$$

With $u = \frac{3}{2}\frac{k}{\mu}T$ – appropriate for a monatomic ideal gas – we obtain [38]

$$\frac{3}{2}kT = \frac{\int \frac{\mu}{2} C^2 f \mathrm{d}c}{\int f \mathrm{d}c}$$

so that T is the mean kinetic energy of the atoms. This may be considered as the *kinetic definition* of temperature, or the *kinetic temperature.*

If $\psi = \mu, \mu c_i, \frac{\mu}{2}c^2$ is introduced into the equations of transfer, one obtains the conservation laws of mass, momentum and energy

$$\frac{\partial \rho}{\partial t} + \frac{\partial \rho v_i}{\partial x_i} = 0$$

$$\frac{\partial \rho v_j}{\partial t} + \frac{\partial(\rho v_j v_i + \int \mu C_j C_i f \mathrm{d}c)}{\partial x_i} = 0$$

$$\frac{\partial \rho(u + \frac{1}{2}v^2)}{\partial t} + \frac{\partial\left(\rho(u + \frac{1}{2}v^2)v_i + \int \mu C_j C_i f \mathrm{d}c\, v_i + \int \frac{\mu}{2} C^2 C_i f \mathrm{d}c\right)}{\partial x_i} = 0.$$

Comparison with the corresponding macroscopic laws, cf. Chap. 3, identifies stress and heat flux of a gas as

$$t_{ij} = -\int \mu C_j C_i f \mathrm{d}c \quad \text{and} \quad q_i = \int \frac{\mu}{2} C^2 C_i f \mathrm{d}c\,.$$

Thus the stress is properly called a momentum flux.

Insert. 4.4

For special choices of ψ, viz. $\psi = \mu$, $\psi = \mu c_i$, $\psi = \frac{1}{2}\mu c^2$, one obtains the conservation laws of mass, momentum and energy from the generic equation, cf. Insert 4.4. In those cases both source terms vanish. For any other choice of ψ the collision term is not generally equal to zero. However, there is an important choice of ψ for which a conclusion can be drawn, although the source does not vanish. That is the case when the production has a sign. A sharp look at the source, – in the suggestive form in which I have written it – will perhaps allow the attentive reader to identify that particular ψ all by himself. Certainly this was no difficulty for

[38] The additive energy constant is routinely ignored in the kinetic theory.

Boltzmann. He chose $\psi = -k \ln\frac{f}{b}$, where k and b are positive constants to be determined. With that choice we have

$$\text{collision source} = \frac{k}{4}\int \ln\frac{f'f'^1}{f\,f^1}(f'f'^1 - f\,f^1)\sigma g \sin\theta \mathrm{d}\theta \mathrm{d}\varphi \mathrm{d}\mathbf{c}^1\,\mathrm{d}\mathbf{c}$$

and that is obviously non-negative, since $\ln\frac{f'f'^1}{f\,f^1}$ and $(f'f'^1 - f\,f^1)$ always have the same sign. In equilibrium, where f is given by the Maxwellian distribution, both expressions vanish so that there is no source. Both properties suggest that

$$S = -k\int f \ln\frac{f}{b}\,\mathrm{d}\mathbf{c}\mathrm{d}\mathbf{x}$$

is a candidate for being considered as the entropy of the kinetic theory of gases. If k is the Boltzmann constant, S *is* the entropy. Indeed, if we insert the Maxwellian – the equilibrium distribution – we obtain

$$S_{equ}(T,p) = S_{equ}(T_R, p_R) + m\left(\frac{5}{2}\frac{k}{\mu}\ln\frac{T}{T_R} - \frac{k}{\mu}\ln\frac{p}{p_R}\right)$$

which agrees with the entropy of a monatomic gas calculated by Clausius, see Chap. 3.

Entropy flux

The interpretation of the quantity $-k\int f \ln\frac{f}{b}\mathrm{d}\mathbf{c}$ as entropy density is not complete unless we relate its rate of change, or its flux, to heat or heating, so as to recognize the status of Clausius's 2nd law $\frac{\mathrm{d}S}{\mathrm{d}t} \geq \frac{\dot{Q}}{T}$ within the kinetic theory. Let us consider that:

If indeed $-k\int f \ln\frac{f}{b}\mathrm{d}\mathbf{c}\mathrm{d}\mathbf{x}$ is the entropy, the non-convective entropy flux should be given by

$$\Phi_i = -k\int C_i f \ln\frac{f}{b}\mathrm{d}\mathbf{c}.$$

We calculate that expression from the Grad 13-moment approximation[39]

[39] All this is terribly anachronistic but it belongs here. Grad proposed the moment approximation of the distribution function in 1949! H. Grad: "On the kinetic theory of rarefied gases." Communications of Pure and Applied Mathematics 2 (1949).

$$f_G = f_{equ}\left(1+\underbrace{\frac{1}{2\rho\frac{k}{\mu}T}t_{\langle ij\rangle}\left(\frac{1}{\frac{k}{\mu}T}C_iC_j-\delta_{ij}\right)-\frac{1}{\rho(\frac{k}{\mu}T)^2}q_iC_i\left(1-\frac{1}{5}\frac{1}{\frac{k}{\mu}T}C^2\right)}_{\varphi}\right),$$

which is the most popular – and most rational – approximate near-equilibrium distribution function available. Insertion provides, if second order terms in φ are ignored

$$\rho s = \rho s_{equ} - \frac{t_{\langle ij\rangle}t_{\langle ij\rangle}}{4\rho\frac{k}{\mu}T^2} - \frac{q_iq_i}{5\rho\left(\frac{k}{\mu}\right)^2T^3} \quad \text{and} \quad \Phi_i = \frac{q_i}{T} + \frac{2}{5}\frac{t_{\langle ij\rangle}q_j}{\rho\frac{k}{\mu}T^2}.$$

Thus s contains non-equilibrium terms and $\Phi_i = \frac{q_i}{T}$ – the Duhem expression for the entropy flux, cf. Chap. 3 – holds only, if non-linear terms are neglected.

Insert. 4.5

Thus Boltzmann had given a kinetic interpretation for the entropy, an interpretation in terms of the distribution function f and its logarithm. That interpretation, however, is in no way intuitively appealing or suggestive, and as such it does not provide the insight into the strategy of nature which I have promised; *not yet anyway.*

In order to find a plausible interpretation, the integral for S has to be discretized and extrapolated in the manner described in Insert 4.6. It is the very nature of extrapolations that there are elements of arbitrariness in them; they are not just corollaries. In the present case – in the reformulation of the integral for S – I have emphasized the speculative nature of the extrapolating steps by introducing them with a bold-face **if**.

The discretization stipulates that the element $d\boldsymbol{x}d\boldsymbol{c}$ of the $(\boldsymbol{x},\boldsymbol{c})$-space has a finite number $P_{d\boldsymbol{x}d\boldsymbol{c}}$ (say) of occupiable points $(\boldsymbol{x},\boldsymbol{c})$ – occupiable by atoms – and that $P_{d\boldsymbol{x}d\boldsymbol{c}}$ is proportional to the volume $d\boldsymbol{x}d\boldsymbol{c}$ of the element with a quantity Y as the factor of proportionality. Thus $^1/_Y$ is the volume of the smallest element, i.e. a *cell,* which contains only one point. In this manner the $(\boldsymbol{x},\boldsymbol{c})$-space is *quantized* and indeed, Boltzmann's procedure in this context foreshadows quantization, although at this stage it may be considered merely as a calculational tool rather than a physical argument. And it was so considered by Boltzmann when he says: … *it seems needless to emphasize that* [for this calculation] *we are not concerned with a real physical problem.* And further on: … *this assumption is nothing more than an auxiliary tool.*[40]

[40] L. Boltzmann (1872) loc.cit.

If the occupancy N_{xc} of all points, or cells, in d$\boldsymbol{x}$d$\boldsymbol{c}$ is equal, Boltzmann obtained by a suitable choice of b viz. $b = eY$, cf. Insert 4.6

$$S = k \ln \frac{1}{\prod_{xc}^{P} N_{xc}!},$$

where P is the total number of cells – of occupiable points – in the $(\boldsymbol{x},\boldsymbol{c})$-space.

This is still not an easily interpretable expression, but it is close to one. Indeed, **if** we multiply the factor $N!$ into the argument of the logarithm, we may write

$$S = k \ln W, \qquad \text{where} \qquad W = \frac{N!}{\prod_{xc}^{P} N_{xc}!}.$$

And that expression *is* interpretable, because W – by the rules of combinatorics – is the number of realizations, often called *microstates,* of the distribution $\{N_{xc}\}$ of N atoms. [The combinatorial rule is relevant here, **if** the interchange of two atoms at different points $(\boldsymbol{x},\boldsymbol{c})$ leads to different realizations.]

Reformulation of $S = -k \int f \ln \frac{f}{b} \mathrm{d}\boldsymbol{c}\mathrm{d}\boldsymbol{x}$

Let there be $P_{\mathrm{d}x\mathrm{d}c}$ occupiable points in the element d$\boldsymbol{x}$d$\boldsymbol{c}$ and let $P_{\mathrm{d}x\mathrm{d}c} = Y\mathrm{d}\boldsymbol{x}\mathrm{d}\boldsymbol{c}$. Let further each point in d$\boldsymbol{x}$d$\boldsymbol{c}$ be occupied by the same number N_{xc} of atoms, cf. figure, so that we have $N_{xc} P_{\mathrm{d}x\mathrm{d}c} = f \,\mathrm{d}\boldsymbol{x}\mathrm{d}\boldsymbol{c}$. Then the contribution of d$\boldsymbol{x}$d$\boldsymbol{c}$ to S may be written as

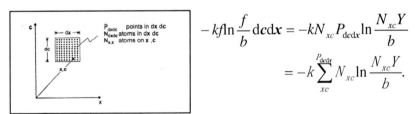

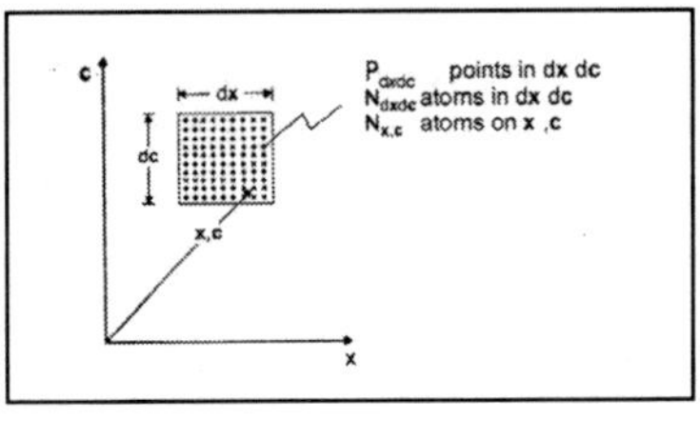

$$-kf\ln\frac{f}{b}\mathrm{d}\boldsymbol{c}\mathrm{d}\boldsymbol{x} = -kN_{xc}P_{\mathrm{d}c\mathrm{d}x}\ln\frac{N_{xc}Y}{b}$$

$$= -k\sum_{xc}^{P_{\mathrm{d}c\mathrm{d}x}} N_{xc}\ln\frac{N_{xc}Y}{b}.$$

Fig. 4.4 An element of $(\boldsymbol{x},\boldsymbol{c})$-space

The sum is really a sum over $P_{\mathrm{d}x\mathrm{d}c}$ equal terms. b may be chosen arbitrarily and we choose $b = eY$, where e is the Euler number so that

We shall see later, cf. Chaps. 6 and 7 that it was S.N. Bose who took the cells seriously, and gave them a value and a physical interpretation.

$$-kf\ln\frac{f}{b}\,\mathrm{d}c\mathrm{d}\boldsymbol{x} = -k\sum_{xc}^{P_{\mathrm{d}c\mathrm{d}x}}(N_{xc}\ln N_{xc} - N_{xc})$$

$$= k\ln\frac{1}{\prod_{xc}^{P_{\mathrm{d}c\mathrm{d}x}} N_{xc}!}.$$

The last step makes use of the Stirling formula $\ln a! = a\ln a - a$, which can be applied, **if** a – here N_{xc} – is much larger than 1. Therefore the total entropy reads

$$S = k\ln\frac{1}{\prod_{xc}^{P} N_{xc}!},$$

where P is the total number of occupiable points in the $(\boldsymbol{x},\boldsymbol{c})$-space.

Insert 4.6

A first extrapolation of the formula for S is that we may now drop the requirement that the values N_{xc} within the element $\mathrm{d}\boldsymbol{x}\mathrm{d}\boldsymbol{c}$ are all equal. This may be a constraint appropriate to the kinetic theory of gases, – where there is only *one* value $f(\boldsymbol{x},\boldsymbol{c},t)$ characterizing the gas in the element – but it has no status in the new statistical interpretation of S. In particular, it is now conceivable that all atoms may be found in the same cell, so that they all have the same position and the same velocity; in that case the entropy is obviously zero, since there is only *one* realization for that distribution.

With $S = k \ln W$ we have a beautifully simple and convincing possibility of interpreting the entropy, or rather of understanding why it grows: The idea is that each *realization* of the gas of N atoms is *a priori* considered to occur equally frequently, or to be equally probable. That means that the realization where all atoms sit in the same place and have the same velocity is just as probable as the realization that has the first N_1 atoms sitting in one place $(\boldsymbol{x},\boldsymbol{c})$ and all the remaining $N - N_1$ atoms sitting in another place, etc. In the former case W is equal to 1 and in the latter it equals $\frac{N!}{N_1!(N-N_1)!}$. In the course of the irregular thermal motion the realization is perpetually changing, and it is then eminently reasonable that the gas – as time goes on – moves to a *distribution* with more possible realizations and eventually to the distribution with most realizations, i.e. with a maximum entropy. And there it remains; we say that equilibrium is reached.

So this is what I have called the *strategy of nature,* discovered and identified by Boltzmann. To be sure, it is not much of a strategy, because it consists of letting things happen and of permitting blind chance to take its course. However, $S = k\ln W$ is easily the second most important formula of physics, next to $E = mc^2$ – or at a par with it. It emphasizes the random component inherent in thermodynamic processes and it implies – as we shall see later – entropic forces of considerable strength, when we attempt to thwart the random walk of the atoms that leads to more probable distributions.

However, the formula $S = k \ln W$ is not only interpretable, it can also be extrapolated away from monatomic gases to any system of many identical units, like the links in a polymer chain, or solute molecules in a solution, or money in a population, or animals in a habitat. Therefore $S = k \ln W$ with the appropriate W has a universal significance which reaches far beyond its origin in the kinetic theory of gases.

Actually $S = k \ln W$ was nowhere quite written by Boltzmann in this form, certainly not in his paper of 1872[41]. However, it is clear from an article of 1877[42] that the relation between S and W was clear to him. In the first volume of Boltzmann's book on the kinetic theory[43] he revisits the argument of that report; it is there – on pp. 40 through 42 –, where he comes closest to writing $S = k \ln W$. The formula is engraved on Boltzmann's tombstone, erected in the 1930's after the full significance had been recognized, cf. Fig. 4.5. From the quotation in the figure we see that Boltzmann fully appreciated the nature of irreversibility as a trend to distributions of greater probability.

Since a given system of bodies can never by itself pass to an equally probable state, but only to a more probable one, … it is impossible to construct a *perpetuum mobile* which periodically returns to the original state.[44]

Fig. 4.5. Boltzmann's tombstone on Vienna's central cemetery

[41] L. Boltzmann: (1872) loc.cit.

[42] L. Boltzmann: „Über die Beziehung zwischen dem zweiten Hauptsatze der mechanischen Wärmetheorie und der Wahrscheinlichkeitsrechnung respektive den Sätzen über das Wärmegleichgewicht". [On the relation between the second law of the mechanical theory of heat and probability calculus, or the theories on the equilibrium of heat.] Sitzungsberichte der Wiener Akademie, Band 76, 11. Oktober 1877.

[43] L. Boltzmann: "Vorlesungen über Gastheorie I und II". [Lectures on gas theory] Verlag Metzger und Wittig, Leipzig (1895) and (1898).

[44] L. Boltzmann: „Der zweite Hauptsatz der mechanischen Wärmetheorie". [The second law of the mechanical theory of heat] Lecture given at a ceremony of the Kaiserliche Akademie der Wissenschaften on May, 29th, 1886. See also: E. Broda: "Ludwig Boltzmann. Populäre Schriften". Verlag Vieweg Braunschweig (1979) p. 26.

Boltzmann's lecture on the second law[45] closes with the words: *Among what I said maybe much is untrue but I am convinced of everything.* Lucky Boltzmann who could say that! As it was, all four bold-faced **ifs** on the forgoing pages – all seemingly essential to Boltzmann's eventual interpretation of entropy – are rejected with an emphatic **not so!** by modern physics:

- Neither is N_{xc} equal for all $(\boldsymbol{x},\boldsymbol{c})$ in $\mathrm{d}\boldsymbol{x}\mathrm{d}\boldsymbol{c}$,
- nor is it true that all $N_{xc} >> 1$,
- nor does the interchange of identical atoms lead to a new realization,
- nor is the arbitrary addition of $N!$ quite so innocuous as it might seem.

And yet, $S = k\ln W$, or the statistical probabilistic interpretation stands more firmly than ever. The formula was so plausible that it *had to be true*, irrespective of its theoretical foundation and, indeed, the formula survived – albeit with a different W – although its foundation was later changed considerably, see Chap. 6.

Reversibility and Recurrence

If Clausius met with disbelief, criticism and rejection after the formulation of the second law, the extent of that adversity was as nothing compared with what Boltzmann had to endure after he had found a positive entropy source in the kinetic theory of gases. And it did not help that Boltzmann himself at the beginning thought – and said – that his interpretation was purely mechanical. That attitude represented a challenge for the mechanicians who brought forth two quite reasonable objections

the **reversibility objection** and the **recurrence objection.**

The discussion of these objections turned out to be quite fruitful, although it was carried out with some acrimony – particularly the discussion of the recurrence objection. It was in those controversies that Boltzmann came to hammer out the statistical interpretation of entropy, i.e. the realization that S equals $k \cdot \ln W$, which we have anticipated above. That interpretation is infinitely more fundamental than the formal inequality for the entropy in the kinetic theory which gave rise to it.

The reversibility objection was raised by Loschmidt: If a system of atoms ran its course to more probable distributions and was then stopped and all its velocities were inverted, it should run backwards toward the less

[45] L. Boltzmann: (1886) loc. cit. p. 46.

probable distributions. This had to be so, because the equations of mechanics are invariant under a replacement of time t by $-t$. Therefore Loschmidt thought that a motion of the system with decreasing entropy should occur **just as often** as one with increasing entropy. In his reply Boltzmann did not dispute, of course, the reversibility of the atomic motions. He tried, however, to make the objection irrelevant in a probabilistic sense by emphasizing the importance of initial conditions. Let us consider this:

By the argument that we have used above, all realizations, or microstates occur equally frequently, and therefore we expect to see the distribution evolve in the direction in which it can be realized by more microstates, – irrespective of initial conditions; initial conditions are never mentioned in the context. This cannot be strictly so, however, because indeed Loschmidt's inverted initial conditions are among the possible ones, and they lead to less probable distributions, i.e. those with less possible realizations. So, Boltzmann[46] argues that, among all conceivable initial conditions, there are only a few that lead to less probable distributions among many that lead to more probable ones. Therefore, when an initial condition is picked at random, we nearly always pick one that leads to entropy growth and almost never one that lets the entropy decrease. Therefore the increase of entropy should occur **more often** than a decrease.

Some of Boltzmann's contemporaries were unconvinced; for them the argument about initial conditions was begging the question, and they thought that it merely rephrased the a priori assumption of equal probability of all microstates. However, the reasoning seems to have convinced those scientists who were prepared to be convinced. Gibbs was one of them. He phrases the conclusion succinctly by saying that an entropy decrease seems (!) not to be impossible but merely improbable, cf. Fig. 4.6.

Kelvin[47] had expressed the reversibility objection even before Loschmidt and he tried to invalidate it himself. After all, it contradicted Kelvin's own conviction of the universal tendency of dissipation and energy degradation, which he had detected in nature. He thinks that the inversion of velocities can never be made exact and that therefore any prevention of degradation is short-lived, – all the shorter, the more atoms are involved.

---

[46] L. Boltzmann: „Über die Beziehung eines allgemeinen mechanischen Satzes zum zweiten Hauptsatz der Wärmetheorie". [On the relation of a general mechanical theorem and the second law of thermodynamics] Sitzungsberichte der Akademie der Wissenschaften Wien (II) 75 (1877).

[47] W. Thomson: „The kinetic theory of energy dissipation" Proceedings of the Royal Society of Edinburgh 8 (1874) pp. 325–334.

... the impossibility of an uncompensated decrease of entropy seems to be reduced to an improbability.[48]

Fig. 4.6. Josiah Willard Gibbs

One of the more distinguished person who remained unconvinced for a long time was Planck. He must have felt that he was too distinguished to enter the fray himself. Planck's assistant, Ernst Friedrich Ferdinand Zermelo (1871–1953), however, was eagerly snapping at Boltzmann's heels.[49] Neither Boltzmann nor the majority of physicists since his time have appreciated Zermelo's role much; most present-day physics students think that he was ambitious and brash, – and not too intelligent; they are usually taught to think that Zermelo's objections are easily refuted. And yet, Zermelo went on to become an eminent mathematician, one of the founders of axiomatic set theory. Therefore we may rely on his capacity for logical thought.[50] And it ought to be recognized that his criticism moved Boltzmann toward a clearer formulation of the probabilistic nature of entropy and, perhaps, even to a better understanding of his own theory.

Zermelo had a new argument, because Jules Henri Poincaré (1854–1912) had proved[51] that a mechanical system of atoms, which interact with forces that are functions of their positions, must return – or *almost* return – to its

---

[48] J.W. Gibbs: "On the equilibrium of heterogeneous substances." Transactions of the Connecticut Academy 3 (1876) p. 229.

[49] To those who know the chain of command in German universities – particularly in the 19th century – it is inconceivable that Zermelo entered into a major discussion with a celebrity like Boltzmann without the approval and encouragement of his mentor Planck. Actually Planck was notoriously slow to accept new ideas, including his own, cf. Chap. 7.

[50] Later Zermelo even helped to make statistical mechanics known among physicists by editing a German translation of Gibbs's "Elementary principles of statistical mechanics", see below.

[51] H. Poincaré: „Sur le problème des trois corps et les équations de dynamique'' [On the three-body problem and the dynamical equations] Acta mathematica 13 (1890) pp.1–270. See also : H. Poincaré: "Le mécanisme et l'expérience'' [Mechanics and experience] Revue Métaphysique Morale 1 (1893) pp. 534–537.

initial position. Clearly therefore, the entropy which, after all, is a function of the atomic positions, cannot grow monotonically. This became known as the recurrence objection. Actually, Zermelo thought that the fault lay in mechanics, because he considered irreversibility to be too well established to be doubted. But he could not bring himself to accept any of Boltzmann's probabilistic arguments.[52]

In the controversy Boltzmann tried at first to get away with the observation that it would take a long time for a recurrence to occur. Zermelo agreed, but declared the fact irrelevant. The publicly conducted discussion[53,54,55,56] then focussed on Boltzmann's assertion that – at any one time – there were more initial conditions leading to entropy growth than to entropy decrease. Zermelo could not understand that assumption, and he ridiculed it. In fact, however, something possibly profound came out of the discussion. Boltzmann conceded the point – without ever admitting it in so many words (!) – when he speculated that

> ... in the universe, which is nearly everywhere in an equilibrium, and therefore dead, there must be relatively small regions of the dimensions of our star space (call them worlds) ... which, during the relatively short periods of eons, deviate from equilibrium and among these [there must be] equally many in which the probability of states increases and decreases. ... A creature that lives in such a period of time and in such a world will denote the direction of time toward less probable states differently than the reverse direction: The former as the past, the beginning, the latter as the future, the end. With that convention the small regions, worlds, will "initially" always find themselves in an improbable state.

Thus, over all *worlds* the number of initial conditions for growth and decay of entropy may indeed be equal, although in some single *world* they are not. It seems that Boltzmann believed that the universe as a whole is essentially in equilibrium, but with occasional fluctuations of the size and

---

52 Ten years later Zermelo must have reconsidered this position. In 1906 he translated Gibbs's memoir on statistical mechanics into German, and surely he would not have undertaken the task if he had still thought statistical or probabilistic arguments to be unimportant. Zermelo's translation helped to make Gibbs's statistical mechanics known in Europe.

53 E. Zermelo: "Über einen Satz der Dynamik und die mechanische Wärmelehre" [On a theorem of dynamics and the mechanical theory of heat] Annalen der Physik 57 (1896) pp. 485–494.

54 L. Boltzmann: "Entgegnung auf die wärmetheoretischen Betrachtungen des Hrn. E. Zermelo" [Reply to the considerations of Mr. E. Zermelo on the theory of heat] Annalen der Physik 57 (1896) pp. 773–784.

55 E. Zermelo: "Über mechanische Erklärungen irreversibler Vorgänge. Eine Antwort auf Hrn. Boltzmanns "Entgegnung". ["On mechanical explanations of irreversible processes. A response to Mr. Boltzmann's "reply"] Annalen der Physik 59 (1896), pp. 392–398.

56 L. Boltzmann: "Zu Hrn. Zermelos Abhandlung "Über die mechanische Erklärung irreversibler Vorgänge" [On Mr. Zermelo's treatise "On the mechanical explanation of irreversible processes"] Annalen der Physik 59 (1896) pp. 793–801.

duration of our own big-bang-world. A fluctuation will grow away from equilibrium for a while and then relax back to equilibrium. In both cases the subjective direction of time – as seen by *a creature* – is toward equilibrium, irrespective of the fact that the growing fluctuation objectively moves away from equilibrium. In order to make that mind-boggling idea more plausible, Boltzmann[57] draws an analogy to the notions of *up* and *down* on the earth: Men in Europe and its antipodes both think that they stand upright, while objectively one of them is upside down. Applied to time, however, the idea does not seem to have gained recognition in present-day physics; it is ignored – at least outside science fiction. Maybe rightly so (?).

Boltzmann tries to anticipate criticism of his daring concept of time and time reversal by saying:

> Surely nobody will consider a speculation of that sort as an important discovery or – as the old philosophers did – as the highest aim of science. It is, however, the question whether it is justified to scorn it as something entirely futile.

Actually we may suspect that Boltzmann was not entirely sincere when he made that disclaimer. Indeed, in the years to come he is on record for repeating his cosmological model several times. After having invented it in the discussion with Zermelo he repeats it, and expands on it in his book on the kinetic theory, and again in his general lecture at the World Fair in St. Louis[58].

All in all, the discussion between Boltzmann and Zermelo – despite considerable acrimony – was conducted on a high level of sophistication which definitely sets it off from the more pedestrian attempts of Maxwell and Kelvin to come to grips with randomness and probability. Those attempts involved the Maxwell demon.

Maxwell Demon

Maxwell invented the demon[59] in the effort to reconcile the irreversibility in the trend toward a uniform temperature with the kinetic theory: ... *a creature with such refined capabilities that it can follow the path of each atom.* It guards a slide valve in a small passage between two parts of a gas with – initially – equal temperatures. The demon opens and closes the valve so that it allows fast atoms from one side to pass, and slow atoms to pass

---

[57] L. Boltzmann: (1898) loc. cit. p.129.

[58] L. Boltzmann: "Über die statistische Mechanik" [On statistical mechanics] Lecture given at a scientific meeting in connection with the World Fair in St. Louis (1904). See also: E. Broda (1979) loc.cit. pp. 206–224.

[59] According to G. Peruzzi (2000) loc.cit. p. 93 f. the demon was first conceived in a letter by Tait to Maxwell in (1867). It appeared in print in Maxwell's "Theory of Heat" Longmans, Green & Co. London (1871).

from the other side. In this manner it creates a temperature difference without work because, indeed, the valve has very little mass.

The Maxwell demon was – and is – much discussed, primarily, I suspect, because it can happily be talked about by people who do not possess the slightest knowledge of mathematics. In the works of Kelvin[60] the notion reached absurd proportions: He invented ... *an army of intelligent Maxwell demons which is stationed at the interface between a cold and a hot gas and ... equipped with clubs, molecular cricket bats, as it were. ... Its mass is several times as big as the molecules ... and the demons must not leave their assigned places except when necessary to execute their orders.*

Enough of that! Brush[61] recommends an article by Klein[62] for the readers who want to familiarize themselves with *the voluminous secondary literature on Maxwell's demon.* But we shall leave the subject as quickly as possible. It has a touch of banality. We might just as well go into some belly-aching over a demon that could improve our chances in a dice game.

Boltzmann and Philosophy

There is a persistent tale that Boltzmann committed suicide in a depressed mood, created by discouragement and lack of recognition of his work. This cannot be true. To be sure, eminent people do not take kindly to criticism, and they become addicted to praise and may need it every hour of every day; but Boltzmann did get that kind of attention: He was a celebrity with an exceptional salary for the time and full recognition by all the people who counted. Even the Zermelo controversy seems to have rankled in his mind only slightly: In his essay "The Journey of a German Professor to Eldorado" Boltzmann reports good-humouredly that Felix Klein tried to push him into writing a review article on statistical mechanics by threatening to ask Zermelo to do it, if Boltzmann continued to delay.

So, no! The *neurasthenic condition* which darkened Boltzmann's life, seems more like the depressing mood that afflicts a certain percentage of the human population *normally* and which is nowadays treated effectively with certain psycho-pharmaca, vulgarly known as *happiness pills.*

It is true though that Boltzmann did not reign supreme in the scientific circles in Vienna; there was also Ernst Mach (1838–1916), a physicist of some note in gas dynamics. Mach was a thorn in Boltzmann's flesh, because he insisted that physics should be restricted to what we can see, hear, feel, and smell, or taste, and that excluded atoms. As late as 1897

---

[60] W. Thomson: (1874) loc.cit.
[61] S.G. Brush: (1976) loc.cit. p. 597.
[62] M.J. Klein: American Science 58 (1970).

Mach maintained that atoms did not exist,[63] and it is therefore clear that he had no appreciation for the kinetic theory of gases. Mach also taught philosophy and his classes were full of students eager to imbibe his tasty intellectual philosophical concoction. Boltzmann taught hard science and insisted that his students master a good deal of mathematics; consequently there were few students. That situation irritated Boltzmann, and he decided to teach philosophy himself.

He brought to the task a healthy contempt of philosophers. After Mach had retired, Boltzmann taught *Naturphilosophie* in Vienna. And in his inaugural lecture[64] he gave an account of the failure of his efforts to learn something about philosophy:

> So as to go to the deepest depths I picked up Hegel; but what an unclear, senseless torrent of words I was to find there! My bad luck conducted me from Hegel to Schopenhauer ... and even in Kant there were many things that I could grasp so little that, judging by the sharpness of his mind in other respects, I almost suspected that he was pulling the reader's leg, or even deceiving him.

For a lecture to the philosophical society of Vienna he proposed the title:

> Proof that Schopenhauer is a stupid, ignorant philosophaster, scribbling nonsense and dispensing hollow verbiage that fundamentally and forever rots people's brains.

When the organizers objected, he pointed out – to no avail – that he was merely quoting Schopenhauer, who had written these exact same words about Hegel. Boltzmann had to change the title to a tame one: *On a Thesis of Schopenhauer*,[65] but he got his own back by explaining the controversy in detail to the audience: Apparently Schopenhauer wrote that sentence about Hegel in a fit of pique, when Hegel had failed to support him for an appointment to an academic position. In contrast to that Boltzmann's intended title had been chosen out of an *objective evaluation* of Schopenhauer's work, – or so he says.

It is thus clear that Boltzmann was not an optimal choice for a teacher of conventional philosophy. His disdain for philosophy, that *doctrine of clap-trap and idle whim* was expressed frequently, with or without provocation. It is a good thing, perhaps, that Boltzmann did not also apply himself to the

---

[63] I recommend an excellent account of Boltzmann's professional work and psyche by D. Lindley: "Boltzmann's atom." The Free Press, New York, London (2001). Lindley starts his Introduction with the apodictic quotation from Mach: "I don´t believe that atoms exist."

[64] L. Boltzmann: "Eine Antrittsvorlesung zur Naturphilosophie" [Inaugural lecture on natural philosophy] Reprinted in the journal "Zeit" December 11, 1903 See also: E. Broda: loc.cit.

[65] L. Boltzmann: "Über eine These von Schopenhauer" Lecture to the Philosophical Society of Vienna, given on January 21, 1905. See also: E. Broda: loc.cit.

teaching of theology. Because indeed, his ideas in that field are again quite unconventional as the following paragraph shows.[66]

> .....only a madman denies the existence of God. However, it is true that all our mental images of God are only inadequate anthropomorphisms, so that the God whom we imagine does not exist in the shape in which we imagine him. Therefore, if someone says that he is convinced of God's existence and someone else says that he does not believe in God, maybe both think exactly the same...

Boltzmann sincerely admired Darwin's discoveries, however. Indeed, there is not a single public lecture in which he did not advertise Darwin's work. That work represents the type of natural philosophy that appealed to Boltzmann. And it is true that there is some congeniality between the two scientists in their emphasis upon the underlying randomness of either thermodynamic processes or biological evolution: The vast majority of all mutations are detrimental to the progeny, just as the vast majority of collisions in a gas lead to more disorder. In contrast to a gas, however, the small minority of advantageous mutative events is assisted by *natural selection* so that nature can create order in living organisms. Natural selection in this view plays the role of the infamous Maxwell demon, see above.[67]

Despite his partisanship for Darwin's ideas Boltzmann professes to *see nothing in his convictions that runs counter to religion.*[68]

In the last ten years of his life Boltzmann did not really do any original research, nor did he follow the research of others. Planck's radiation theory of 1900, and Einstein's works on photons, on $E = mc^2$, and on the Brownian movement – all in 1905 – passed by him. In the end his *neurasthenia* caught up with him in a summer vacation. He sent his family to the beach and hanged himself in the pension on the crossbar of a window.

---

[66] L. Boltzmann: "Über die Frage nach der objektiven Existenz der Vorgänge in der unbelebten Natur" [On the question of the objective existence of events in the inanimate nature] Sitzungsberichte der kaiserlichen Akademie der Wissenschaften in Wien. Mathematisch-naturwissenschaftliche Klasse; Bd. CVI. Abt. II (1897) p. 83 ff.

[67] Boltzmann does not seem to have argued like that. I read about this idea in one of Asimov's scientific essays. I. Asimov: "The modern demonology" in "Asimov on Physics." Avon Publishers of Bard, New York (1979).

[68] Some church leaders see this differently. So also Pope Benedikt XVI. Says he in his inaugural speech on April 24th, 2005: ... *each being is a thought of God and not the product of a blind evolutionary process.* The catholic church does not like random evolution, nor does George W. Bush, 43rd president of the United States of America, who advocates that *intelligent design* be taught in the schools of his country.

Kinetic Theory of Rubber

We have already remarked that the formula $S = k\ln W$ can be extrapolated away from monatomic gases, where it was discovered. One such extrapolation – an important one, and a particularly plausible one – occurred in the 1930's when chemists started to understand polymers and to use their understanding to develop a thermal equation of state for rubber. The *kinetic theory of rubber* is a masterpiece of thermodynamics and statistical thermodynamics, and it laid the foundation for an important modern branch of physics and technology: Polymer science.

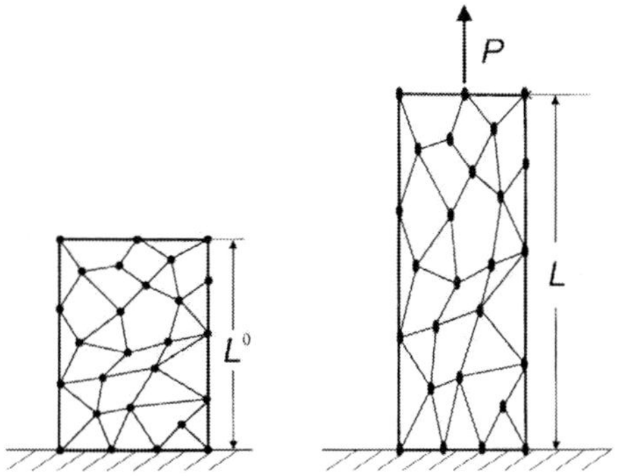

Fig. 4.7. A rubber bar in the un-stretched and stretched configurations

At the base of the theory is the Gibbs equation, see Chap. 3. In the above form the term $-p\mathrm{d}V$ represents the work done on the gas. That term must be replaced by $P\mathrm{d}L$ for a rubber bar of length L under the uni-axial load P, which depends on L and T, cf. Fig 4.7. Therefore the appropriate form of the Gibbs equation for a bar reads

$$T\mathrm{d}S = \mathrm{d}U - P\mathrm{d}L\,.$$

The Gibbs equation obviously implies

$$P = \frac{\partial U}{\partial L} - T\frac{\partial S}{\partial L}\,,$$

so that we may say that the load has an *energetic* and *entropic* part.

The integrability condition implied by the Gibbs equation reads

$$\frac{\partial U}{\partial L} = P - T\frac{\partial P}{\partial T} \quad \text{and hence follows} \quad \frac{\partial S}{\partial L} = \frac{\partial P}{\partial T}\,.$$

Therefore the entropic part of the load may be identified as the slope of the tangent of the easily measured (P,T)-curve of the bar for a fixed length L. The energetic part is determined from the ordinate intercept of that tangent, cf. Fig. 4.8.

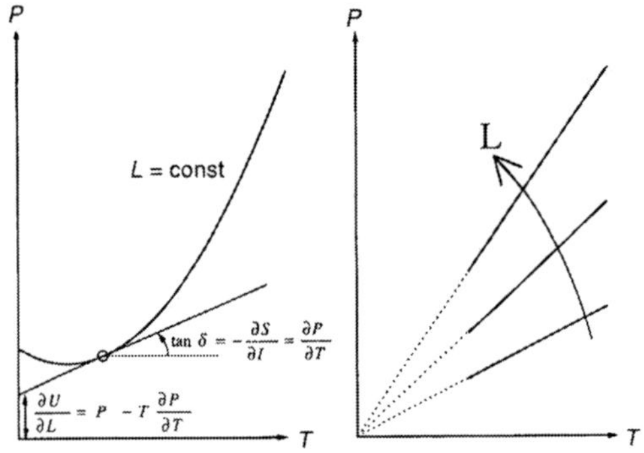

Fig. 4.8. *Left*: (load, temperature)-curve for a generic material *Right*: ditto for rubber. Tangent identifies entropic and energetic parts of force

When the (P,T)-curve is measured for rubber it turns out to be a straight line through the origin of the (P,T)-diagram. Therefore in rubber $\frac{\partial S}{\partial L}$ is independent of T, and U does not depend on L. We obtain

$$P = -T\frac{\partial S}{\partial L} \qquad \text{(for rubber)}\,,$$

a relation that is sometimes expressed by saying that rubber elasticity is *entropic,* or that the elastic force of rubber is *entropy induced*; energy plays no role in rubber elasticity.

This was first noticed by Kurt H. Meyer and Cesare Ferri [69] and they describe their discovery by saying: *L´origine de la contraction* [du caoutchouc] *se trouve dans l´orientation par la traction des chaînes polypréniques. A cette orientation s´opposent les mouvements thermiques qui provoquent finalement le retour des chaînes orientées à des positions désordinées (variation de l´entropie).*[70]

---

[69] K.H. Meyer & C. Ferri: "Sur l´élasticité du caoutchouc". Helvetica Chimica Acta 29, p. 570 (1935).

[70] The cause for the contraction of rubber lies in the orientation imparted to the polymer chains by the traction. The orientation is opposed by the thermal motion which eventually causes the return of the oriented chains to disordered positions (change of entropy).

Apart from rubber, and some synthetic polymers, entropic elasticity occurs only in gases. Indeed, different as gases and rubber may be in appearance, thermodynamically those materials are virtually identical. A joker with an original turn of mind has once commented on this similarity by saying that *rubbers are the ideal gases among the solids.*[71]

It is clear then that we need S as a function of L, if we wish to calculate the thermal equation of state $P(T,L)$ of rubber. We know that $S=k\ln W$ holds and for the calculation of W we need a model for the *chaînes desordineés,* the unordered polymer chains. Werner Kuhn (1899–1963)[72] has provided such a model by imagining the rubber molecules as chains of N independently oriented links of length b with an end-to-end distance r. Fig. 4.9 shows such a molecule and also its one-dimensional caricature, where $N_\pm$ links point to the right or left. Obviously for that simplified model – which we use here – we must have

$$\begin{aligned} N_+ + N_- &= N \\ N_+ b + N_- b &= r \end{aligned} \qquad \text{hence} \qquad N_\pm = \frac{N}{2}\left(1 \pm \frac{r}{Nb}\right).$$

The pair of numbers $\{N_+, N_-\}$ is called the *distribution of links*, and the number of possible realizations of this distribution is

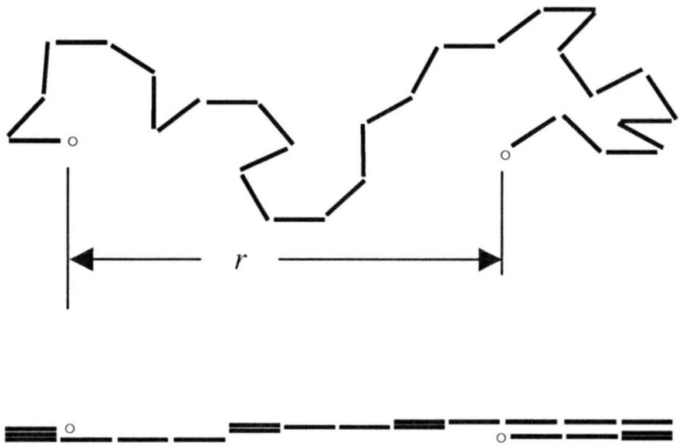

Fig. 4.9. Model for rubber molecule and its one-dimensional caricature

[71] I. Müller, W. Weiss: "Entropy and energy – a universal competition" Springer, Heidelberg (2005). In Chap. 5 of that book the analogy between thermodynamic properties of rubber and gases is highlighted by a juxtaposition.

[72] W. Kuhn: "Über die Gestalt fadenförmiger Moleküle in Lösungen" [On the shape of filiform molecules in solution] Kolloidzeitschrift 68, p. 2 (1934).

$$W = \frac{N!}{N_+!N_-!} = \frac{N!}{\left[\frac{N}{2}\left(1+\frac{r}{Nb}\right)\right]!\left[\frac{N}{2}\left(1-\frac{r}{Nb}\right)\right]!}.$$

Thus W and $S_{mol} = k\ln W$, the entropy of a molecule, are functions of the end-to-end distance r. That function may be simplified by use of the Stirling formula and by an expansion of the logarithm, viz.

$$\ln a! = a\ln a - a \qquad \text{and} \qquad \ln\left(1 \pm \frac{r}{Nb}\right) \approx \pm\frac{r}{Nb} - \frac{1}{2}\left(\frac{r}{Nb}\right)^2.$$

The former is true for large values of a, and we apply it to N as well as to $N_\pm$. The approximation of the logarithm is good for $\frac{r}{Nb} << 1$, i.e. for a strong degree of folding of the molecular chain. We obtain

$$S_{mol} = Nk\left(\ln 2 - \frac{1}{2}\left(\frac{r}{Nb}\right)^2\right),$$

so that the entropy of the molecule is maximal when its end-to-end distance is small.

The understanding of the rubber molecule does a lot for grasping the notion of entropy and its growth property, more – perhaps – than the understanding of gases. Let us consider:

The basic a priori axiom is: *Equal probability of all realizations or microstates.* Thus each and every microstate occurs just as frequently as any other one during the course of the thermal motion. This means in particular that the fully stretched-out microstate shown in Fig. 4.10 occurs just as frequently as the partially folded microstate of the figure with the end-to-end distance $r < Nb$. This means also, however, that a folded distribution $\{N_+, N_-\}$ with $r < Nb$ occurs more frequently than the fully stretched distribution $\{N,0\}$, because the former can be realized by more microstates, while the latter has only one realization.

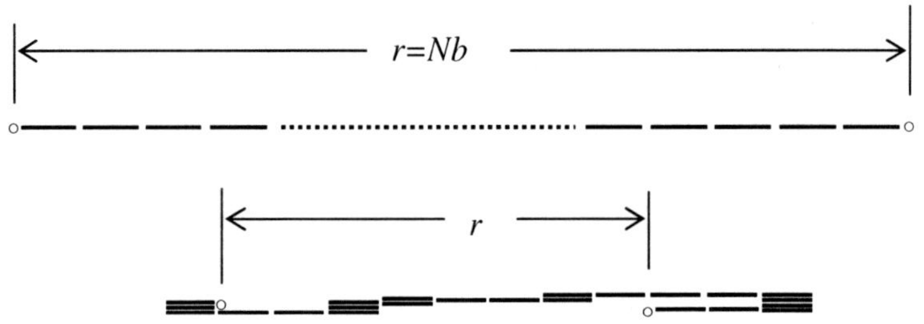

Fig. 4.10. Fully stretched and folded realizations of the chain molecule

Therefore, if the chain molecule starts out straight – with $W = 1$, i.e. $S_{mol} = 0$ – the thermal motion will very quickly mess it up, and kick the molecule into a distribution with many microstates and eventually – with overwhelming probability – into the distribution with most microstates, which we call equilibrium. In equilibrium we therefore have $N_+ = N_- = {}^1/_2$ so that r is zero. During that process the entropy S_{mol} grows from zero to $k \ln \frac{N!}{\frac{N}{2}!\frac{N}{2}!}$. Thus the entropy growth is the result of a random walk of the chain between its microstates.

Of course, we can prevent this growth. If we wish to maintain the straight microstate, – or any r in the interval $0 < r < Nb$ – we need only give the molecule a tug at the ends each time when the thermal motion kicks it. And, if the thermal motion kicks the molecule 10^{12} times per second – a reasonable number – we may apply a constant force at the ends. That is the nature of entropic forces and of entropic elasticity. And that is the nature of the force needed to keep a rubber molecule extended. If $r << Nb$, the entropy is linear in r^2, see above, and the force is proportional to r with the factor of proportionality linear in T, the temperature: The more vigorous the thermal motion is, the bigger is the entropic force. Mechanicians like to speak of the *entropic spring*; its hallmark is an elastic modulus proportional to T.

It is often said that the value of the entropy of a distribution is a measure for the disorder in the arrangement of its particles. This interpretation is most easily understood for the rubber molecule. Indeed, the stretched out, orderly distribution of Fig. 4.10 has zero entropy, because it can only be realized in one single manner. The disordered, folded distribution has positive entropy. And the most disorderly distribution with very many possible realizations has the maximum value of entropy. Therefore the growth of entropy toward equilibrium involves a growth of disorder.

Having said this I like to stress that *order* and *disorder* are not well-defined physical concepts. To be sure, in the present context the notions jibe with our intuition, but they do not always do that. Thus a cubic lattice in an alloy – judged well-ordered on intuitive grounds – has a higher entropy than the more disorderly monoclinic lattice. For that reason the cubic phase is often the high-temperature phase, because for higher temperature the entropy becomes more important in the free energy, cf. Chap. 5. This apparent violation of the equivalence of entropy and disorder can be explained, but the explanation does not employ the notion of crystallographic order or disorder.

A rubber bar consists of a network of rubber molecules all with different length vectors $(\theta_1,\theta_2,\theta_3)$ and different lengths $r = \sqrt{\theta_1^2 + \theta_2^2 + \theta_3^2}$, as shown in Fig. 4.7. Thus the entropies in the un-stretched and stretched states are given, respectively, by

$$S_0 = \int S_{mol} z_0(\theta_1,\theta_2,\theta_3) d\theta_1 d\theta_2 d\theta_3 \text{ and } S = \int S_{mol} z(\theta_1,\theta_2,\theta_3) d\theta_1 d\theta_2 d\theta_3 ,$$

where $z_0(\theta_1,\theta_2,\theta_3)d\theta_1 d\theta_2 d\theta_3$ and $z(\theta_1,\theta_2,\theta_3)d\theta_1 d\theta_2 d\theta_3$ are the numbers of distance vectors in the interval $d\theta_1 d\theta_2 d\theta_3$ at $\theta_1,\theta_2,\theta_3$.

The determination of the functions $z_0(\theta_1,\theta_2,\theta_3)$ and $z(\theta_1,\theta_2,\theta_3)$ is again due to Kuhn in 1936.[73] For the argument he ingeniously employed the inversion of $S_{mol} = k \ln W$: He assumed the number $z_0(\theta_1,\theta_2,\theta_3)d\theta_1 d\theta_2 d\theta_3$ to be proportional to the number $W = \exp\{S_{mol}/k\}$ of realizations of chains with $r = \sqrt{\theta_1^2 + \theta_2^2 + \theta_3^2}$ and obtained

$$z_0(\theta_1,\theta_2,\theta_3) = \frac{n}{\sqrt{2\pi N b^2}^3} \exp\left\{-\frac{\theta_1^2 + \theta_2^2 + \theta_3^2}{2Nb^2}\right\},$$

where n is the total number of chains. As for the number $z(\theta_1,\theta_2,\theta_3)d\theta_1 d\theta_2 d\theta_3$ Kuhn assumed that

$$z(\theta_1,\theta_2,\theta_3) = z_0\left(\frac{1}{\lambda}\theta_1, \sqrt{\lambda}\theta_2, \sqrt{\lambda}\theta_3\right)$$

holds, so that the components of the length vectors are deformed exactly as the edges of the (incompressible) rubber bar, whose deformation in the direction of the force is given by $L = \lambda L_0$.

Thus Kuhn obtained

$$S = S_0 - \frac{1}{2} nk\left(\lambda^2 + \frac{2}{\lambda} - \frac{3}{2}\right) \quad \text{and} \quad P = \frac{nkT}{L_0}\left(\lambda - \frac{1}{\lambda^2}\right).$$

The latter formula represents the thermal equation of state of a rubber bar which gives the load as a function of the temperature T and the stretch $\lambda = {}^{L}/_{L_0}$. The (P,λ)-relation is obviously non-linear.

This formula marks the beginning of polymer science[74] and the field of non-linear elasticity. Its derivation provides a deep insight into the

[73] W. Kuhn: "Beziehungen zwischen Molekülgröße, statistischer Molekülgestalt und elastischen Eigenschaften hochpolymerer Stoffe" [Relations between molecular size, statistical molecular shape and elastic properties of high polymers] Kolloidzeitschrift 76, p. 258 (1936).

[74] Modern representations of the field may be found in the monograph by P.J. Flory: "Principles of Polymer Chemistry." Cornell University Press, Ithaca (1953). The book has gone through many editions and re-printings in later years.

thermodynamic *mechanism* of polymer elasticity. That is its lasting significance, although as a quantitative description of rubber it is less than perfect, particularly for bi-axial loading.[75]

Gibbs's Statistical Mechanics

Our students are mildly bored when we use ideal gases to explain the power of statistical arguments. And indeed, the development of the atomic or statistical theory of gases in equilibrium is inextricably entwined with old, old conjectures, observations, and measurements of ideal gas properties. Thus the thermal equation of state $p = p(v,T)$ for ideal gases was known before Daniel Bernoulli explained the phenomenon of pressure in terms of moving atoms, cf. Insert 4.1. And Bernoulli's argument implied, that the internal energy of an ideal gas depends on T only, which was – much later – conjectured by Clausius and experimentally confirmed by Joule and Kelvin. In a manner of speaking everything about gases was known to the students – one way or another – before its molecular interpretation was discovered; hence the boredom.

This is not so for rubber! In that case statistical arguments have given us a thermal equation of state $P = P(L,T)$ that had *not* been known before. So, in my experience this is the point where the students perk up – those of them who are capable of such a reaction – and they demand a *da capo*: Let us maximize entropy, they might say, given by

$$S = k \ln W \quad \text{with} \quad W = \frac{N!}{\prod_{xc} N_{xc}!} ,$$

and calculate the thermal equations of state of a liquid (say), or of a metal! This is not a bad proposition but, alas, it is impractical and we cannot satisfy this reasonable request. Let us consider:

In a liquid the N atoms $\alpha = 1,2,\ldots N$ interact, – each one with all others. Therefore there is a kinetic *and* a potential energy. The latter is the sum over all pair-potentials $\Phi(|x_\beta - x_\alpha|)$. We have

$$U = \sum\nolimits_{\alpha=1}^{N} \frac{\mu}{2} c_\alpha^{\,2} + \frac{1}{2} \sum\nolimits_{\alpha,\beta=1}^{N} \Phi\left(|\, x_\alpha - x_\beta|\right) ,$$

or, in terms of N_{xc}

[75] For a discussion of rubber elasticity in some detail and a discussion of the limitations of the *Kinetic Theory of Rubber* the reader is referred to the booklet "Rubber and Rubber Balloons – Paradigms of Thermodynamics" by I. Müller and P. Strehlow, Springer Lecture Notes in Physics, Springer, Heidelberg (2005).

$$U = \sum_{xc} N_{xc} \frac{\mu}{2} c^2 + \frac{1}{2} \sum_{xc,x'c'} N_{xc} N_{x'c'} \Phi(|x - x'|).$$

Thus, when we maximize S under the constraints of fixed N and U, the equilibrium distribution N_{xc}^{equ} must be determined from the system[76]

$$-k \ln N_{xc}^{equ} - \alpha - \beta \left(\frac{\mu}{2} c^2 + \frac{1}{2} \sum_{x'c'} N_{x'c'}^{equ} \Phi(|x - x'|) \right) = 0$$

with as many equations as there are cells in (x,c)-space. These equations cannot be solved analytically for N_{xc}^{equ}. Therefore we cannot calculate equilibrium values for U and S; so we are stymied right at the beginning.

Nor did Gibbs solve the problem with statistical mechanics. However, he shifted the difficulty from the *beginning* of the argument to the *end* by expressing U and S in terms of a single function, the *partition function.* To be sure, the partition function cannot be calculated *either* in terms of the thermodynamic variables, like the volume $V$and the temperature T – except in trivial cases like the gas and the rubber – but it may sometimes be approximated.

Gibbs's statistical thermodynamics represents a daring and ingenious extrapolation of Boltzmann's ideas. Boltzmann and Maxwell had always applied probabilistic arguments to systems of identical elements: atoms in a gas, or dipoles in paramagnetic fluids or – one might add – links in a rubber chain. Gibbs proposed a giant step away from this by suggesting that

> for some purposes, however, it is desirable to take a broader view ... We may imagine a great number of systems of the same nature, but differing in the configurations and velocities which they have at a given instant, and differing not merely infinitesimally, but it may be so as to embrace every conceivable combination of configurations and velocities.[77]

The *great number of systems* was called an *ensemble* by Gibbs. He introduced different kinds of ensembles:

- An ensemble of systems with the same energy, now called microcanonical.
- An ensemble of systems with the same volume and temperature which, *on account of its unique importance in the theory of statistical equilibrium, I have ventured to call canonical.*[78]

---

[76] For gases the energy constraint is *linear* in N_{xc}, because $\Phi(|x - x'|)$ is absent, which makes things easy.

[77] J.W. Gibbs: "Elementary principles in statistical mechanics – developed with especial reference to the rational foundation of thermodynamics." Yale University Press (1902). This memoir is available as a Dover booklet, first published in 1960. My page numbers refer to the Dover publication.

[78] J.W. Gibbs: Ibidem p. XI.

- a *grand* ensemble ... composed of *h petits* ensembles[79] appropriate for mixtures of *h* constituents.

How does the ensemble-idea help? In order to see that let us concentrate on the canonical ensemble – of total energy ε and total entropy σ – of v liquids, each in a volume V, with particle number N, and all in *thermal contact,* so that they have the same temperature. Among the v imagined liquids let there be $v_{x_1....c_N}$ in the state $x_l...c_N$ with energy $U(x_1...c_N)$ such that

$$\varepsilon = \sum_{x_1...c_N} U(x_1...c_N)\, v_{x_1....c_N} .$$

The summation extends over all $x_j \in V$ and all velocities.

In a big step of extrapolation away from Boltzmann's entropy of a *gas,* Gibbs writes the entropy σ of the *ensemble* as

$$\sigma = k \ln W \text{ with } W = \frac{v!}{\prod\limits_{x_1...c_N} v_{x_1...c_N}!}$$

such that it represents the number of realizations of the distribution $v_{x_1....c_N}$.

In order to find the equilibrium distribution $v^{equ}{}_{x_1....c_N}$ he maximizes σ, cf. Insert 4.7. Thus he was able to calculate the mean energy $U = \varepsilon/v$ and the mean entropy $S = \sigma/v$ of a single liquid as

$$U = kT^2 \frac{\partial \ln P}{\partial T} \quad \text{and} \quad S = k\frac{\partial}{\partial T}(T \ln P)$$

both in terms of the partition function P with $P = \sum\limits_{x_1...c_N} \exp\left(-\frac{U(x_1...c_N)}{kT}\right)$.

Canonical ensemble

What interests us are the thermal and caloric equations of state of a single liquid of N atoms in a volume V with energy

$$U(x_1...c_N) = \sum_{xc} N_{xc} \frac{\mu}{2} c^2 + \frac{1}{2} \sum_{xc,x'c'} N_{xc} N_{x'c'} \Phi(|x - x'|) .$$

Instead we consider an ensemble of v such liquids with a fixed total energy ε. Among the v liquids let there be $v_{x_1....c_N}$ in the state $x_l...c_N$ such that – ignoring the interaction between the liquids – we have

[79] J.W. Gibbs: Ibidem p. 190.

$$\varepsilon = \sum_{x_1 \ldots c_N} U(x_1 \ldots c_N)\, \nu_{x_1 \ldots c_N}.$$

The entropy σ of the ensemble is calculated as

$$\sigma = k \ln W \quad \text{with} \quad W = \frac{\nu!}{\prod\limits_{x_1 \ldots c_N} \nu_{x_1 \ldots c_N}!}.$$

Summation and product extend over all positions $x_i \in V$ and all velocities. The equilibrium distribution $\nu_{x_1 \ldots c_N}$ is *the one* that maximizes σ under the constraints of constant ε and ν, namely the *canonical distribution*

$$\nu_{x_1 \ldots c_N} = \nu \frac{\exp(-\beta U(x_1 \ldots c_N))}{\sum\limits_{x_1 \ldots c_N} \exp(-\beta U(x_1 \ldots c_N))}.$$

Hence follows for the energy and the equilibrium entropy of the ensemble

$$\varepsilon = \nu \frac{\partial \ln P}{\partial \beta} \quad \text{and} \quad \sigma = \nu k \left(\beta \frac{\varepsilon}{\nu} + \ln P \right),$$

where $P = \sum\limits_{x_1 \ldots c_N} \exp(-\beta U(x_1 \ldots c_N))$ is the partition function. β is the Lagrange multiplier that takes care of the energy constraint.

By equipartition the energy of each atom must be equal to $\frac{3}{2} kT$ and that helps us to identify β as $\frac{1}{kT}$, because we have

$$\varepsilon_{kin} = \sum_{x_1 \ldots c_N} \nu^{equ} \left(\frac{\mu}{2} c_1^{\,2} + \ldots + \frac{\mu}{2} c_N^{\,2} \right)_{x_1 \ldots c_N} = \nu \cdot N \frac{3}{2} \frac{1}{\beta}.$$

Thus the mean energy and entropy of a single liquid is

$$U = \frac{\varepsilon}{\nu} = kT^2 \frac{\partial \ln P}{\partial T} \quad \text{with} \quad P = \sum_{x_1 \ldots c_N} \exp\left(-\frac{U(x_1 \ldots c_N)}{kT} \right)$$

$$S = \frac{\sigma}{\nu} = k \frac{\partial}{\partial T}(T \ln P) \quad \text{hence the free energy} \quad F = U - TS:$$

$$F = -kT \ln P.$$

Insert 4.7

Thus Gibbs arrived at a final result, after a fashion, even though it is quite impossible for liquids – and most other non-trivial systems – to evaluate the sum in the partition function and obtain $P(V,T)$ explicitly.

However, the problem is reduced to the evaluation of a multiple sum. In that form it represents a challenge for mathematicians, and one may think of making intelligent approximations. In fact

- For liquids J.E. Mayer and M.G. Mayer developed a cumbersome but effective *cluster method* to approximate the thermal equation of state of a *real gas*[80]
- Lars Onsager was able to evaluate the partition function exactly for the *Ising model* of a ferromagnet, although I believe that the success was restricted to a two-dimensional case,
- Recently Oliver Kaster[81] has approximated the partition function of a shape memory alloy and was able to simulate the austenitic ↔ martensitic phase transition that is typical for such alloys.

So Gibbs's idea proved to be quite useful. Conceptually, however, there are problems: We may very well conceive of ensembles, of course. But in actual fact we have a single liquid – never an ensemble. So, how do we argue in order to get the ensemble out of our minds and concentrate on the single liquid? The conventional idea is that the ensemble does no more than provide the individual liquids with a temperature. From that thought it is a simple conceptual step to forget the ensemble entirely, and replace it by a *heat bath* for the one and only liquid in our laboratory.

Gibbs did not address such lingering misgivings nor do most books on statistical mechanics.[82] A notable exception is Schrödinger in a written account of thoughtful seminars.[83] Says he:

> ....here the ν identical systems are mental copies of the one system under consideration – of the one macroscopic device that is actually erected on our laboratory table. Now what on earth could it mean, physically, to distribute a given amount of energy ε over these ν mental copies? The idea is in my view, that you can, of course, imagine that you really had ν copies of your system, that they really were in "weak interaction" with each other, but isolated from the rest of the world. Fixing your attention on one

---

80 J.E. Mayer, M.G. Mayer: "Statistical Mechanics." John Wiley & Sons, New York (1940) Chap. 13.

81 O. Kastner: "Zweidimensionale molekular-dynamische Untersuchung des Austenit ↔ Martensit Phasenübergangs in Formgedächtnislegierungen." [Two-dimensional molecular dynamics of the austenite↔martensite phase transition in shape memory alloys] Dissertation TU Berlin, Shaker Verlag (2003).

82 In modern books on the subject it is not uncommon to have the partition function appear on the first half-page, and the rest of the book is given to its evaluation in special cases. That is what is known as the *deductive approach*, or *understanding by doing*.

83 E. Schrödinger: "Statistical thermodynamics. A course of seminar lectures." Cambridge at the University Press (1948).

of them, you find it in a peculiar kind of "heat bath" which consists of the $\nu - 1$ others.

The only treatment of a proper and realistic *ensemble,* known to me, is due to Maxwell in his paper "On Boltzmann's theorem on the average distribution of energy in a system of material points" [84] Maxwell considers a gas of $\nu \cdot N$ atoms and – in imagination – he splits it into ν gases of N atoms. Then he proceeds to determine the distribution[85]

$$\nu^{equ}_{x_1 \ldots c_N} = \frac{1}{V^N} \frac{1}{\sqrt{2\pi\mu kT}^{3N}} \exp\left(-\frac{\mu c_1^2 + \cdots + \mu c_N^2}{2kT} \right),$$

which is the canonical distribution for the case. For N=1 – one single atom – Maxwell thus recovers the Maxwell distribution, which he had derived originally in two different manners, see above. Maxwell is acknowledged by Gibbs – along with Clausius and Boltzmann – as *one of the principal founders of statistical mechanics.*

Another question which Gibbs took in his stride – without much ado – concerns the *mean value over the ensemble*: What is the significance of that mean value for the single liquid under consideration? The answer is given by the *ergodic hypothesis.* This implies that the number $\nu^{equ}_{x_1 \ldots c_N}$ calculated for the ensemble of ν liquids is also the frequency of the state $x_1 \ldots c_N$ in a single liquid, if that liquid is observed ν times at sufficiently large intervals. The hypothesis is often expressed by saying

ensemble average = expectation value for single liquid[86]

so that the average over the imagined ensemble is immediately relevant for the one and only system under consideration. Obviously, the prescription for the calculation of the time average – or expectation value – can only be relevant for equilibria.

In the wording of arguments and in the formulae I have so far concentrated on liquids. This was for definiteness and suggestiveness only. Statistical mechanics of other bodies follows the same lines. One of the more amazing applications[87] is a *single hydrogen atom,* a proton with one electron which may occupy $2n^2$ orbits (n = 1,2,...) with energies

---

[84] J.C. Maxwell: (1879) loc.cit.

[85] This paper of 1879 thus contains Maxwell's third derivation of the Maxwell distribution. We have reviewed the other two derivations above.
Maxwell's third derivation is now a popular exercise for physics students, because it provides them with the opportunity to acquaint themselves with volumes and surface areas of spheres in many dimensions.

[86] A trivial illustration is this: Suppose that on an aerial photo of a city you identify the fraction of cars which drive with 50 km/h. Next, consider that you drive a car yourself for some long time *randomly* through the city and register the fraction of seconds that your speed is 50 km/h. The ergodic hypothesis implies that the two fractions are equal.
Mathematicians have tried to prove the ergodic hypothesis and their efforts have led to a branch of set theory, the *ergodic theory*. That theory, however, offers little to the physicist.

[87] I found this simple problem in the book by J.D. Fast: "Entropie. Die Bedeutung des Entropiebegriffs und seine Anwendung in Wissenschaft und Technik." [Entropy. The

$$E_n = \frac{2\pi^2 e^4 \mu}{(4\pi\varepsilon_0 h)^2}\left(1 - \frac{1}{n^2}\right) = 2.171 \cdot 10^{-18}\,\mathrm{J}\left(1 - \frac{1}{n^2}\right)$$

according to Bohr's model of atomic structure.[88] In the jargon developed in statistical mechanics we place the atom in a heat bath of temperature T and form its partition function

$$P = \sum_{n=1}^{\infty} 2n^2 \exp\left(-\frac{E_n}{kT}\right).$$

Hence follows the entropy and the free energy of the atom

$$S = k\frac{\partial}{\partial T}\left(T\ln\left(\Sigma_{n=1}^{\infty} 2n^2 \exp\left(\frac{E_n}{kT}\right)\right)\right) \quad \text{and} \quad F = -kT\ln\left(\Sigma_{n=1}^{\infty} 2n^2 \exp\left(\frac{E_n}{kT}\right)\right).$$

For any *normal* earthly temperature only the first term with $n = 1$ contributes appreciably to the sum so that we have

$$S = k\ln 2 \quad \text{and} \quad F = -kT\ln 2.$$

The 2 in these equations represents the two possibilities in which the electron may realize the energy zero: *spin up* and *spin down.*

Writing this I am reminded of an exchange between two eminent thermodynamicists at a conference, which I attended as a young man. One, a Nobel prize winner – call him P – emphatically opposed the other one – let him be called T – for having applied statistical mechanics to a single atom. The discussion culminated in this dialogue:

> P: Your application is not permissible and, if you had read my book carefully, you would know it.
>
> T: I read your book more carefully than you wrote it, and …

The rest of the answer was lost in an outbreak of hilarity in the audience.

Other Extrapolations. Information

The interpretations of entropy as a number of realizations of a distribution and as a measure of order and disorder have led to extrapolations of the concept to fields other than gases. We have already discussed the case of rubber properties and we shall later discuss the power of the *entropy of*

---

meaning of the concept of entropy and its application in science and technology] Philips's Technische Bibliothek (1960). Also available in Dutch, English, French, and Spanish.

[88] The energy of the ground state is set equal to zero. e and μ are the electric charge and mass of the electron respectively; $h = 6.625 \cdot 10^{-34}$ Js is the Planck constant, and ε_0 is the vacuum di-electricity.

mixing in mixtures. Both applications belong to main-stream thermodynamics. However, there are also fairly esoteric extrapolations, popular among physicists affecting sensitivity for the unusual – and there are many of those.

Sometimes such arguments come along as challenges, like when it is pointed out that a great piece of literature – usually *Hamlet*, or *Faust,* no less (!) – is obviously highly ordered in comparison to a random distribution of its words, or letters. It should therefore have a small entropy, and so Shakespeare, or Goethe must have defeated the universal tendency of entropy to increase. In this case the challenge is: How did the poets do that, and where is the inevitable overall increase of entropy to offset the decrease effected by the dramas? No serious answer is available!

And then, there is information theory, invented in 1948 by Claude Elwood Shannon (1916–2001). Shannon[89] put a number on a message which somehow represents its informational value, cf. Insert 4.8. The expression for the calculation of the number can look – under certain circumstances – like Boltzmann's entropy $S = k \ln W$ with $W = \frac{N!}{\prod_{xc}^{P} N_{xc}!}$.

And so Shannon called his number the *entropy of the message.* There is a story about this which is reported by Denbigh[90]:

> When Shannon had invented his quantity and consulted von Neumann on what to call it, von Neumann replied: *Call it entropy. It is already in use under that name and besides, it will give you a great edge in debates because nobody knows what entropy is anyway.*

No doubt Shannon and von Neumann thought that this was a funny joke, but it is not, – it merely exposes Shannon and von Neumann as intellectual snobs. Indeed, it may sound philistine, but a scientist *must* be clear, – as clear as he can be –, and avoid wanton obfuscation at all cost. And if von Neumann had a problem with entropy, he had no right to compound that problem for others – students and teachers alike – by suggesting that entropy had anything to do with information.

Shannon's information

If a message consists of a single "*sign*" a which naturally occurs with the probability $p(a)$, Shannon calls its information value – or simply *information* –

$$Inf = \log_2 \frac{1}{p(a)} \text{ bit.}$$

[89] C.E. Shannon: "A mathematical theory of communication." Bell Systems Technology Journal 27, (1948), pp. 379–423, 623–657.

[90] K. Denbigh: "How subjective is entropy?". In: "Maxwell's demon, entropy, information, computing." H.S. Leff, A.F. Rex (eds.) Rrinceton University Press (1990) pp. 109–115.

The smaller the probability of the "*sign*" the more information we gain by receiving it.

1 bit is the unit of information; it stands for **b**inary **i**ndissoluble information uni**t**. The name stems from simple cases when the message a has the probability $(^1/_2)^n$ so that it can be identified by n successive alternatives, *binary decisions,* with the probability $^1/_2$ each.

An example occurs when we draw a card from a stack of 32 with {7, 8, 9, 10, knave, queen, king, ace}. We may then give out messages about our card like this: "black" $p = {}^1/_2$, or "spades" $p = {}^1/_4$, or "spades unnumbered" $p = {}^1/_8$, or "spades with queen or king" $p = {}^1/_{16}$, or "queen of spades" $p = {}^1/_{32}$. The corresponding informations come out as 1 bit for "black" through 5 bit for "queen of spades". They are *higher* for the less probable "sign" and, when the "*sign*" is least probable, as the queen of spades is, the information is complete, i.e. the card is fully identified. The predilection for the *dual* logarithm is due to the fact that we want integers as information for this simple case, – or Shannon did.

The logarithm itself is chosen so that information is additive, when a message consists of several (independent) signs $a_1, a_2, \ldots, a_n$ (say) with probability $p(a_1)$ $p(a_2)\ldots p(a_n)$. In that case we have

$$Inf = \left(\sum_{i=1}^{n} \log_2 \frac{1}{p(a_i)} \right) \text{bit},$$

and if the sign a_i occurs N_i times in the message – with $\sum_{i=1}^{n} N_i = N$ – we obviously obtain

$$Inf = \left(\sum_{i=1}^{n} N_i \log_2 \frac{1}{p(a_i)} \right) \text{bit} \cdot$$

This is the expression called *entropy* by Shannon.

If the probability $p(a_i)$ of sign a_i is equal to the relative frequency N_i/N of its occurrence – as may perhaps happen in very long messages – we obtain

$$Inf = -\left(\sum_{i=1}^{n} N_i \log_2 \frac{N_i}{N} \right) \text{bit} \quad \text{or} \quad Inf = \left(\log_2 \frac{N!}{\sum_{i=1}^{n} N_i!} \right) \text{bit},$$

where the Stirling formula has been used for the last step. Thus the analogy to Boltzmann's entropy is complete to within a multiplicative constant.

If we wish, we can now assign an entropy to the message which Shakespeare sent us when he wrote *Hamlet*: We look up the probability of each letter a_i of the English alphabet, count how often they occur in *Hamlet* and calculate *Inf.* People do that and we may suppose that they know why.

Insert 4.8

For level-headed physicists entropy – or order and disorder – is nothing by itself. It has to be seen and discussed in conjunction with temperature and heat, and energy and work. And, if there is to be an extrapolation of entropy to a foreign field, it must be accompanied by the appropriate extrapolations of temperature and heat and work. Lacking this, such an extrapolation is merely at the level of the following graffito, which is supposed to illustrate the progress of western culture to more and more disorder, i.e. higher entropy:

Hamlet: to be or not to be
Camus: to be is to do
Sartre: to do is to be
Sinatra: do be do be do be do

Ingenious as this joke may be, it provides no more than amusement.

5 Chemical Potentials

It is fairly seldom that we find resources in the form in which we need them, which is as pure substances or, at least, strongly enriched in the desired substance. The best known example is water: While there is some sweet water available on the earth, salt water is predominant, and that cannot be drunk, nor can it be used in our machines for cooling (say), or washing. Similarly, natural gas and mineral oil must be refined before use, and ore must be smelted down in the smelting furnace. Smelting was, of course, known to the ancients – although it was not always done efficiently – and so was distillation of sea water which provided both, sweet water and pure salt in one step, the former after re-condensation. Actually, in ancient times there was perhaps less scarcity of sweet water than today, but – just like today – there was a large demand for hard liquor that had to be distilled from wine, or from other fermented fruit or vegetable juices.

The ancient distillers did a good enough job since time immemorial, but still their processes of separation and enrichment were haphazard and not optimal, since the relevant thermodynamic laws were not known.

The same was largely true for chemical reactions, when two constituents combine to form a third one (say), or when the constituents of a compound have to be separated. Sometimes heating is needed to stimulate the reaction and on other occasions the reaction occurs spontaneously or even explosively. The chemists – or alchemists – of early modern times knew a lot about this, but nothing systematic, because chemical thermodynamics – and chemical kinetics – did not yet exist.

Nowadays it is an idle question which is more important, the thermodynamics of energy conversion or chemical thermodynamics. Both are essential for the survival of an ever growing humanity, and both mutually support each other, since power stations need fuel and refineries need power. Certainly, however, chemical thermodynamics – the thermodynamics of mixtures, solutions and alloys – came late and it emerged in bits and pieces throughout the last quarter of the 19th century, although Gibbs had formulated the comprehensive theory in one great memoir as early as 1876 through 1878.

Josiah Willard Gibbs (1839–1903)

Gibbs led a quiet, secluded life in the United States, which during the 19th century was as far from the beaten track as Russia.[1] As a postdoctoral fellow Gibbs had had a six year period of study in France and Germany, before he became a professor of mathematical physics at Yale University, where he stayed all his life. His masterpiece "On the equilibrium of heterogeneous substances" was published in the "Transactions of the Connecticut Academy of Sciences"[2] by reluctant editors, who knew nothing of thermodynamics and who may have been put off by the size of the manuscript – 316 pages! The paper carries Clausius's triumphant slogan about the energy and entropy of the universe as a motto in the heading, see Chap. 3, but it extends Clausius's work quite considerably.

The publication was not entirely ignored. In fact, in 1880 the American Academy of Arts and Sciences in Boston awarded Gibbs the Rumford medal – a legacy of the long-dead Graf Rumford. However, Gibbs remained largely unknown where it mattered at the time, in Europe.

Friedrich Wilhelm Ostwald (1853–1932), one of the founders of physical chemistry, explains the initial neglect of Gibbs's work: Only partly, he says, is this due to the small circulation of the Connecticut Transactions; indeed, he has identified what he calls an *intrinsic handicap* of the work: … *the form of the paper by its abstract style and its difficult representation demands a higher than usual attentiveness of the reader*. And it *is* true that Gibbs wrote overlong sentences, because he strove for maximal generality and total un-ambiguity, and that effort proved to be counterproductive to clarity of style. However, it is also true that the concepts in the theory of mixtures, with which Gibbs had to deal, are somewhat further removed from everyday experience – and bred-in perspicuity – than those occurring in single liquids and gases.

Ostwald translated Gibbs's work into German in 1892, and in 1899 le Chatelier translated it into French. Then it turned out that Gibbs had anticipated much of the work of European researchers of the previous decades, and that he had in fact gone far beyond their results in some cases. Ostwald encourages researchers to study Gibbs's work because … *apart from the vast number of fruitful results which the work has already provided, there are still hidden treasures.* Gibbs revised Ostwald's translation but … *lacked the time to make annotations, whereas the translator* [Ostwald] *lacked the courage.*[3]

---

[1] I. Asimov: "Biographies …" loc.cit.

[2] J.W. Gibbs: Vol III, part 1 (1876), part 2 (1878).

[3] So Ostwald in the foreword of his translation: "Thermodynamische Studien von J. Willard Gibbs" [Thermodynamic studies by J. Willard Gibbs] Verlag W. Engelmann, Leipzig (1892).

Those translations made Gibbs known. His work came to be universally recognized, and in 1901 he received the Copley medal of the Royal Society of London. In 1950 – nearly fifty years after his death – he was elected a member of the Hall of Fame for Great Americans.

The greatest achievement, perhaps, of Gibbs is the discovery of the chemical potentials of the constituents of a mixture. The chemical potential of a constituent is representative for the presence of that constituent in the mixture in much the same way as temperature is representative for the presence of *heat*. I shall explain as we go along.

While evolution has provided us, the human race, with a good sensitivity for temperature, it has done less well with chemical potentials. To be sure, our senses of smell and taste can discern foreign admixtures to air or water, but such observations are at a low level of distinctness. Therefore the thermodynamic laws of mixtures have to be learned *intellectually* – rather than intuitively – and Gibbs taught us how this is best done.

Because of that it seems impossible to explain Gibbs's work – and to do it justice – without going into some technicalities. Nor is it possible to relegate all the more technical points into *Inserts*. Therefore I am afraid that parts of this chapter may read more like pages out of a textbook than I should have liked.

Entropy of Mixing. Gibbs Paradox

Chemical thermodynamics deals with mixtures – or solutions, or alloys – and the first person in modern times who laid down the laws of mixing, was John Dalton again, the re-discoverer of the atom, see Chap. 4. Dalton's law, as we now understand it, has two parts.

The first one is valid for all mixtures, or solutions, and it states that, in equilibrium, the pressure p of the mixture and the densities of mass, energy and entropy of the mixture are sums of the respective partial quantities appropriate for the constituents. If we have v constituents, indexed by α = 1,2,…v, we may thus write

$$p = \sum_{\alpha=1}^{v} p_\alpha,\ \rho = \sum_{\alpha=1}^{v} \rho_\alpha(T, p_\beta),\ \rho u = \sum_{\alpha=1}^{v} \rho_\alpha u_\alpha(T, p_\beta),\ \rho s = \sum_{\alpha=1}^{v} \rho_\alpha s_\alpha(T, p_\beta).$$

The second part of Dalton's law refers to ideal gases: If we are looking at a mixture of ideal gases, the partial quantities ρ_α, u_α, and s_α depend on T and on only *their own* p_α, and, moreover, the dependence is the same as in a single gas, i.e. cf. Chap. 3

$$p_\alpha = \rho_\alpha \frac{k}{\mu_\alpha} T, \quad u_\alpha = u_\alpha(T_R) + z_\alpha \frac{k}{\mu_\alpha}(T - T_R), \quad \text{and}$$

$$s_\alpha = s_\alpha(T_R, p_R) + (z_\alpha + 1)\frac{k}{\mu_\alpha} \ln \frac{T}{T_R} - \frac{k}{\mu_\alpha} \ln \frac{p}{p_R}.$$

A typical mixing process is indicated in Fig. 5.1, where ν single constituents under the pressure p and at temperature T are allowed to mix after the opening of the connecting valves. When the mixing is complete, the volume, internal energy and entropy of the mixture may be different from their values before mixing. We write

$$V = \sum_{\alpha=1}^{\nu} V_\alpha + V_{Mix}, \quad U = \sum_{\alpha=1}^{\nu} U_\alpha + U_{Mix}, \quad S = \sum_{\alpha=1}^{\nu} S_\alpha + S_{Mix}$$

and thus we identify the *volume, internal energy* and *entropy of mixing*.

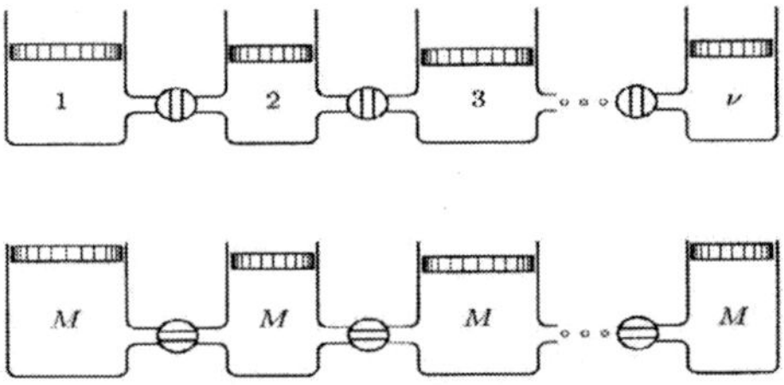

Fig. 5.1. Pure constituents at T, p before mixing (*top*). Homogeneous mixture at T, p (*bottom*). Note that the volume may have changed during the mixing process

For ideal gas mixtures V_{Mix} and U_{Mix} are both zero and S_{Mix} comes out as

$$S_{Mix} = -k \sum_{\alpha=1}^{\nu} N_\alpha \ln \frac{N_\alpha}{N},$$

where N_α is the number of atoms of gas α and $N = \sum_{\alpha=1}^{\nu} N_\alpha$. By Avogadro's law – and, of course, by the thermal equation of state $p_\alpha = \rho_\alpha \frac{k}{\mu_\alpha} T$ – the numbers N_α are independent of the nature of the gases. Therefore the entropy of mixing is the same, irrespective of the gases that are being mixed. This is an observation due to Gibbs and the *Gibbs paradox*[4] is closely related to it: If the *same* gas fills all volumes at the beginning, the situation before and after opening of the valves is the same one, and yet the entropies should differ, since the entropy of mixing does

[4] J.W. Gibbs: loc.cit. pp. 227–229.

not depend on the nature of the gases, but only on their number of atoms or molecules.

The Gibbs paradox persists to this day. The simplicity of the argument makes it mind-boggling. Most physicists think that the paradox is resolved by quantum thermodynamics, but it is not! Not, that is, as it has been described above, namely as a proposition on the equations of state of a mixture and its constituents as formulated by Dalton's law.[5]

Gibbs himself attempted to resolve the paradox by discussing the possibility of un-mixing different gases, and the impossibility of such an un-mixing process in the case of a single gas. It is in this context that Gibbs pronounced his often-quoted dictum: ... *the impossibility of an uncompensated decrease of entropy seems to be reduced to an improbability,* see Fig. 4.6. Gibbs also suggested to imagine mixing of different gases which are more and more alike and declared it noteworthy that the entropy of mixing was independent of the degree of similarity of the gases. None of this really helps with the paradox, as far as I can see, although it provided later scientists with a specious argument. Thus Arnold Alfred Sommerfeld (1868–1951) [6] pointed out that gases are inherently distinct and that there is no way to make them gradually more and more similar. Then Sommerfeld quickly left the subject, giving the impression that he had said something relevant to the Gibbs paradox which, however, is not so, – or not in any way that I can see.

Homogeneity of Gibbs Free Energy for a Single Body

So far, when we have discussed the trend toward equilibrium, or the increase of disorder, or the impending heat death, we might have imagined that equilibrium is a homogeneous state in all variables. The truth is, however, that indeed, temperature T and pressure p[7] are homogeneous in equilibrium, but the mass density is *not*, or not necessarily. What *is* homogeneous are the fields of temperature, pressure and *specific Gibbs free*

---

[5] The easiest way to deal with a paradox is to maintain that it does not exist, or does not exist anymore. The Gibbs paradox is particularly prone to that kind of solution, because it so happens that a superficially similar phenomenon occurs in statistical thermodynamics. That statistical paradox was based on an incorrect way of counting realizations of a distribution, and it *has indeed been resolved* by quantum statistics of an ideal gas, cf. Chap. 6. It is easy to confuse the two phenomena.

[6] A. Sommerfeld: „Vorlesungen über theoretische Physik, Bd. V, Thermodynamik und Statistik" [Lectures on theoretical physics, Vol. V. Thermodynamics and Statistics] Dietrich'sche Verlagsbuchhandlung, Wiesbaden, 1952 p. 76.

[7] Pressure is only homogeneous in equilibrium in the absence of gravitation.

energy $u - Ts + pv$.[8] The specific Gibbs free energy is usually abbreviated by the letter g and it is also known as the *chemical potential*,[9] although that name is perhaps not quite appropriate in a single body.

We proceed to show briefly how, and why, this unlikely combination – at first sight – of u,s,v with T and p comes to play a central role in thermodynamics: We know that the entropy S of a closed body with an impermeable and adiabatic surface *at rest* tends to a maximum, which is reached in equilibrium. The interior of the body may at first be in an arbitrary state of non-equilibrium with turbulent flow (say) and large gradients of temperature and pressure. While the body approaches equilibrium, its mass m and energy $U + E_{kin}$ are constant, because of the properties of the surface. In order to find necessary conditions for equilibrium we must therefore maximize S under the constraints of constant m and $U + E_{kin}$. If we take care of the constraints by Lagrange multipliers λ_m and λ_E, we have to find the conditions for a maximum of

$$\int_V \rho s \,\mathrm{d}V - \lambda_m \int_V \rho \,\mathrm{d}V - \lambda_E \int_V \rho(u + \tfrac{1}{2}v^2)\mathrm{d}V .$$

The specific values s and u of entropy and internal energy are assumed to satisfy the Gibbs equation locally:[10]

$$T\mathrm{d}s = \mathrm{d}u + p\mathrm{d}v \quad \text{or, equivalently} \quad T\,\mathrm{d}(\rho s) = \mathrm{d}(\rho u) - g\mathrm{d}\rho .$$

Since u is a function of T and ρ, the variables in the expression to be maximized are the values of the fields $T(\boldsymbol{x})$, $v_i(\boldsymbol{x})$, and $\rho(\boldsymbol{x})$ at each point $\boldsymbol{x}$. By differentiation we obtain the necessary conditions for *thermodynamic* equilibrium in the form

$$v_i = 0, \qquad \text{and}$$

$$\begin{aligned} &\frac{\partial \rho s}{\partial T} - \lambda_E \frac{\partial \rho u}{\partial T} = 0 \\ &\frac{\partial \rho s}{\partial \rho} - \lambda_m - \lambda_E \frac{\partial \rho u}{\partial \rho} = 0 \end{aligned} \qquad \text{hence with the Gibbs equation}: \qquad \begin{aligned} &\frac{1}{T} = \lambda_E \\ &g = -T\lambda_m \end{aligned} .$$

Therefore in thermodynamic equilibrium the body is at rest throughout V, and T and $g = u - Ts + pv$ are homogeneous. This is what we have set out to show. The homogeneity of the pressure p follows from the momentum

---

[8] $v = {}^1/\rho$ is the specific volume.

[9] On the European continent g is also called the *specific free enthalpy*.

[10] This assumption is known as the *principle of local equilibrium* since – as we recall – the Gibbs equation holds for reversible processes, i.e. a succession of equilibria. Gibbs accepts this principle remarking that it requires the *changes of type and state* of mass elements to be small.

balance because, when the motion has stopped, the condition of *mechanical* equilibrium reads $\frac{\partial p}{\partial x_i} = 0$.

One might be tempted to think that, since u, s, and v – and hence g – are all functions of T and p, the homogeneity of g should be a corollary of the homogeneity of T and p, – and therefore not very exciting. But this is not necessarily so, since $g(T,p)$ may be a different function in different parts of the body. Thus one part may be a liquid, with $g'(T,p)$, and another part may be a vapour with $g''(T,p)$. Both phases have the same temperature, pressure and specific Gibbs free energy in equilibrium, but very different values of u, s, and v, i.e., in particular, very different densities. And since the values of $g'(T,p)$ and $g''(T,p)$ are equal, there is a relation between p and T in phase equilibrium: That relation determines the vapour pressure in phase equilibrium as a function of temperature; it may be called the thermal equation of state of the saturated vapour or the boiling liquid.

Gibbs Phase Rule

A very similar argument provides the equilibrium conditions for a mixture. To be sure, in a mixture the local Gibbs equation cannot read

$$T\mathrm{d}(\rho s) = \mathrm{d}(\rho u) - g\mathrm{d}\rho\,,$$

as it does in a single body, because s and u may generally depend on the densities ρ_α of all constituents rather than only on ρ. Accordingly, one may write

$$T\mathrm{d}(\rho s) = \mathrm{d}(\rho u) - \sum_{\alpha=1}^{\nu} g_\alpha \mathrm{d}\rho_\alpha\,;$$

the g_α's may be thought of as partial Gibbs free energies, but Gibbs called them *potentials* and nowadays they are called *chemical potentials.*[11] Obviously they are functions of T and ρ_β ($\beta = 1,2\ldots\nu$). Let us consider their equilibrium properties.

Thermodynamic equilibrium means – as in the previous section – a maximum of S, now under the constraints

$$m_\alpha = \int_V \rho_\alpha \mathrm{d}V \ (\alpha = 1,2\ldots\nu), \quad \text{and} \quad U + E_{kin} = \int_V \left(\rho u + \sum_{\alpha=1}^{\nu} \frac{\rho_\alpha}{2} v_\alpha^{\,2}\right)\mathrm{d}V$$

in a volume with an adiabatic impermeable surface at rest.

[11] The canonical symbol for the chemical potential of constituent α, introduced by Gibbs, is μ_α. I choose g_α instead, since μ_α already denotes the molecular mass. Moreover, the symbol g_α emphasizes the fact that the chemical potential g_α is the specific Gibbs free energy of constituent α in a mixture.

As before we take care of the constraints by Lagrange multipliers λ_m^{α} and λ_E and obtain as necessary conditions for thermodynamic equilibrium

$$v_e^{\alpha} = 0, \quad \text{and} \quad \frac{1}{T} = \lambda_E, \quad \text{and} \quad g_\alpha = -T\lambda_m^{\alpha}.$$

Thus in thermodynamic equilibrium all constituents are at rest, and T, and all g_α ($\alpha = 1,2,\ldots\nu$) are homogeneous throughout V. The pressure p is also homogeneous; as before, this is a condition of mechanical equilibrium.

And once again – just like in the previous section – if the body in V is all in one phase, liquid (say), the homogeneity of T and g_α means that all densities ρ_α are homogeneous. However, if there are f spatially separated phases, indexed by $h = 1,2\ldots f$, the homogeneity of g_α implies

$$g_\alpha^{\,h}(T,\rho_\alpha^{\,h}) = g_\alpha^{\,f}(T,\rho_\alpha^{\,f}) \qquad (\alpha = 1,2,\ldots\nu), \quad (h = 1,2,\ldots f-1)$$

so that the chemical potentials of all constituents have equal values in all phases. This condition is known as the *Gibbs phase rule.*

Since the pressure p is also equal in all phases, so that $p = p(T, \rho_\alpha^{\,h})$ holds for all h, the Gibbs phase rule provides $\nu(f\text{-}1)$ conditions on $f(\nu-1)+2$ variables. That leaves us with $F = \nu - f + 2$ independent variables, or *degrees of freedom* in equilibrium.[12] In particular, in a single body the coexistence of three phases determines T and p uniquely, so that there can only be a triple *point* in a (p,T)-diagram. Or, two phases in a single body can coexist along a *line* in the (p,T)-diagram, e.g. the vapour pressure curve, see above, Inserts 3.1 and 3.7. Further examples will follow below.

Law of Mass Action

If a single-phase body within the impermeable adiabatic surface at rest is already at rest itself and homogeneous in all fields T and ρ_α, the Gibbs equation may be written – upon multiplication by V – as

$$T\,\mathrm{d}S = \mathrm{d}U - \sum_{\alpha=1}^{\nu} g_\alpha \mathrm{d}m_\alpha .$$

While such a body is in mechanical and thermodynamic equilibrium, it may not be in equilibrium *chemically.* In chemical reactions, with the stoichiometric coefficients $\gamma_\alpha^{\,a}$, the masses m_α can change in time according to the mass balance equations[13]

[12] Sometimes this corollary of the Gibbs phase rule is itself known by that name.

[13] Often, or usually, there are several reactions proceeding at the same time; they are labelled here by the index a, ($a = 1,2\ldots n$). n is the number of independent reactions. There is some arbitrariness in the choice of independent reactions, be we shall not go into that.

$$m_\alpha(t) = m_\alpha(0) + \sum_{a=1}^{n} \gamma_\alpha{}^a \mu_\alpha R^a(t),$$

so that the *extents* R^a of the reactions determine the masses of all constituents during the process. And in equilibrium the masses m_α assume *the* values that maximize S under the constraint of constant U. We use a Lagrange multiplier and maximize S-$\lambda_E U$, which is a function of T and R^a. Thus we obtain necessary conditions of chemical equilibrium, viz.

$$\frac{\partial S}{\partial T} - \lambda_E \frac{\partial U}{\partial T} = 0 \qquad \text{hence} \qquad \frac{1}{T} = \lambda_E$$

$$\sum_{\alpha=1}^{\nu}\left(\frac{\partial S}{\partial m_\alpha} - \lambda_E \frac{\partial U}{\partial m_\alpha}\right)\gamma_\alpha{}^a \mu_\alpha = 0 \quad \text{hence} \quad \boxed{\sum_{\alpha=1}^{\nu} g_\alpha \gamma_\alpha{}^a \mu_\alpha = 0,\ (a = 1,2\ldots n).}$$

The framed relation is called the *law of mass action.* It provides as many relations on the equilibrium values of m_α as there are independent reactions.

Gibbs's fundamental equation

In a body with homogeneous fields of T and ρ_β the local Gibbs equation $T\mathrm{d}(\rho s) = \mathrm{d}(\rho u) - \sum_{\alpha=1}^{\nu} g_\alpha \mathrm{d}\rho_\alpha$ holds in all points and, if we consider slow changes of volume V – reversible ones, so that the homogeneity is not disturbed –, we obtain by multiplication by V

$$T\mathrm{d}S = \mathrm{d}U - \rho\Big(u - Ts - \sum_{\alpha=1}^{\nu} g_\alpha \frac{\rho_\alpha}{\rho}\Big)\mathrm{d}V - \sum_{\alpha=1}^{\nu} g_\alpha \mathrm{d}m_\alpha\,.$$

In a closed body, where $\mathrm{d}m_\alpha = 0$, $(\alpha = 1,2\ldots\nu)$ holds, we should have $T\mathrm{d}S = \mathrm{d}U + p\mathrm{d}V$ and this requirement identifies p so that we may write

$$u - Ts + \frac{p}{\rho} = \sum_{\alpha=1}^{\nu} g_\alpha \frac{\rho_\alpha}{\rho} \quad \text{and hence} \quad \underline{T\mathrm{d}S = \mathrm{d}U + p\,\mathrm{d}V - \sum_{\alpha=1}^{\nu} g_\alpha \mathrm{d}m_\alpha}\,.$$

Alternatively for the whole homogeneous body we have

$$G = \sum_{\alpha=1}^{\nu} g_\alpha m_\alpha \qquad \text{hence} \qquad \underline{\mathrm{d}G = -S\mathrm{d}T + V\mathrm{d}p - \sum_{\alpha=1}^{\nu} g_\alpha \mathrm{d}m_\alpha}\,.$$

The first one of these relations is called the *Gibbs-Duhem relation* and the underlined differential forms are two versions of the *Gibbs fundamental equation*; they accommodate all changes in a homogeneous body, including those of volume and of all masses m_α. However, the last two equations imply

$$\sum_{\alpha=1}^{\nu} m_\alpha \mathrm{d}g_\alpha = -S\,\mathrm{d}T + V\,\mathrm{d}p\,,$$

so that $g_\alpha(T,p,m_\beta)$ can only depend on such combinations of m_β that are invariant under multiplication of the body by any factor; they may depend on the concentrations $c_\beta = \rho_\beta/\rho$ for instance, or on the mol fractions $X_\beta = N_\beta/N$.

If we know all chemical potentials $g_\alpha(T,p,m_\beta)$ as functions of all variables, we may use the Gibbs-Duhem relation to determine the Gibbs free energy $G(T,p,m_\beta)$ of the mixture and hence, by differentiation, $S(T,p,m_\beta)$, $V(T,p,m_\beta)$, and finally $U(T,p,m_\beta)$.

The integrability conditions implied by Gibbs's fundamental equation viz.

$$\frac{\partial g_\alpha}{\partial m_\varepsilon} = \frac{\partial g_\varepsilon}{\partial m_\alpha}, \quad \frac{\partial g_\alpha}{\partial T} = -\frac{\partial S}{\partial m_\alpha}, \quad \frac{\partial g_\alpha}{\partial p} = \frac{\partial V}{\partial m_\alpha}$$

help in the determination of the chemical potentials $g_\alpha(T,p,m_\beta)$.

Insert 5.1

Semi-Permeable Membranes

The above framed relations, – the Gibbs phase rule, and the law of mass action – are given in a somewhat synthetic form, because they are expressed in terms of the chemical potentials g_α. What we may want, however, are predictions about the masses m_α in chemical equilibrium, or the mass densities ρ_α^h of the constituents in phase equilibrium. For that purpose it is obviously necessary to know the functional form of $g_\alpha(T,p,m_\beta)$. In general there is no other way to determine these functions than to measure them. So, how can chemical potentials be measured?

An important, though often impractical, conceptual tool of thermodynamics of mixtures is the semi-permeable membrane. This is a wall that lets particles of some constituents pass, while it is impermeable for others. One may ask what is continuous at the wall, and one may be tempted to answer, perhaps, that it is the partial densities ρ_α of those constituents that can pass, or their partial pressures p_α. However, we know already that the answer is different: In general it is neither of the two; rather it is the chemical potentials $g_\alpha(T,p,m_\beta)$.

This knowledge gives us the possibility – in principle – to measure the chemical potentials: Let a wall be permeable for only one constituent γ (say). Then we can imagine a situation in which we have that constituent in pure form on side I of the wall at a pressure p^{I}, while there is an arbitrary mixture – including γ – on side II under the pressure p^{II}. We thus have in thermodynamic equilibrium

$$g_\gamma(T,p^{\mathrm{I}}) = g_\gamma(T,p^{\mathrm{II}},m_\beta^{\mathrm{II}}) .$$

The Gibbs free energy $g_\gamma\ (T,p^{\mathrm{I}}) = u_\gamma(T,p^{\mathrm{I}}) - Ts_\gamma(T,p^{\mathrm{I}}) + pv_\gamma(T,p^{\mathrm{I}})$ of the single, or pure constituent γ can be calculated – to within a linear function of T – because $u_\gamma(T,p)$, and $s_\gamma(T,p)$, and $v_\gamma(T,p)$ can be measured and calculated, the former two to within an additive constant each, see Chap. 3.[14] Thus a value of $g_\gamma(T,p,m_\beta)$ can be determined for one given (ν+2)-tupel $(T,p^{\mathrm{II}},m_\beta^{\mathrm{II}})$. Changing these variable we may – in a laborious process indeed – experimentally determine the whole function $g_\gamma(T,p,m_\beta)$.

In real life this is impossible for two reasons: First of all, measurements like these would be extremely time-consuming, and expensive to the degree of total impracticality. Secondly, in reality we do not have semi-permeable walls for all substances and all types of mixtures or solutions. Indeed, we have them for precious few only.

But still, *imagining* that we had semi-permeable membranes for every substance and every mixture, we can conceive of a hypothetical *definition* of the chemical potential g_γ as *the* quantity that is continuous at a γ-permeable membrane. In that sense the kinship of chemical potentials and temperature is put in evidence: Temperature measures how hot a body is and the chemical potential g_γ measures how much of constituent γ is in the body. Both measurements are made from outside, by contact.

On Definition and Measurement of Chemical Potentials

However, Gibbs's definition of chemical potentials has nothing to do with semi-permeable membranes. He writes[15]

> Definition. – Let us suppose that an infinitely small mass of a substance is added to a homogeneous mass, while entropy and volume are unchanged; then the quotient of the increase of energy and the increase of mass is the *potential* of this substance for the mass under consideration.

Obviously this definition is read off from the fundamental equation

$$T\mathrm{d}S = \mathrm{d}U + p\,\mathrm{d}V - \sum_{\alpha=1}^{\nu} g_\alpha \mathrm{d}m_\alpha$$

and Gibbs blithely ignores the fact that the *increase of energy* is unknown before we have calculated it from the knowledge of the chemical potentials $g_\gamma(T,p,m_\beta)$.

This is the same type of logical somersault, which also *defines* temperature as $(\frac{\partial U}{\partial S})_V$, and which ignores the fact that $U(S,V)$ is unknown before we

---

[14] All it takes for that is (p,V,T)-measurements and measurements of $c_v(T,v_0)$ for *one* v_0.

[15] J.W. Gibbs: loc.cit. p. 149.

have calculated it from measurements that involve temperature measurements. I have done my best to discredit this procedure before, cf. Chap. 3.

Having said this and having seen that the implementation of semi-permeable membranes – although logically sound – is strongly hypothetical, we are left with the problem of how to determine the chemical potentials. There is no easy answer and no pat solution; rather there is a thorny process of guessing and patching and extrapolating away from ideal gas mixtures.

Indeed, for ideal gases we know everything from Dalton's law, see above. In particular we know the Gibbs free energy explicitly as

$$G=\sum_{\alpha=1}^{\nu} m_\alpha\left[u_\alpha(T_R)+z_\alpha\frac{k}{\mu_\alpha}(T-T_R)-T\left(s_\alpha(T_R,p_R)+(z_\alpha+1)\frac{k}{\mu_\alpha}\ln\frac{T}{T_R}-\frac{k}{\mu_\alpha}\ln\frac{p_\alpha}{p_R}\right)+\frac{k}{\mu_\alpha}T\right]$$

$$G=\sum_{\alpha=1}^{\nu} m_\alpha\left[u_\alpha(T_R)+z_\alpha\frac{k}{\mu_\alpha}(T-T_R)-T\left(s_\alpha(T_R,p_R)+(z_\alpha+1)\frac{k}{\mu_\alpha}\ln\frac{T}{T_R}-\frac{k}{\mu_\alpha}\ln\frac{p}{p_R}\right)+\frac{k}{\mu_\alpha}T+\frac{k}{\mu_\alpha}T\ln X_\alpha\right]$$

The last term represents the entropy of mixing, see above. By the fundamental equation we thus obtain the prototype of all chemical potentials, viz.

$$g_\alpha(T,p,m_\beta)=\frac{\partial G}{\partial m_\alpha}=g_\alpha(T,p)+\frac{k}{\mu_\alpha}T\ln X_\alpha,$$

where $g_\alpha(T,p)$ is the specific Gibbs free energy of the single ideal gas α at T and p; $X_\alpha - N_\alpha/N$ is the mol fraction of constituent α. So, in this special case of ideal gases we *may indeed* use the Gibbs definition, because we do know the functional form of $G(T,p,m_\alpha)$, which generally, we do not know.

And yet, this specific form has become the prototypical expression for chemical potentials, considered applicable sometimes even for solutions and alloys. To be sure, in those cases $g_\alpha(T,p)$ are the Gibbs free energies of the single liquids or solids, respectively, rather than of the single gases. Originally that extrapolation was a wild guess, made by van't Hoff and born out of frustration, perhaps. When the guess turned out to give reasonable results occasionally, – often for dilute solutions – the expression was admitted, and nowadays, if valid, it is said to define an *ideal mixture;* such a mixture may be gaseous, liquid, or solid.

But, even when our mixture, or solution, or alloy is not ideal, the ideal-gas-expression still serves as a reference: The departure from ideality is represented by correction factors γ_α or φ_α and we write

$$g_\alpha(T,p,m_\beta)=g_\alpha(T,p)+\frac{k}{\mu_\alpha}T\ln(\gamma_\alpha X_\alpha),\qquad \text{or}$$

$$g_\alpha(T,p,m_\beta)=g_\alpha(T,p_\alpha(T))+\frac{k}{\mu_\alpha}T\ln\left(\frac{p}{p_\alpha(T)}\varphi_\alpha X_\alpha\right).$$

The former is primarily used for liquid solutions, because the *activity coefficient* $\gamma_\alpha(T,p,m_\beta)$, if it is different from 1, represents the deviation from an ideal solution. The latter expression is mostly used for vapours, because the *fugacity coefficient* $\varphi_\alpha(T,p,m_\beta)$, if it is different from 1, represents the deviation of the vapour from a mixture of ideal gases; $p_\alpha(T)$ is the vapour pressure of the single constituent α.

We shall not go further into this matter. Suffice it to say that an army of chemical engineers are busy determining activity coefficients and fugacity coefficients, and they lay down their results in books of tables. Their tools are varied. They use semi-permeable membranes whenever they exist, otherwise they use temperature measurements of incipient boiling and condensation, and occasionally they use the integrability conditions for the chemical potentials, mentioned in Insert 5.1. Their task is important, but their life is hard. It is worlds removed from the lofty positions of the theoreticians who think that they have understood thermodynamics when they have understood the properties of monatomic gases.[16]

Osmosis

Although good semi-permeable membranes are rare, there are some efficient ones, for water particularly. Wilhelm Pfeffer (1845–1920), a botanist, experimented with them. He invented the Pfeffer tube which is sealed with a water-permeable membrane [17] at one end and stuck – with that end – into a water reservoir, cf. Fig. 5.2. The water level will then be equal in tube and reservoir. Afterwards some salt is dissolved in the water of the tube; the membrane is impermeable for the sodium ion Na^+ and the chloride Cl^- into which the salt dissociates upon solution. One observes that the solution in the tube rises, because water pushes its way into the tube in a process called *osmosis.*[18] For reasonable data, viz.

2 litre reservoir, $1\ \text{cm}^2$ tube diameter, 1 g salt, $T = 298$ K, $p = 1$ atm

the solution in the tube rises to a height of nearly 10 m (!).[19]

---

[16] These practical people have their own pride in their work though, and rightly so: They like to ridicule the theoreticians as suffering from *argonitis.*

[17] A ferro cyan copper membrane.

[18] The Greek word *osmos* means *to push.*

[19] The Pfeffer tube is nowadays a popular show piece in high-school laboratories. The solution does usually not reach its full height during the lab session.

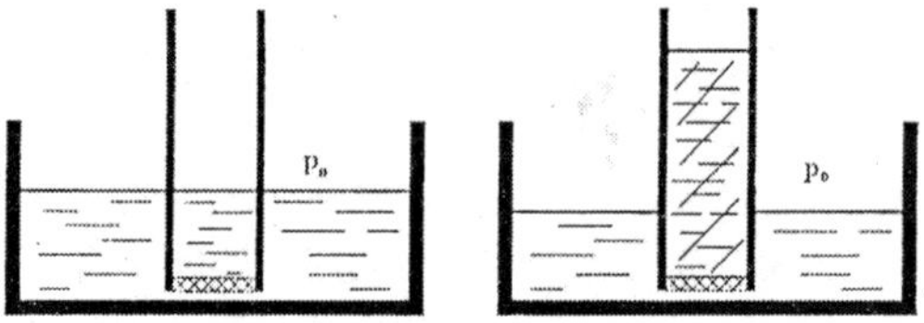

Fig. 5.2. Pfeffer tube

After equilibrium is established, the membrane has to support a considerable pressure difference, the *osmotic pressure* $P = p^{II} - p^{I}$.

Pfeffer reported his experiments in 1877, just in the middle of the two-year-period when Gibbs published the two parts of his great paper. Had Pfeffer known Gibbs's work, he could have written a formula for the calculation of the pressure p^{II} on top of the membrane, namely

$$g_{\text{Water}}(T,p^{I}) = g_{\text{Water}}(T,p^{II},m_{\text{Na+}},m_{\text{Cl-}},m_{\text{Water}}{}^{II})$$

and, of course, he would have had to know the functions g_{Water} in order to calculate p^{II} or, in fact, to calculate the osmotic pressure $P = p^{II} - p^{I}$.

As it was, Pfeffer did no calculations at all, nor did he present any formulae. However, he knew how to measure the osmotic pressure and he noticed that – given the mass of the solute – the pressure decreased with the size of the dissolved molecules. Being a botanist he dissolved organic macro-molecules, like proteins, and he was thus the first person to make some reasonably reliable measurements on the size of giant molecules.[20]

It is not by accident that it was a botanist who concerned himself with semi-permeable membranes. Plants and animals make extensive use of osmotic phenomena in order to transport substances, often water, through cell boundaries, and life would be impossible without them.

Thus the roots of trees lie in the ground water and their surface membranes are permeable for the water. The water can therefore dilute the nutritious sap inside the roots and, at the same time, push it upwards through the ducts that lead from the roots to the tree tops. It has been estimated that in a tree this osmotic effect can overcome a height difference of 100 m.

In animals and humans the cell boundaries are also permeable for water and the osmotic pressure across the membranes of blood cells amounts to 7.7 bar (!). Therefore the cells would burst, if we injected a patient with pure water. The fluid in the *drips* fixed to hospital beds is a salt solution – 8.8g per litre water – which balances the osmotic pressure in the cell by exerting itself a counter-pressure of 7.7 bar. The solution is known as the *physiological salt solution*; physicians say that it is *isotonic* to the contents of the cell.

[20] I. Asimov: "Biographies...." loc.cit p. 441.

Dilute solutions are analogous to ideal gases in some respect. At least that was the hypothesis made by Jacobus Henricus van't Hoff (1852–1911), a chemist of note and physical chemist, who was the first Nobel prize winner in chemistry in 1901. Van't Hoff assumed that the molecules of $\nu - 1$ solutes move freely in a solution much in the same way as gas molecules move through empty space. Thus the osmotic pressure of a solution on a semi-permeable membrane – permeable for the solvent ν – should be given by

$$P = \sum_{\alpha=1}^{\nu-1} \rho_\alpha \frac{k}{\mu_\alpha} T ,$$

as if it were the pressure of a mixture of ideal gases. That relation is known as *van't Hoff's law.*

Van't Hoff's suggestion met with heavy disapproval among more conservative chemists; but then he produced experimental evidence and it turned out that the law was sometimes true. Van't Hoff published it in 1886 and, of course, he had been anticipated – at least partly – by Gibbs. Indeed the continuity of the chemical potential of the solvent ν across the semi-permeable membrane, and the assumption of an ideal solution reads, according to Gibbs, see above

$$g_\nu(T, p') = g_\nu(T, p'') + \frac{k}{\mu_\nu} T \ln X_\nu .$$

If the single solvent is incompressible, with ρ_ν as density, $g_\nu(T,p)$ is a linear function of p with $^1/\rho_\nu$ as coefficient, and if the solution is dilute, we have

$$\ln X_\nu \approx - \sum_{\alpha=1}^{\nu-1} \frac{N_\alpha}{N_\nu^S} ,$$

where N_ν^S is the number of solvent molecules in the solution. Thus one obtains for the osmotic pressure

$$P = p^{II} - p^{I} = \frac{\rho_\nu}{\rho_\nu^S} \sum_{\alpha=1}^{\nu-1} \rho_\alpha \frac{k}{\mu_\alpha} T .$$

The ratio of ρ_ν and ρ_ν^S, the density of the solvent in the solution, is very nearly equal to 1 in a dilute solution, so that van't Hoff's law emerges from Gibbs's thermodynamics, at least approximately.

Having said this, I must qualify: One can easily become over-enthusiastic ascribing discoveries to Gibbs. It is true that Gibbs had the general rule about the continuity of the chemical potential. Also he had the form of the chemical potential in a mixture of ideal gases. But he did not conceive of ideal mixtures other than mixtures of ideal gases so that he could not get as far as van't Hoff's law for dilute solutions.

Van't Hoff's extrapolation of ideal gas properties to solutions must have seemed a wild guess to himself and his contemporaries, and it seemed quite properly to be a dubious assumption to the chemical establishment. But it was also a *lucky* guess and the question is why? The answer, or at least a good motivation, can be found in Boltzmann's molecular interpretation of entropy, cf. Insert 5.2.

Entropy of mixing in a solution

We have seen that the specific term $\frac{k}{\mu_\nu} T \ln X_\nu$ comes from the entropy of mixing of ideal gases, namely

$$S_{Mix} = -k \sum_{\alpha=1}^{\nu} N_\alpha \ln \frac{N_\alpha}{N}.$$

But then the entropy has a molecular interpretation, see Chap. 4, and we may consider S_{Mix} in the present case as $k \ln W$, where W is the increase in the number of realizations during the mixing process. Assuming a homogeneous distribution $\{N_{\boldsymbol{x}}\}$ of particles at position $\boldsymbol{x}$ in V after mixing, and homogeneous distributions $\{N_{\boldsymbol{x}}^{\alpha}\}$ in V_α before mixing we have

$$S_{Mix} = \left(k \ln \frac{N!}{\prod_{x \in V} N_x!} - \sum_{\alpha=1}^{\nu} \ln \frac{N^\alpha!}{\prod_{x \in V_\alpha} N_x^\alpha!} \right), \quad \text{with } N = \frac{N}{XV} \text{ and } N_x^\alpha = \frac{N^\alpha}{XV_\alpha},$$

where X is the factor of proportionality between the number of positions in V and V itself.[21] It follows by use of the Stirling formula that we have

$$S_{Mix} = -k \ln \sum_{\alpha=1}^{\nu} N_\alpha \ln \frac{V_\alpha}{V}.$$

V/V_α is equal to N/N_α in gases but not necessarily in liquids, unless the particles of all constituents are equal in size. With this *proviso* Boltzmann's interpretation of entropy supports the entropy of mixing of ideal mixtures.

Insert 5.2

Raoult's Law

Francois Marie Raoult (1830–1901) was one of the founders of physical chemistry. He observed experimentally that – in liquid-vapour phase equilibrium of a mixture – the partial pressure of a vapour constituent is

[21] Recall this kind of *quantization* in Boltzmann's arguments, see above. Since X drops out at the end, the argument may be considered as a calculational auxiliary.

proportional to the mol fraction of that constituent in the solution. Obviously, if this is true, we must have [22]

$$p_\alpha'' = X_\alpha' \, p_\alpha(T) \, ,$$

where $p_\alpha(T)$ is the saturation vapour pressure of the single constituent α.

Therefore *carbonated* mineral water – water with CO_2 in solution – is kept in the bottle under CO_2-pressure; upon opening the bottle we hear the hiss when the gas escapes and we see the CO_2-bubbles that are released by the water under the lowered CO_2-pressure.

If the vapour is an ideal gas mixture under the pressure p, we have $p_\alpha'' = X_\alpha'' \, p$ and thus we obtain *Raoult's law*

$$X_\alpha'' \, p = X_\alpha' p_\alpha \, (T) \qquad (\alpha = 1, 2 \ldots \nu) \, .$$

Raoult found this law in 1886 and he was lucky indeed to find it at all, because there are few solutions which satisfy this law. The exploitation of Gibbs's phase rule for two phases, viz.

$$g_\alpha''(T, p, m_\beta'') = g_\alpha'(T, p, m_\beta')$$

reveals the conditions under which the law is valid:

- the solution must be ideal,[23]
- the liquid constituents must be incompressible,
- the vapour must be a mixture of ideal gases,
- the vapour densities must be much smaller than liquid densities.

However, when Raoult's law *is* valid and when it is applied to a binary system, the two equations allow the calculation of X_1' and X_1'' – hence $X_2' = 1 - X_1'$ and $X_2'' = 1 - X_1''$ – as functions of p, when T is prescribed. Usually these functions are plotted inversely as $p(X_1'; T)$ and $p(X_1''; T)$. The analytic form of Raoult's law then reads

$$p = p_2(T) + (p_1(T) - p_2(T)) X_1' \quad \text{and} \quad p = \frac{p_2(T)}{1 - \left(1 - \frac{p_2(T)}{p_1(T)}\right) X_1''}$$

and the graphs are shown in Fig. 5.3_{left}. That figure represents the prototype of all *(p, X_1)-phase-pressure-diagrams* with separate boiling and condensation lines and the two-phase-region in-between. The diagram is drawn for the case that constituent 1 is the *high-boiling liquid* and constituent 2 is the *low-boiler*: As single liquids they boil at high and low temperatures respectively.

---

[22] As on some occasions before we characterize the liquid by a prime and the vapour by a double-prime.

[23] CO_2 dissolved in water does *not* form an ideal solution. Therefore the above discussion of mineral water must be taken *with a grain of salt.*

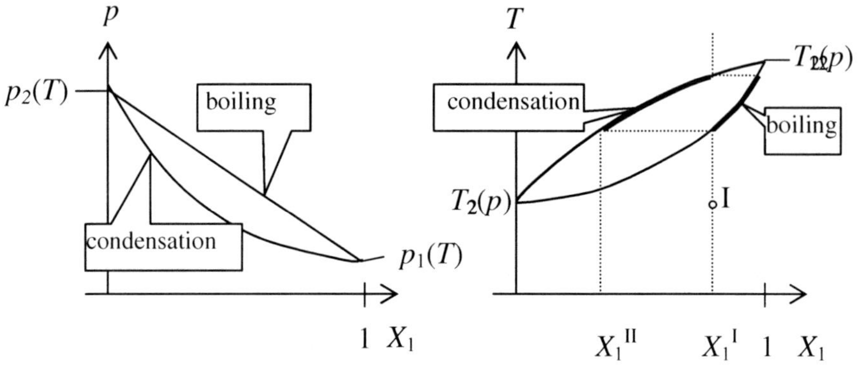

Fig. 5.3. *Left*: Phase-pressure-diagram. *Right*: Phase-diagram

If the equations are solved for T – at fixed p –, we obtain the curves $T(X_1';p)$ and $T(X_1'';p)$, which may be plotted in the (T,X_1)-*phase-diagram*, albeit not in analytic form, since the vapour pressure functions $p_\alpha(T)$ are not known analytically. Fig. 5.3_{right} shows a (T,X_1)-phase-diagram qualitatively

Diagrams of this type are important tools for the chemical engineer and for the metallurgist, because they provide them with the knowledge needed for enriching solutions or alloys in one of their constituents, or even to separate the constituents.[24] Let us consider this:

We start at point I in Fig. 5.3_{right} with a feed-stock solution of mol fraction X_1^{I} – as it was found or provided – and at low temperature, where the liquid prevails. Then we increase T until the boiling line is reached. The vapour that is formed there has the mol fraction X_1^{II}, i.e. it is enriched in constituent 2. Consequently the boiling liquid grows richer in constituent 1. At the new composition the solution needs to be hotter for boiling and at the higher temperature the new vapour is not quite so rich in constituent 2 as the old one, but still richer than $X_2^{\text{I}} = 1 - X_1^{\text{I}}$. When the process of evaporation continues, the state of the remaining solution moves upwards along the boiling line and the state of the vapour moves upwards along the condensation line until X_1^{I} is reached in the vapour, and the solution is all used up. Further heating will only make the vapour hotter at constant X_1.

The clever chemical engineer interrupts the process at an intermediate point and comes away with a 2-rich vapour and a 1-rich liquid. Both may serve his purpose.

If we wish to separate both constituents completely, the feed-stock solution must be fed into a *rectifying column* consisting of many levels of

[24] Metallurgists are dealing with alloys, and solid-melt equilibria. The thermodynamics of solutions and alloys is nearly identical despite the different appearances of those substances. To be sure, neither melts nor solids are much affected by pressure and therefore metallurgists prefer the (T,X_1)-diagram over the (p,X_1)-diagram.

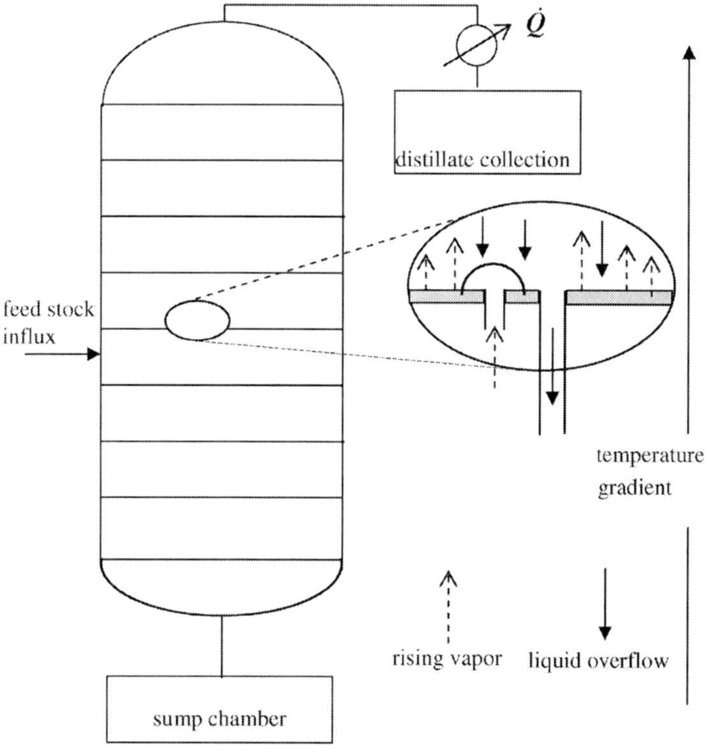

Fig. 5.4. Schematic view of a rectifying column

boiling liquid, cf. Fig. 5.4.[25] The vapour rising from the feed level is led through the liquid solution on top and there it condenses partially, primarily of course the high-boiling constituent. After passing through several – or many – such levels, the vapour arrives at the top, where it contains essentially only the low-boiling constituent. That vapour is condensed in a cooler which it leaves as a virtually pure liquid constituent, the distillate. Similarly the liquid solution, enriched in the high-boiling constituent by the partial vapour condensation, overflows the rim of its level and drops into the solution of the next lower level, enriching it in the high-boiling constituent beyond the degree of enrichment that was the result of the evaporation. After several such steps the liquid at the bottom level becomes nearly pure in the high-boiling constituent and is led out. In the stationary process the liquid at each level is boiling at the temperature appropriate to its composition.

[25] In the jargon of chemical engineering *to rectify* means *to purify,* or to separate into constituents as pure as possible. The process in a rectifying column is also called suggestively *distillation by reverse circulation.*

Rectifying columns are up to 30m high, 5m in diameter and may contain 30 levels. Unfortunately the method does not work well for complex multi-constituent solutions like mineral oil. For such solutions one has to be content with obtaining certain *fractions* like benzine, petroleum, or heavy benzine, etc. which are not pure substances, but pure enough for efficient use in automobiles (say).

The rectifying column represented in Fig. 5.4 and similar modern designs are developments of engineers working in the chemical industry and trying hard to optimise the process for output and energy consumption. The process itself of rectification by distillation, however, is age-old. So old in fact, that no inventor can be identified. To be sure, whoever the inventor was, he was not concerned with mineral oil. Rather he worked in order to satisfy the pressing need – of himself and others – for high percentage *hard liquor*, such as brandy, whiskey, gin, rum, grappa and the likes. This requires separation of alcohol from water by boiling fermented fruit juices or grain mash, and then condensing it. The process was – and is – carried on in distilleries, vulgarly known as *stills*.

Alternatives of the Growth of Entropy

One of the sections in Gibbs's memoir is entitle: "On the quantities ψ,χ and ζ" [26] and in that section Gibbs explains what happens to a body when its surface is not adiabatic and at rest. We proceed to discuss that point.

We know from Clausius that the entropy of a body with an adiabatic surface ∂V grows, and if the body reaches an equilibrium, the entropy is maximal. That is the case, for instance, when the adiabatic surface is at rest, so that the energy $U + E_{\text{kin}}$ is constant. The question arises, however, what happens when the surface is *not* adiabatic, or when it is *not* at rest, or both. The easy answer is, that in such cases generally equilibrium will not be approached.

However, that is too pat for an answer. There *are* special boundary conditions – other than adiabaticity and rest – for which equilibrium *can* be approached and some of them may be characterized as follows:

- Homogeneous and constant temperature T_0 on ∂V and body at rest there,
- adiabatic boundary ∂V and homogeneous and constant pressure there,
- homogeneous and constant temperatures T_0 and pressure p_0 on ∂V.

We refer to Chap. 4 and recall the equations of balance of energy and entropy

---

[26] J.W. Gibbs: loc.cit. p. 144.

Energy:[27] $$\frac{\mathrm{d}(U+E_{kin})}{\mathrm{d}t} = -\int_{\partial V} q_i n_i \mathrm{d}A - \int_{\partial V} p v_i n_i \mathrm{d}A$$

Entropy: $$\frac{\mathrm{d}S}{\mathrm{d}t} \geq \int_{\partial V} \frac{q_i n_i}{T} \mathrm{d}A.$$

It is then fairly obvious that under the three stipulated sets of conditions we obtain

- $\frac{\mathrm{d}(U+E_{kin}-T_o S)}{\mathrm{d}t} \leq 0,$
- $\frac{\mathrm{d}(U+E_{kin}+p_o V)}{\mathrm{d}t} = 0$ and $\frac{\mathrm{d}S}{\mathrm{d}t} \geq 0,$
- $\frac{\mathrm{d}(U+E_{kin}-T_o S+p_o V)}{\mathrm{d}t} \leq 0.$

This means that

- $U + E_{kin} - T_o S \rightarrow$ minimum for T_0 constant and ∂V at rest,
- $S \rightarrow$ maximum for an adiabatic surface,
- $U + E_{kin} - T_o S + p_0 V \rightarrow$ minimum for T_0 constant and p_0 constant.

The first and last conditions are alternatives of the growth of entropy, appropriate for the stipulated conditions. The validity of these trends toward equilibrium is independent of how far the body is away from equilibrium; indeed, initially the process in ∂V may be characterized by turbulent flow fields and strong gradients of temperature and pressure. At the end, however, when equilibrium is near, we know that E_{kin} is negligible and the fields of temperature and pressure are very nearly homogeneous, apart from being constant. That is the situation considered by Gibbs.

Indeed, Gibbs uses a method akin to the method of *virtual displacement* known in mechanics. The kinetic energy never occurs and temperature and pressure are always equal to their boundary values. Therefore he concludes:

- *Free energy* $F = U - TS$ is minimal in equilibrium compared to its values in other states with the same T and V.
- Entropy S is maximal in equilibrium compared to its values in other states with the same p and *enthalpy* $H = U + pV$.
- *Gibbs free energy* $G = U - TS + pV$ is minimal in equilibrium compared to its values in other states with the same T and p.

Free energy, enthalpy and Gibbs free energy are the quantities ψ, χ and ζ in Gibbs's work. He does not name these quantities apart from calling ψ and ζ *force functions* under the appropriate conditions of constant (T,V) and (T,p) respectively. I have introduced the now common names and chosen

[27] The working term is simplified here, because we do not account for viscous stresses.

the symbols F, H, and G which are most often used in the modern literature.[28]

The question is, of course, what it is that can change when T and p are already equal to the constant boundary values. One possibility is that the masses m_α of a chemically reacting mixture can change and at constant T and p they will change so as to minimize G; see above, where we have derived the law of mass action. Another possibility is that different phases in a body can readjust themselves – at constant T and V – so as to minimize F and to make the chemical potentials homogeneous.

Entropy and Energy in Competition

The knowledge, that the free energy

$$F = \text{energy} - T \cdot \text{entropy}$$

tends to a minimum as equilibrium is approached, is more than the result of some formal rearrangement of equations and inequalities. Indeed, the knowledge provides a deep insight into the driving forces of nature. Obviously, a decrease of energy and an increase of entropy are both conducive to making the free energy small. If T is small, such that the entropic part of F is negligible, the free energy tends to a minimum because the energy does. And, if T is large, so that the entropic part of F dominates, the free energy becomes minimal, because entropy tends to a maximum. Those are the extremes; at intermediate temperatures it is neither energy that reaches a minimum, nor entropy that reaches a maximum. Both quantities have to compromise and the result of the compromise is the minimum of the free energy.

The Pfeffer tube provides an instructive example for that situation, cf. Fig. 5.2. The energy – gravitational potential energy in this case – tends to adjust the levels of liquid in tube and reservoir to be equal; that is the situation where the energy is minimal. The entropy, on the other hand, tends to *pull* all the water from the reservoir into the tube, because that means maximal entropy of mixing of water and salt. Neither energy nor entropy succeed; they compromise and as a result some water remains in the reservoir, – less for a higher temperature.

The phenomenon is also interesting for another aspect: Obviously it is essentially the water that *pays the cost*, as it were, because its potential

[28] It is not uncommon though to see the free energy be denoted by ψ, as in Gibbs´ work; others prefer the letter A for *available* free energy. The letter H for enthalpy stands for *heat content* which is the literal translation of the Greek word *enthalpos*: en *inside* + thalos *heat*. This is a good name, since the enthalpy comes closest among all thermodynamic quantities to what the layman calls *heat*. The G for the Gibbs free energy is, of course, in honour of Gibbs himself.

energy rises considerably; and it is the salt that *profits* because its entropy increases with the larger volume of the solution in the tube. We conclude that nature does not allow the constituents of a mixture to be selfish: The system *as a whole* profits by decreasing its free energy.

Even closer to home is the case of our atmosphere: The potential energy of the air–molecules would be best served, if all of them lay at rest on the surface; but the entropy would be best off, if all molecules were spread evenly throughout infinite space. The compromise of minimal *free* energy in this case provides earth with a thin layer of thin air. If the earth were hotter, like the planet mercury, that atmosphere would have left us, and if it were smaller, like mars, the atmosphere would be even thinner.[29]

Considerations like these help to create an intuitive feeling for the significance of Gibbs's *force functions*.

Phase Diagrams

Let the Gibbs free energy G of a binary mixture with a fixed mass $m = m_1 + m_2$ at some fixed values of T and p be represented – as a function of m_1 – by the convex graph of Fig. 5.5_{left}. It follows from the relations of Insert 5.1 that the graph begins and ends at $g_2(T,p)$ and $g_1(T,p)$ respectively as indicated in the figure. Moreover, if we draw the tangent at some point $G(T,p,m_1^*)$, the intercepts of that tangent with the vertical lines $m_1 = 0$ and $m_1 = m$ represent the chemical potentials $g_2(T,p,m_2^*)$ and $g_1(T,p,m_1^*)$, respectively, cf. figure.

Now, let there be *two* such graphs, corresponding to two phases ′ and ″ (say). These are shown in Fig. 5.5_{right} for a (T,p)-pair for which they intersect. If the two phases are to be in phase equilibrium, the Gibbs phase rule requires that the chemical potentials g_α' and g_α'' $(\alpha = 1,2)$ be equal. That requirement provides an easy graphical method for the determination of m_1' and m_1'' in phase equilibrium: Indeed, m_1' and m_1'' are the abscissae of the point of contact of the *common tangent* of the graphs G' and G'', see Fig. 5.5_{right}.

For fixed p and changing T the common tangent shifts, since the end points $g_2(T,p)$ and $g_1(T,p)$ of both phases change in their own ways. At high temperatures the Gibbs free energy G'' of the vapour phase is everywhere below G' so that the body minimizes its Gibbs free energy by being in the vapour phase. Similarly, at low temperature we have $G' < G''$, irrespective of the value of m_1 and the liquid phase prevails, since *it* has the smaller Gibbs free energy. More interesting is the case where G' and G'' intersect so that two phases can coexist with the masses m_1' and m_1'' corresponding to

[29] These and other examples have been worked out by Müller and Weiss in a recent book. I. Müller, W. Weiss: "Entropy and energy – a universal competition." Springer, Heidelberg (2005).

the end point of the common tangent. For $m_1' < m_1 < m_1''$ the Gibbs free energy has values on that tangent, because those values are lower than the values of either phase.

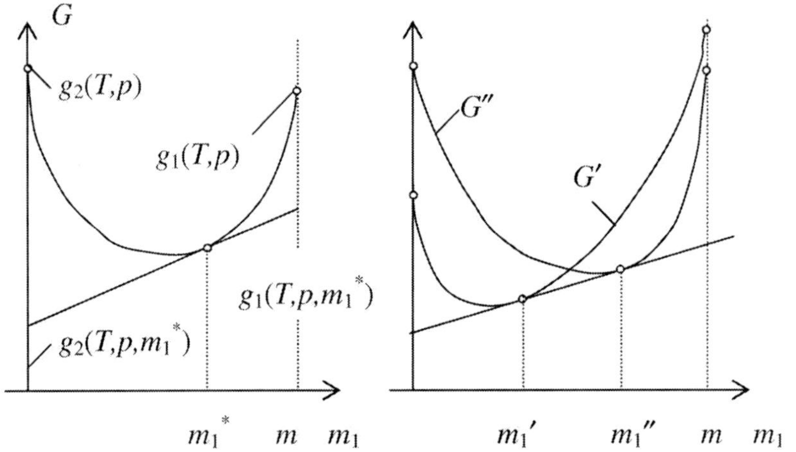

Fig. 5.5. *Left*: Gibbs free energy and chemical potentials. *Right*: Common tangent for phase equilibrium

All this is described – not in an optimal fashion – by Gibbs, who has a large chapter on "Geometric visualization".[30] After Gibbs it has become common practice to project the common tangents for different temperatures onto the corresponding isotherms in a (T,m_1)-diagram, or a (T,X_1)-diagram. The end points of those projections are then connected and form boiling and condensation curves like those of Fig. 5.3_{right}.

The convex graphs of Fig. 5.5 are appropriate for ideal solutions, or ideal alloys, where S_{Mix} is the only non-zero mixing quantity. When, on the other hand, U_{Mix} and V_{Mix} are non-zero, they combine in the Gibbs free energy to $H_{Mix} = U_{Mix} + pV_{Mix}$, the *heat of mixing*. The heat of mixing can be both positive and negative. It is due to the fact that unequal next neighbours among molecules are respectively either unfavourable or favourable energetically. In the latter case the mixing process must be accompanied by cooling, if the temperature is to be maintained. The former case requires heating lest the mixture cool off during mixing; that case is the interesting one, because the Gibbs free energy can become non-convex, if the heat of mixing is big enough.

It makes sense to consider the special case that only the liquid phase is affected by the heat of mixing, while the vapour – whose molecules are far apart – is ideal. In such a case we have Gibbs free energies G' and G'' of the type shown in Fig. 5.6_{left}. That figure corresponds to a fixed pair of pressure

---

[30] J.W. Gibbs: loc.cit. pp. 172–187.

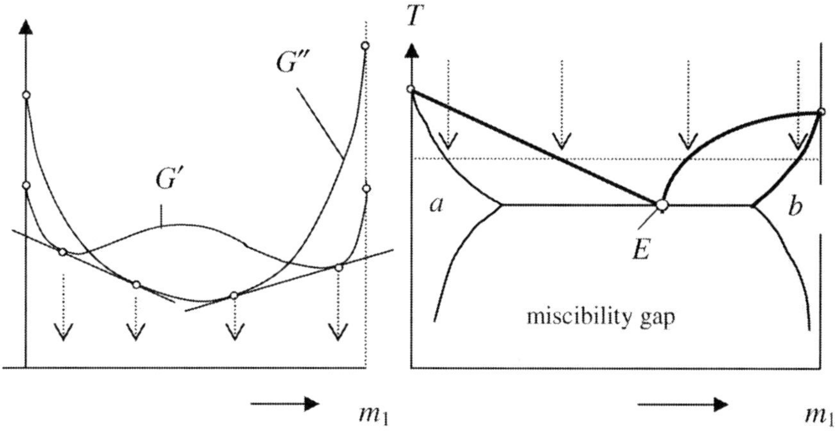

Fig. 5.6. *Left*: Gibbs free energies of vapour and liquid. *Right*: Phase diagram with a miscibility gap

and temperature and obviously there is the possibility now for *two* common tangents. When the temperature drops, the graph G' comes down with respect to G'' and there is the limiting case when the two tangents grow together and a three-phase-equilibrium exists between two liquids and a vapour. At still lower temperatures a common tangent can connect the convex branches of the liquid graph so that two liquids coexist, solutions *a* and *b,* which are 2-rich and 1-rich respectively and which do not mix: There is a miscibility gap in the liquid phase.

The phase diagram is constructed as before by projecting the common tangents into a (T,m_1)-diagram onto the appropriate isotherm. When the end-points of the projections are connected we obtain a diagram of the type shown in Fig. 5.6_{right}. The point denoted by *E* is called the *eutectic point*. For alloys the eutectic composition has the lowest melting point and that is what eutectic means in Greek: *eutektos = easy melting*.

In solutions the existence of the miscibility gap helps to separate the constituents and since one solution is invariably lighter than the other one, it floats on top and may be scooped off – like the fat from the milk.

Gibbs did not draw phase diagrams, but he did know the theory. Thus the three-phase equilibrium at the eutectic point in a binary solution allows $F = 2 - 3 + 2 = 1$ degree of freedom, see above. This means that the eutectic points form a *line* in a three-dimensional (p,T,X_1)-diagram.

Phase diagrams of the type shown in Figs. 5.5 and 5.6 – and more complex ones – for solutions and alloys are being measured and refined to this day. The first person to make this part of Gibbs's work explicit, and draw conclusions from it, was Hendrik Willem Bakhuis Roozeboom (1854–1907). He learned about Gibbs from van der Waals and made many experiments to confirm the phase rule. The modern physics and chemistry of alloys started with his work.

Law of Mass Action for Ideal Mixtures

We recall the law of mass action as derived by Gibbs and combine it with the form of the chemical potentials valid for ideal solutions. Thus we obtain the law of mass action for an ideal solution

$$\prod_{\alpha=1}^{\nu} X_\alpha^{\gamma_\alpha^a} = \exp\left[-\frac{\sum_{\alpha=1}^{\nu}\gamma_\alpha^a \mu_\alpha g_\alpha(T,p)}{kT}\right] \qquad (a = 1,2\ldots n)$$

The right-hand side is independent of composition and it is therefore often called the chemical *constant* K^a, although it depends on T and p. Of course, there are n such constants, one each for every independent reaction. The left-hand side – when written out – is the quotient of mol fractions of resultants and reactants with the appropriate exponents. It is customary to consider the stoichiometric coefficients of the resultants as positive, and those of the reactants as negative. Also the *most negative* one among the stoichiometric coefficients is often set equal to –1.[31] Thus for the reaction

$$C + \tfrac{1}{2}O_2 \rightarrow CO \text{ or } -C - \tfrac{1}{2}O_2 + CO = 0$$

the law of mass action has the form – with $K(T,p)$ as the chemical constant –

$$\frac{X_{CO}}{X_C\sqrt{X_{O_2}}} = K(T,p), \qquad \text{where} \qquad X_C + X_{O_2} + X_{CO} = 1,$$

so that a decrease of the mol fraction X_{O_2} will increase the output of CO.

Early chemists like Torbern Olof Bergman (1735–1784) or Claude Louis Comte de Berthollet (1748–1822)[32] were somewhat confused about *mass action* (sic), i.e. the shift of the chemical equilibrium upon the addition of mass of a constituent to a mixture. Bergman had conceived of *affinities* between substances, such that substance A reacted with B but not with C, if

[31] There is a certain arbitrariness here. The convention mentioned in the text is not universally accepted. Some people prefer to set the smallest absolute value among the γ_α^a equal to 1, thus avoiding fractional coefficients. Both conventions – and still others – are perfectly good, but they must not be mixed. And before we use tables – for chemical constants, or heats of reaction (say) – we must know which one of the conventions was employed in the compilation of the table.

[32] Berthollet was yet another scientist ennobled and made a senator by Napoléon, who generally knew how to create loyal followers. In the case of Berthollet, however, he made a mistake, because the chemist later voted for the deposition of the emperor. That gained him a peerage under the returning Bourbons, cf. I. Asimov: "Biographies…." loc.cit.

the affinity between A and B was great, while the affinity between A and C was small. Bergman prepared tables of affinities that were much used at his time. But then Berthollet observed that A and C *would* react after all, if only C was present in sufficiently great quantity. Thus we see how the somewhat strange name of *mass action* came about. Berthollet wrote a book about his findings,[33] in which he showed deep insight into the nature of chemical reactions, but did not quite arrive at the proper form of the law of mass action.

The correct formulation of the law came in a paper by Cato Maximilian Guldberg (1836–1902) and Peter Waage, both professors of chemistry at the university of Christiania – now Oslo – in Norway, and brothers in law. The paper was written in Norwegian and remained unnoticed by most chemists. Even the French translation in 1867 did not help and it was only the German translation in 1879 that made the work known. So for once, Gibbs lost his priority for the law of mass action; he whose memoir made so many other people lose *their* priority in subsequent years.

The argument of Guldberg and Waage is extremely simple: They argued very sensibly that a reaction can occur only, when the molecules of all reactants meet at one point in the numbers required by the stoichiometric equation. They considered – again plausibly – the probability for a molecule of reactant α to be at a certain point as being proportional to X_α. Therefore the probability for the forward reaction *reactant* → *resultant* should be given by

$$P_{\rightarrow} = C_{\rightarrow} \prod_{\alpha=1}^{\nu_-} X_\alpha^{|\gamma_\alpha|} ,$$

where ν_- is the number of reactants and $C_{\rightarrow}$ is a factor of proportionality. Accordingly the probability for the backward reaction *resultant* → *reactant* should be

$$P_{\leftarrow} = C_{\leftarrow} \prod_{\alpha=\nu_-+1}^{\nu} X_\alpha^{\gamma_\alpha} ,$$

In equilibrium both probabilities ought to be equal and so Guldberg and Waage came to the condition of chemical equilibrium in the form

$$\prod_{\alpha=1}^{\nu} X_\alpha^{\gamma_\alpha} = \frac{C_{\leftarrow}}{C_{\rightarrow}} .$$

The nature of the right-hand side – and its dependence on T and p – could not be determined in this simple manner. The law is truly a law of *mass* action and not, as it were, of pressure action, or temperature action.

And so, although Gibbs was anticipated by Guldberg and Waage with respect to mass action, his discovery went beyond that of the Norwegians, because he knew the structure of the right hand side, viz. of $K^{\alpha}(T,p)$:

---

[33] C.L. Berthollet: "Essay de statique chimique" (1803).

$$K^a(T,p) = \exp\left(-\frac{\sum_{\alpha=1}^{\nu} \gamma_\alpha{}^a \mu_\alpha g_\alpha(T,p)}{kT} \right).$$

We have argued before that $g(T,p)$ can be determined by (p,V,T)-measurements and by measurements of heat capacities $C_V(T,V_0)$ for *one* V_0. Actually we showed that such measurements leave us with unknown additive constants in U and S. Therefore $K^a(T,p)$ contains a linear function of T of the type

$$\Delta h_R^a - T\Delta s_R^a = \sum_{\alpha=1}^{\nu} \gamma_\alpha^a \mu_\alpha h_\alpha(T_R, p_R) - T\sum_{\alpha=1}^{\nu} \gamma_\alpha^a \mu_\alpha s_\alpha(T_R, p_R)$$

with unknown coefficients, the specific *heat of reaction* Δh_R^a and Δs_R^a, the specific *entropy of reaction.*

It is worth mentioning, perhaps, that those constants do nowhere play a role in thermodynamics, *except when it comes to chemical reactions.* Indeed, when a constituent vanishes or emerges, then energy and entropy of the constituent vanish and emerge along with the mass, and that includes the additive constant terms in energy and entropy.

The heat of reaction a, viz. Δh_R^a can be measured by measuring how much heating or cooling the reaction requires, if temperature and pressure are to be maintained. And after Δh_R^a has been obtained in that way, the entropy Δs_R^a of reaction results from a quantitative analysis of the reaction products. A systematic experimental campaign was needed for that and that was not Gibbs's thing.

Anyway, Gibbs had done enough. And we have not even considered his contribution to the thermodynamics of solids, where elastic stresses take over the role of the single pressure in fluids in determining the working term in the first law. Nor have we considered Gibbs's work on thermodynamic stability or the large second part of his memoir which is entitled "Theory of Capillarity", where Gibbs deals with surface effects and treats droplets, bubbles and inclusions. These are all important contributions to thermodynamics, but they represent collateral tributaries in the history of the field rather than the main stream.

Measurements of the heat of reaction had already been made by Lavoisier. And Germain Henri Hess (1802–1850) measured enough of them to pronounce Hess's rule in 1840 which states that the heats of reaction in successive reactions must be added. This rule helped to determine values for reactions which are difficult to investigate directly. After Gibbs's work the Hess rule became a corollary of that work.

Heats of reaction are usually measured in *calorimetric bombs*, i.e. strong chambers, capable of enduring high pressures at constant volume. Pierre Eugène Marcelin Berthelot (1827–1907) measured hundreds of heats of reaction, while Hans Peter Jörgen Thomsen (1826–1909) measured thousands of them. So we may assume that heats of reaction were available to chemists at large. This is not to say, however, that the significance of the quantity was universally recognized. Berthelot in particular was confused about the role of heats of reaction. He considered them the sole driving force for a reaction, such that only exothermic [34] reactions – those with a negative Δu[35] – could proceed spontaneously. The idea is plausible and, indeed, it is very often true. When it is not true, it is because the entropy of reaction interferes: Its growth Δs during the reaction may be so big that – at the prevailing temperature – it may offset a positive value Δu and still allow the necessary decrease Δf of free energy.

A well-known example is the reaction

$$H_2 + I_2 \rightarrow 2\,HI$$

which is endothermic with $\Delta\bar{h}_R = 25\frac{\text{kJ}}{\text{mol}}$ and yet – at about 450 °C – hydrogen iodide makes up $^4/_5$ of all molecules in equilibrium. Indeed, the tables provide the value $\Delta\bar{s}_R = 166\frac{\text{J}}{\text{mol}}$, so that the entropy moves upwards by the formation of HI, while the free energy moves downwards. That is the desired direction for both of them. So, how can H_2 and I_2 survive at all, albeit in the small proportion of 20%? The answer lies in the T- and p-dependent part of $\Delta\bar{h}$ and $\Delta\bar{s}$. We shall consider a similar situation below when we deal with the ammonia synthesis.

It was Helmholtz who pointed out Berthelot's misunderstanding about the decisive role of the heat of reaction in 1882[36] and – we know it, but he did not – he had been anticipated by Gibbs. Most scientists, however, learned about the delicate balance between Δu and Δs from Helmholtz and that is why the free energy $F = U - T{\cdot}S$ is known as the Helmholtz free energy in English speaking countries.

Le Châtelier (1850–1936), the most eminent chemist at the turn of the century, did not indulge in speculation. He simply reported what he had observed when, in 1888, he pronounced the *principle of least constraint* or simply *le Châtelier's principle*: *Every change of one of the factors of an equilibrium* [e.g. pressure or temperature] *brings about a rearrangement of*

---

[34] It was Berthelot who coined the words endothermic and exothermic.

[35] In a calorimetric bomb the reactions proceed at constant volume. Therefore the heat of reaction is equal to Δu; for a reaction that happens at constant pressure the heat of reaction is $\Delta h = \Delta u + p\Delta v$, because some of the energy change is converted into work.

[36] H. Helmholtz; "Die Thermodynamik chemischer Vorgänge" [Thermodynamics of chemical processes] Sitzungsberichte der preussischen Akademie der Wissenschaften. Berlin (1882).

the system – actually of its constituents – *in such a direction as to decrease the original change.* For instance: An endothermic reaction – one with a positive heat of reaction – proceeds further at increased temperature, so that the temperature in the end does not rise quite as far as it would have done without the reaction. Similarly: A volume-increasing reaction is helped along by a pressure decrease, so that the eventual pressure drop is smaller than without the reaction.

Le Châtelier, when he translated Gibbs's work into French in 1899, must have had mixed emotions when he saw that his principle had been proved by Gibbs ten years before he himself stated it. However, there was a consolation: Gibbs's proof was valid only for ideal gas mixtures, whereas le Châtelier's statement claims general validity.

Ostwald, the German translator of Gibbs, had said that he undertook the task because he believed in *hidden treasures* in Gibbs's work. He was right, and le Châtelier and later Haber and Bergius were chemists who uncovered and lifted the treasures. Of course, Roozeboom had been another one, see above.

Fritz Haber (1868–1934)

Fritz Haber was a chemist who knew Gibbs's work well enough to make use of it in the production of ammonia NH_3 from the nitrogen N_2 of the air. The overall stoichiometric formula reads

$$H_2 + \tfrac{1}{3} N_2 \rightarrow \tfrac{2}{3} NH_3$$

and the heat and entropy of reaction are[37]

$$\Delta\bar{h}_R = -30.8 \tfrac{\text{kJ}}{\text{mol}} \text{ and } \Delta\bar{s}_R = -59.5 \tfrac{\text{J}}{\text{molK}} .$$

Thus at normal temperature T_R and pressure p_R the energy, or enthalpy, drops and it is therefore conducive to the formation of ammonia. The entropy drops also and that fact is bad for ammonia. But the energetic term dominates, since $\Delta\bar{h}_R - T_R \Delta\bar{s}_R .= -13.1 \tfrac{\text{kJ}}{\text{mol}}$ holds, so that the Gibbs free energy favours ammonia, and that is what counts, or it should be; see Fig. 5.7 which shows Gibbs free energies as a function of the extent of reaction. For the reference state T_R = 298 K and p_R = 1atm the minimum of G lies very close to 100% ammonia.

---

[37] Chemists – at least the an-organic types – like to use molar quantities $\bar{a}_\alpha$ which are related to the mass-specific quantities a_α by $\bar{a}_\alpha = a_\alpha M_\alpha$, where $M_\alpha = M^r{}_\alpha \frac{\text{g}}{\text{mol}}$ is the molar mass and $M^r{}_\alpha = \frac{\mu_\alpha}{\mu_o}$ is the relative molecular mass.

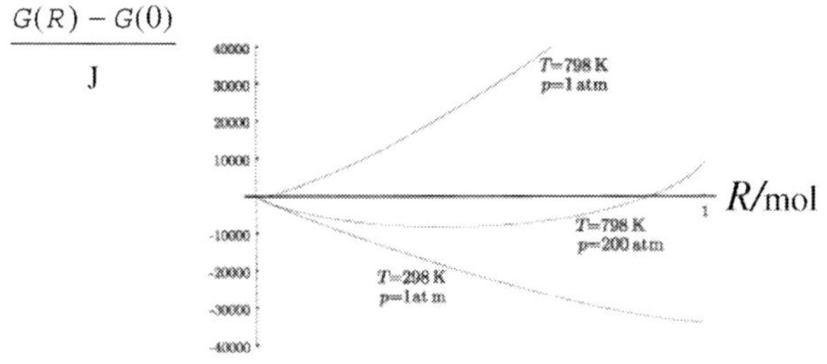

Fig. 5.7. Gibbs free energies for the ammonia synthesis as functions of the extent of reaction

And yet, nothing happens when hydrogen and nitrogen are mixed. No ammonia is formed and the mixture is perfectly stable, or rather metastable, since the strong chemical bonds between the atoms in H_2 and N_2 must first be severed or weakened before ammonia can be formed.

For that purpose Haber used perforated iron sheets whose surface catalyses the dissociations $H_2 \rightarrow 2H$ and $N_2 \rightarrow 2N$ at high temperature, say 500°C. Unfortunately, such a high temperature emphasizes the negative value of $\Delta\overline{s}_R$ so that the minimum of the Gibbs free energy lies on the side of hydrogen and nitrogen, cf. Fig 5.7, and once again, nothing happens. But then Haber knew what to do: The stoichiometric equation shows that the reaction, if it proceeds all the way, cuts the number of molecules by half and, since all constituents are ideal gases, the volume is halved as well. Therefore, by le Châtelier's principle and Gibbs's formulae, a high pressure should assist the reaction. Haber put the mixture under 200 atm and achieved a good output of ammonia,[38] cf. Fig. 5.7.

Ammonia can easily be converted into nitrates which the world craves for the production of fertilizers and explosives. Before Haber the main source of nitrates were the guano fields on the west coast of South America, over which Chile, Peru and Bolivia fought the *guano war*. Chile won and Bolivia lost her access to the sea.

The Haber-Bosch synthesis was developed in 1908, just in time for the first world war. It was clear that in case of war Germany would be cut off from guano imports by a British naval blockade and therefore a huge ammonia plant was built in Saxony. Its output supplied the German army throughout the four years of war easily. The country ran out of men and food and morale, but never of explosives. Haber received the Nobel prize in

[38] The process is known as the *Haber-Bosch synthesis* after Haber, of course, and Karl Bosch (1874–1940), who suggested a good strong material for the pressure vessel. The apparatus is exhibited on the campus of the University of Karlsruhe, where it rusts away on an unkempt lot.

1918 and, after the war he was made director of the Kaiser Wilhelm Institute for physical chemistry.[39]

Haber's patriotism led him to propagate and direct the use of chlorine and mustard gas as a means of warfare on the western front. Chlorine was first. On April 22, 1915 it was used at Ypres against Canadian troops. The troops fled and the result was an unprecedented five-mile gap in the front. The strategic effect, however, was nil, since the German general staff had not really believed that the project would work, and was not prepared for an offensive.[40]

Only a little later Haber became a tragic figure. He was Jewish and, when Hitler came to power, he was stripped of all posts and driven into exile. He was not alone in that, of course, but while others were made welcome in Britain by an international initiative of scientists led by Ernest Rutherford, Haber was *not*, because of his poison gas activity. He left for Italy but died *en route*.

Haber continued in what he saw as his patriotic duty after the disastrous war. He attempted to isolate gold from sea water in the hope to help Germany repay the huge war indemnity demanded under the Versailles peace treaty. In this effort Haber failed. However, he could have saved himself the effort because in the end the indemnity was never paid.

Fig. 5.8. Fritz Haber

The impact of chemistry on war and warfare confirmed itself in the second world war. That war was largely fought by mobile troops with mechanized transport, and by tanks and airplanes, and the biggest logistical problem was the supply of fuel. Germany has no natural mineral oil but a lot of coal, – both brown coal and pit coal. And again, just in time, it became possible to convert both types of coal into benzine. That was the invention

[39] It is a sign of the schizophrenia of German politics between the two world wars that the Kaiser Wilhelm Institute retained its name in the time of the Weimar republic and during the national-socialist rule, although the monarchy was thoroughly discredited. It took another world war to shake the name loose. The institute was renamed Max Planck Institute in 1946, cf. M. Planck: "Physikalische Abhandlungen und Vortraege" [Papers and lectures on physics] Vieweg, Braunschweig (1958). Foreword by M.von Laue.

[40] According to I. Asimov: "Biographies." loc.cit

of Friedrich Karl Rudolf Bergius (1884–1949),[41] who had studied catalytic high-pressure chemistry under Nernst and Haber. He developed the *Bergin process* to combine coal and hydrogen at high pressure and high temperature. Huge hydrogenation plants were built in Germany to supply the Wehrmacht, the German armed forces. Strangely enough the Allied Bomber Command overlooked the strategic importance of these vulnerable plants – 54 of them – until well into 1944. Then they were bombed and destroyed in May 1944.[42]

Fuel became very scarce indeed after that, and soon the vehicles of the German army were converted for the use of *wood-gas*, a comparatively low-tech application of mass action: Wood was burned with an insufficient air supply in a barrel-shaped furnace – that was loaded into the trunk –, and the resulting carbon monoxide was fed into the motor. I remember from my childhood that, half-way up even moderate hills, the drivers had to stop and stoke before they could proceed. Obviously this would not do for airplanes.

Socio-thermodynamics

On several occasions in previous chapters I have hinted at the usefulness of thermodynamic concepts in *remote areas*, i.e. fields that have little or nothing to do with thermodynamics at first sight. Those hints would be wanton remarks unless I corroborated them somehow, in order to acquaint the reader with the spirit of extrapolation away from thermodynamics proper. To be sure, most such subjects belong more to the future of thermodynamics rather than to its history. They are struggling to be taken seriously, and to obtain admission into the field. But anyway, let us consider the non-trivial proposition which has been called *socio-thermodynamics*. It extends the concepts described above for the construction of phase diagrams in binary solutions to a mixed population of hawks and doves with a choice of different contest strategies.

We let ourselves be motivated by an often discussed model of game theory[43] for a mixed population of hawks and doves who compete for the

---

[41] Bergius shared the 1931 Nobel prize with Karl Bosch, Haber's colleague and assistant in the Haber-Bosch synthesis of ammonia.

[42] According to A. Galland: "Die Ersten und die Letzten, die Jagdflieger im Zweiten Weltkrieg." [The first and the last, fighter pilots in World War II] Verlag Schneekluth, Augsburg (1953).
Adolf Galland was himself a highly decorated fighter pilot before he was given an office job; he became the last inspector of the Luftwaffe in the war and then the first inspector of the after-war Luftwaffe in 1956.

[43] J. Maynard-Smith, G.R. Price: "The logic of animal conflict." Nature 246 (1973).
P.D. Straffin: "Game Theory and Strategy." New Mathematical Library. The Mathematical Association of America 36 (1993).

same resource, whose value, or price, is denoted by τ. Prices are out of control for the birds, but they must be taken into account by them. Indeed, in their competition the birds may assume different strategies A or B which we define as follows.

Strategy A

If two hawks meet over the resource, they fight until one is injured. The winner gains the value τ, while the loser, being injured, needs time for healing his wounds. Let that time be such that the hawk must buy 2 resources, worth 2τ to feed himself during convalescence. Two doves do not fight. They merely engage in a symbolic conflict, posturing and threatening, but not actually fighting. One of them will eventually win the resource – always with the value τ – but on average both lose time such that after every dove-dove encounter they need to catch up by buying part of a resource, worth 0.2τ. If a hawk meets a dove, the dove walks away, while the hawk wins the resource; there is no injury, nor is any time lost.

Assuming that winning and losing the fights or the posturing game is equally probable, we conclude that the elementary expectation values for the gain per encounter are given by the arithmetic mean values of the gains in winning and losing, i.e.

$$e_A^{HH} = 0.5\,(\tau - 2\,\tau) \;= -\,0.5\,\tau$$
$$e_A^{HD} = \tau$$
$$e_A^{DH} = 0$$
$$e_A^{DD} = 0.5\,\tau - 0.2\,\tau = 0.3\,\tau$$

for the four possible encounters HH, HD, DH, and DD.

Note that both, the fighting of the hawks and the posturing of the doves, are irrational acts, or luxuries. Indeed both species would do better, if they cut down in these activities, or abandoned them altogether. Also the meekness of the doves confronted with a hawk may be regarded as overcautious. Such observations have let to the formulation of strategy B.

Strategy B

The hawks adjust the severity of the fighting – and thus the gravity of the injury – to the prevailing price τ. If the price of the resource is higher than 1, they fight less, so that the time of convalescence in case of a defeat is shorter and the value to be bought during convalescence is reduced from 2τ to $2\tau(1\text{-}0.2(\tau - 1))$. Likewise the

---

The issue in these presentations is the proof that a mixed population of two species may be evolutionarily stable, if the species follow the proper contest strategy. In the present account of socio-thermodynamics the objective is different: No evolution is allowed but two different strategies may be chosen which both depend on the price of the contested resource.

doves adjust the duration of the posturing, so that the payment for lost time is reduced from 0.2τ to $0.2\tau\,(1-0.3(\tau-1))$. But that is not all: To be sure, in strategy B the doves will still not fight when they find themselves competing with a hawk, but they will try to grab the resource and run. Let them be successful 4 out of 10 times. However, if unsuccessful, they risk injury from the enraged hawk and may need a period of convalescence at the cost $2\tau\,(1+0.5(\tau-1))$.

Thus the elementary expectation values for gains under strategy B may be written as

$$\begin{aligned}
e_B^{HH} &= 0.5(\tau - 2\,\tau(1-0.2(\tau-1))) &&= (0.2\,\tau - 0.7)\,\tau \\
e_B^{HD} &= 0.6\,\tau \\
e_B^{DH} &= 0.4\,\tau - 0.6\cdot 2\,\tau\,(1+0.5(\tau-1)) &&= -(0.6\,\tau + 0.2)\,\tau \\
e_B^{DD} &= 0.5\,\tau - 0.2\,\tau\,(1-0.3(\tau-1)) &&= (0.06\,\tau + 0.24)\,\tau\,.
\end{aligned}$$

The assignment of numbers is always a problem in game theory. Here the numbers have been chosen so as to fit a conceivable idea of the behaviour of the species. Let us consider this:

The grab-and-run policy is clearly not a wise one for the doves, because they get punished for it. So, why do they adopt that policy? We may explain that by assuming, that doves are no wiser than people, who start a war with the expectation of a quick gain and then meet disaster. This has happened often enough in history.

Note that for $\tau > 1$ the intra-species penalties for either fighting or posturing become smaller, because we have assumed that these activities are reduced when their execution becomes more expensive. However, the interspecies penalty – the injury of the doves – increases, because the hawks will exert more violence against the impertinent doves when the stolen resource is more valuable.

$\tau = 1$ is a reference price in which both strategies coincide, except for the grab-and-run feature of strategy B. Penalties for either fighting or posturing should never turn into rewards for whatever permissible value of τ. This condition imposes a constraint on the permissible values of τ: $0 < \tau < 4.33$.[44]

Now, let z_H and $z_D = 1 - z_H$ be the fractions of hawks and doves, and let all hawks and doves either employ strategy A or B. Therefore the gain expectations e_i^H and e_i^D (i = A,B) of a hawk and a dove per encounter with another bird may be written as

$$e_i^H = z_H\, e_i^{HH} + (1 - z_H)\, e_i^{HD} \qquad \text{and} \qquad e_i^D = z_H\, e_i^{DH} + (1 - z_H)\, e_i^{DD}$$

in terms of the elementary expectation values. And the gain expectations e_i for strategy i per bird and per encounter reads

[44] Such a constraint could be avoided, if we allowed non-linear penalty reductions which, for simplicity, we do not.

$$e_i = z_H\, e_i^{H} + (1 - z_H)\, e_i^{D} \text{ or explicitly}$$

$$e_i = z_H^{2}(e_i^{HH} + e_i^{DD} - e_i^{HD} - e_i^{DH}) + z_H(e_i^{HD} + e_i^{DH} - 2\, e_i^{DD}) + e_i^{DD}.$$

Specifically we have

$$e_A = -1{,}2\, \tau\; z_H^{2} + 0.4\, \tau\, z_H + 0.3$$
$$e_B = 0.86\, \tau\, (\tau - 1)\, z_H^{2} - (0.72\tau + 0.08)\tau\, z_H + (0.06\, \tau + 0.24)\tau.$$

The graphs of these functions are parabolae which – for some values of τ – are plotted in Fig. 5.9.a–e.

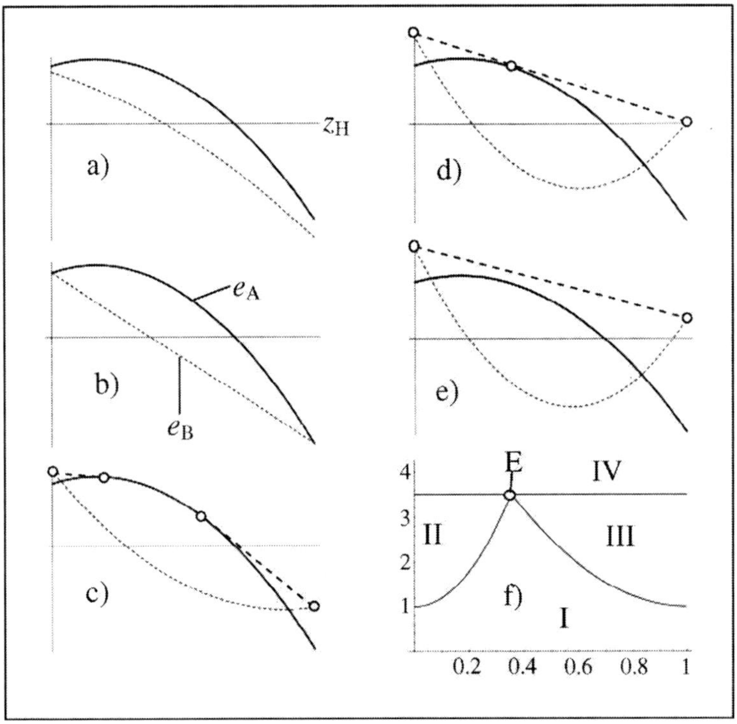

Fig. 5.9. Expectation values as functions of z_H for some values of the price τ. Concavification. Strategy diagram

The interpretation of those graphs is contingent on the reasonable assumption that the population chooses *the* strategy that provides the maximal gain expectation. Obviously for $\tau = 0.6$ and $\tau = 1$ that strategy is strategy A. At that price level the hawks and doves will therefore all choose strategy A irrespective of the hawk fraction z_H in the population.

For higher price levels the situation is more subtle, because the graph $max[e_A, e_B]$ is not concave. This provides the possibility of concavification, cf. Fig. 5.9.c–e. There are intervals of z_H where the concave envelope of $max[e_A, e_B]$ lies higher than that graph itself. The population then has the possibility to increase the expected gain by un-mixing; it *segregates* into homogeneous *colonies* with hawk fractions corresponding to the end-points of the concavifying straight lines, which are dashed in the figures. In Figs. 5.9c,d the adopted strategies are A and B and the species are mixed in the colony with strategy A, whereas the colony with strategy B is pure-dove or pure-hawk, depending on whether the extant overall hawk fraction lies below the left, or right tangent respectively. For $\tau > 3.505$ the concave envelope connects the end-points of the parabolae e_B so that hawks and doves are fully segregated in two colonies, both employing strategy B.

Mutatis mutandis all this is strongly reminiscent of the considerations of phase diagrams of solutions or alloys with a miscibility gap, see above at Fig. 5.6. To be sure, there we *minimized* Gibbs free energies whereas here we *maximize* gain. Accordingly in solutions we *convexify* the graph $max[G',G'']$ whereas her we *concavify* the graph $max[e_A, e_B]$, but those are superficial differences. And just as we constructed phase diagrams before, we may now construct a strategy diagram by projecting the concavifying lines unto the appropriate horizontal line in a (price, hawk fraction)-diagram, cf. Fig. 5.9f. We recognize four regions in that diagram.

- I: Full integration of species employing strategy A.
- II: Colony of pure doves with strategy B and integrated colony of hawks and doves with strategy A. Partial segregation.
- III: Colony of pure hawks with strategy B and integrated colony with strategy A. Partial segregation.
- IV: Colonies of pure doves and pure hawks. Full segregation.

The curves separating the regions II and III from region I can easily be calculated:

$$\tau = 20\, z_H^{\,2} + 1 \qquad \text{and} \qquad \tau = 6\, z_H^{\,2} - 12\, z_H + 7$$

respectively. Those two curves intersect in the eutectic point E, so called in analogy to thermodynamics.

Although the analogy between our sociological model and thermodynamics of solutions is fairly striking, there are differences. In particular, the present strategy diagram lacks the lateral regions, denoted by *a* and *b* in Fig. 5.6. This is due to the fact that we have not accounted for an entropy of mixing in the present case. For socio-thermodynamics in full – including the entropy of mixing – I refer to my recent article "Socio-thermodynamics – integration and segregation in a population." [45] In that paper the analogy is fully developed, including first and second laws of socio-thermodynamics, and with the proper interpretations of working and heating etc.[46]

---

[45] I. Müller: Continuum Mechanics and Thermodynamics 14 (2002) pp. 389–404.

[46] The simplified presentation given above follows a paper by J. Kalisch, I. Müller: "Strategic and evolutionary equilibria in a population of hawks and doves." Rendiconti del Circolo Matematico in Palermo, Serie II, Supplemento 78 (2006), pp. 163–171.

The upshot of the present investigation is that, if integration of species – or, perhaps, ethnic groups – is desired and segregation is to be avoided, political leaders should provide for low prices, if they can. In good times integration is no problem, but in bad times segregation is likely to occur. We all know that. But here is a mathematical representation of the fact with – conceivably – the possibility for a quantification of parameters.

The analogy of segregation in a population and the miscibility gap in solutions and alloys has been noticed before by Jürgen Mimkes, a metallurgist.[47] His approach is more phenomenological than mine, without a model from game theory. Mimkes has studied the integration and segregation of protestants and catholics in Northern Ireland, and he came to interesting conclusions about mixed marriages.

It is interesting to note that socio-thermodynamics is only accessible to chemical engineers and metallurgists. These are the only people who know phase diagrams and their usefulness. It cannot be expected, in our society, that sociologists will appreciate the potential of these ideas. They have never seen a phase diagram in their lives.

That paper also includes evolutionary processes, which make the hawk fraction change so that the population may eventually reach the evolutionarily stable strategy appropriate to the price level τ.

[47] J. Mimkes: "Binary alloys as a model for a multicultural society." Journal of Thermal Analysis 43 (1995).

6 Third Law of Thermodynamics

In cold bodies the atoms find potential energy barriers difficult to surmount, because the thermal motion is weak. That is the reason for liquefaction and solidification when the intermolecular van der Waals forces overwhelm the free-flying gas atoms. If the temperature tends to zero, no barriers – however small – can be overcome so that a body must assume the state of lowest energy. No other state can be realized and therefore the entropy must be zero. That is what the *third law of thermodynamics* says.

On the other hand cold bodies have slow atoms and slow atoms have large de Broglie wave lengths so that the quantum mechanical wave character may create macroscopic effects. This is the reason for gas-degeneracy which is, however, often disguised by the van der Waals forces.

In particular, in cold *mixtures* even the smallest malus for the formation of unequal next neighbours prevents the existence of such unequal pairs and should lead to un-mixing. This is in fact observed in a cold mixture of liquid He^3 and He^4. In the process of un-mixing the mixture sheds its entropy of mixing. Obviously it must do so, if the entropy is to vanish.

Let us consider low-temperature phenomena in this chapter and let us record the history of low-temperature thermodynamics and, in particular, of the science of cryogenics, whose objective it is to reach low temperatures. The field is currently an active field of research and lower and lower temperatures are being reached.

Capitulation of Entropy

It may happen – actually it happens more often than not – that a chemical reaction is constrained. This means that, at a given pressure p, the reactants persist at temperatures where, according to the law of mass action, they should long have been converted into resultants; the Gibbs free energy g is lower for the resultants than for the reactants, and yet the resultants do nor form. We may say that the mixture of reactants is *under-cooled*, or *over-heated* depending on the case. As we have understood on the occasion of the ammonia synthesis, the phenomenon is due to energetic barriers which must be overcome – or bypassed – before the reaction can occur. The bypass may be achieved by an appropriate catalyst.

An analogous behaviour occurs in phase transitions,[1] mostly in solids: It may happen that there exist different crystalline lattice structures in the same substance, one stable and one meta-stable, i.e. as good as stable or, anyway, persisting nearly indefinitely. Hermann Walter Nernst (1864–1941) studied such cases, particularly for low and lowest temperatures.

Take tin for example. Tin, or pewter, as *white tin* is a perfectly good metal at room temperature – with a tetragonal lattice structure – popular for tin plates, pewter cups, organ pipes, or toy soldiers.[2] Kept at 13.2°C and 1atm, white tin crumbles into the unattractive cubic *grey tin* in a few hours. However, if it is not given the time, white tin is meta-stable below 13.2°C and may persist virtually forever.[3]

It is for a pressure of 1atm that the phase equilibrium occurs at 13.2°C. At other pressures that temperature is different and we denote it by $T_{w\leftrightarrow g}(p)$; its value is known for all p. At that temperature $\Delta g = g_w - g_g$ vanishes, and below we have $g_w > g_g$, so that grey tin is the stable phase. Δg may be considered as the frustrated driving force for the transition and it is sometimes called the *affinity* of the transition. It depends on T and p and has two parts

$$\Delta g(T,p) = \Delta h(T,p) - T \cdot \Delta s(T,p),$$

an energetic and an entropic one.

$\Delta h(T,p)$ is the latent heat of the transition and $\Delta s(T,p)$ is the entropy change.[4] For any given p the latent heat $\Delta h(T,p)$ can be measured as a function of T by encouraging the transition catalytically, e.g. by *doping* white tin with a small amount of grey tin. And $\Delta s(T,p)$ may be calculated by integration of $c_p(T,p)/T$ of both variants, white and grey, between $T = 0$, – or as low as possible – and the extant T. Thus we have

$$\Delta g(T,p) = \Delta h(T,p) - T \cdot \left\{ s_w(0,p) + \int_0^T \frac{c_p^w(\tau,p)}{\tau} d\tau - s_g(0,p) - \int_0^T \frac{c_p^g(\tau,p)}{\tau} d\tau \right\}$$

---

[1] From the point of view of thermodynamics phase transitions are much like chemical reactions, although the phenomena differ in appearance. One might go so far as to say that phase transitions *are* chemical reactions of a particularly simple type.

[2] In ancient times tin was much in demand because, alloyed to copper, it provided bronze, the relatively hard material used for weapons, tools, and beads and baubles in the bronze age (sic).

[3] Not so, however, when it coexists with previously formed traces of grey tin. If that is the case, tin appliances are affected by the *tin disease* at low temperature. A church may lose its organ pipes in a short time, and that loss did in fact occur during a cold winter night in St. Petersburg in the 19th century.

[4] Note that the heat and entropy of transition depend on T *and* p, if the transition occurs in the under-cooled range. If it occurs at the equilibrium point, both quantities depend only on *one* variable, since $T = T_{w\leftrightarrow g}(p)$ holds at that point.

Inspection shows that for $T \to 0$ the affinity tends to the latent heat. This would even be true, if the specific heats $c_p(T,p)$ were constant for $T \to 0$. In reality, in Nernst's time – between the 19th and the 20th century – there was already ample evidence that all specific heats tend to zero polynomially, with $T \to 0$, e.g. as $(a \cdot T^3)$ for non-conductors, or as $(a \cdot T^3 + b \cdot T)$ for conductors. Given this observation, the integrals in $\Delta s(T,p)$ themselves tend to zero, and the curly bracket reduces to $s_w(0,p) - s_g(0,p)$. This difference may be related to the heat of transition $\Delta h(T_{w\leftrightarrow g}(p))$ at the equilibrium point, because in phase equilibrium we have $\Delta g(T_{w\leftrightarrow g}(p)) = 0$, or

$$s_w(T_{w\leftrightarrow g}(p)) - s_g(T_{w\leftrightarrow g}(p)) = \frac{\Delta h(T_{w\leftrightarrow g}(p))}{T_{w\leftrightarrow g}(p)} \qquad \text{or}$$

$$s_w(0,p) - s_g(0,p) = \frac{\Delta h(T_{w\leftrightarrow g}(p))}{T_{w\leftrightarrow g}(p)} - \int_0^{T_{w\leftrightarrow g}(p)} \frac{c_p^{\,w}(\tau,p) - c_p^{\,g}(\tau,p)}{\tau} d\tau.$$

From some measurements Nernst convinced himself that this expression – which after all is equal to $\Delta s(T,p)$ for $T \to 0$ – is zero, irrespective of the pressure p, and for all transitions.[5] So he came to pronounce his law or *theorem* which we may express by saying that *the entropies of different phases of a crystalline body become equal for* $T \to 0$, *irrespective of the lattice structure. Moreover, they are independent of the pressure p.*

This became known as the **third law of thermodynamics**.

We recall Berthelot, who had assumed the affinity to be given by the heat of transition. And we recall Helmholtz, who had insisted that the contribution of the entropy of the transition must not be neglected. Helmholtz was right, of course, but the third law provides a low-temperature niche for Berthelot: Not only does $T \cdot \Delta s(T,p)$ go to zero, $\Delta s(T,p)$ itself goes to zero. The entropy *capitulates* to low temperature and gives up its efficacy to influence reactions and transitions.

Inaccessibility of Absolute Zero

In 1912 Nernst pointed out that absolute zero could not be reached because of the third law.[6] Indeed, since $s(T,p)$ tends to the same value for $T \to 0$ irrespective of pressure, the graphs for different p's must look qualitatively

---

[5] W. Nernst: "Über die Berechnung chemischer Gleichgewichte aus thermodynamischen Messungen" [On calculations of chemical equilibria from thermodynamic measurements] Königliche Gesellschaft der Wissenschaften Göttingen 1, (1906).

[6] W. Nernst: "Thermodynamik und spezifische Wärme" [Thermodynamics and specific heat]. Berichte der königlichen preußischen Akademie der Wissenschaften (1912).

like those of Fig. 6.1.a. Therefore the usual manner for decreasing temperature, – namely isothermal compression followed by reversible adiabatic expansion – indeed decreases the temperature, *but never to zero*, since the graphs become ever closer for $T \rightarrow 0$.

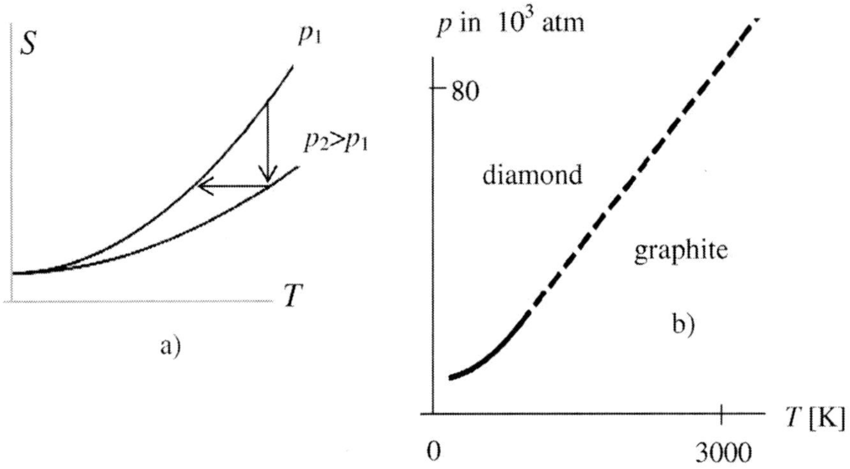

Fig. 6.1. (**a**) Isothermal compression (↓) and adiabatic expansion (←) (**b**) Equilibrium pressure for the transition graphite↔diamond

Having presented that argument, Nernst summarizes the three laws of thermodynamics thus:[7]

> It is impossible to build an engine that produces heat or work from nothing.
> It is impossible to build an engine that produces work from nothing else than the heat of the environment.
> It is impossible to take all heat from a body.

This accumulation of negatives appealed to Nernst and it has appealed to physicists ever since.

Diamond and Graphite

One of the more unlikely cases of coexisting phases occurs in solid carbon and they are known as graphite and diamond. Both are crystalline in different ways: Graphite consists of plane layers of benzene rings tightly bound – inside the layer – in a hexagonal tessellation. And each layer is

[7] W. Nernst: "Die theoretischen und experimentellen Grundlagen des neuen Wärmesatzes." [Theoretical and experimental basis for the new heat theorem] Verlag W. Knapp, Halle (1917), p. 77.

loosely bound to the neighbouring ones. If one rubs graphite against a sheet of paper (say), the uppermost layers are scraped off and leave a mark on the paper. That is why graphite can be used for writing. Hence the name: graphos = to write in Greek. The *lead* inside our pencils consists of graphite mixed with clay. It has the gloss of lead.

And then there is diamond, the hardest material of all; it cannot be scratched or ground except by use of other diamonds and it is unaffected by most chemicals. The Greek word was "adamas" = untameable and that is where, after some distortion, the name diamond comes from. In diamonds the carbon atoms sit in the centre of tetrahedra and are quite tightly bound, although not as tightly as the in-plane atoms in the graphite layers. At normal pressure and temperature graphite is stable and diamond is meta-stable.

All this, of course, was unknown until modern times and, naturally, since diamond was rare and beautiful, and therefore valuable, it was of much interest to chemists and alchemists alike. To investigate its properties, however, it needed a rich patron. Cosimo III, Grand Duke of Tuscany – true to the Medici tradition of patronizing the arts and sciences – provided a good-size sample for scientific investigation. For security he entrusted it to a group of three scientists who could not – try as they might – affect it in any way. Eventually they brought a burning glass to bear, in order to *heat* the stone. It developed a halo and then – it was gone! Naturally the report was met with some scepticism,[8] but nobody was much tempted to repeat the experiment until Lavoisier did so 80 years later. Lavoisier, living up to his reputation, controlled his experimental conditions by using a closed jar. He found that, after the diamond had been burned, the air inside the jar contained an appropriate amount of carbon di-oxide and so he could conclude that diamond is pure carbon.

After the inevitable sceptics had been convinced, there arose a strong desire to reverse the process and make diamond from graphite. Since $\Delta g(T,p) = g_{dia}(T,p) - g_{graph}(T,p)$ is the affinity of the process and since $\left(\frac{\partial g}{\partial p}\right)_T = v$ holds, we have

$$\Delta g(T,p) = \Delta g(T,0) + \int_0^p \Delta v(T,\pi)d\pi \,.$$

Diamond is a lot denser than graphite $-3.5 \,{}^{\mathrm{g}}/_{\mathrm{cm}^3}$ as compared to $2 \,{}^{\mathrm{g}}/_{\mathrm{cm}^3}$ – and therefore we have $\Delta v < 0$ so that $\Delta g(T,p)$ decreases with increasing p. For phase equilibrium $\Delta g(T,p)$ must vanish and thus we obtain an equation for the requisite p as a function of T

[8] According to I. Asimov: "The unlikely twins" in: "The tragedy of the moon" Dell Publishing Co. New York (1972).

$$\int_0^p \Delta v(T,\pi)d\pi = \Delta g(T,0)\ .$$

By the third law $\Delta g(T,0)$ is known – without any unknown constants – from measurements of the latent heat of the transition for $p = 0$ and from measurements of the specific heats $c_p(T,0)$ of both phases starting at $T = 0$, or as low as possible. Also $v(T,p)$ is known for all T as a function of pressure. Of course, it takes a protracted experimental campaign to measure all these values, but the end might justify the means: For every fixed temperature we obtain the pressure that should convert graphite into diamond. Fig. 6.1.b shows the graph.[9]

Inspection of the graph shows that, at room temperature, it should take approximately 15 kbar to obtain diamond, if indeed the transition occurred in equilibrium. However, in both directions the transition is hampered by energetic barriers: In the interesting direction the planar benzene configuration must first be destroyed before diamond can be formed, and in the other direction the tetragonal diamond structure must be weakened before diamond turns to graphite. For both it needs high temperature and therefore the equilibrium graph of Fig. 6.1.b is really relevant only in the upper part. When diamonds were eventually synthesized in 1955, by scientists of the General Electric Company in the USA, it occurred at 2800 K and at a pressure of about 100 kbar.[10]

There had been several false alarms before that time. But the reported results turned out to be either fakes or hoaxes. It is believed that the chemist Henri Moisseau had been hoodwinked by one of his assistants when – in 1893 – he presented a diamond which he believed he had created in his laboratory. Certainly he could never repeat the feat.

Hermann Walter Nernst (1864–1941)

It is difficult to say much in praise of Nernst which was not already said better by Nernst himself, cf. Fig. 6.2. He was a bon-vivant, as much as that is possible for a hard-working professor, operator and administrator. He hunted in the stylised European manner, was a connoisseur of wine and women, an early gentleman automobilist and, quite generally, a person endowed with a healthy self-regard. That by itself is one way to get ahead in the world and Nernst was good at that.

---

[9] J. Wilks: "Der dritte Hauptsatz der Thermodynamik" [The third law of thermodynamics] Vieweg, Braunschweig (1963)

[10] Or 700 tons per square inch in the cute American units.

Nernst reassures us concerning the emergence of further thermodynamic laws:

The 1st law had three discoverers: Mayer, Joule and Helmholtz.

The 2nd law had two discoverers: Carnot and Clausius.

The 3rd law has only one discoverer, namely himself: Nernst.

The 4th law … (?)

Fig. 6.2. Hermann Walter Nernst

He had obtained the patent for an essentially useless electric lamp – the Nernst pin – which nevertheless, to Edison's amazement,[11],[12] he sold to industry for a million marks, a very sizable amount of money indeed at the time. Nernst suggested to Röntgen that he should patent X-rays so as to make money, an idea that had never occurred to Röntgen; nor was he tempted.

Nernst's law, or *theorem* stood on uncertain grounds at first. It is now recognized that, at the beginning,[13] it was a daring proposition with little or no evidence to back it up.[14] To bc sure, the theorem was not presented cautiously, but rather with some fanfare. A somewhat irrelevant differential equation was solved and one solution was preferred arbitrarily over all others, because *a priori* that seemed to Nernst to be the *easiest solution*.[15] However, at the end, just like with his pin, Nernst was lucky. Others collected the evidence, which he had failed to present. By and large, Nernst's proposition was confirmed through painstaking work lasting many years. To be sure, *amorphous* solids had to be excluded somewhere along the way, but that was a secondary qualification, perhaps.

Despite Nernst's proud statement, cf. Fig. 6.2, about being the sole discoverer of the third law, there were really *two* discoverers. Indeed, Planck strengthened the law on the grounds of statistical thermodynamics by demanding that the entropy of all crystalline bodies tend to zero for $T \rightarrow 0$.

---

[11] Thomas Alva Edison (1847–1931), the greatest inventor of all times, owned 1300 patents at the end of his career, among them one for the electric light bulb. He held a poor opinion of the practical skills of professors like Nernst.

[12] I. Asimov: "Biographies …" loc.cit.

[13] W. Nernst: "Über die Berechnung ...." loc.cit. (1906).

[14] See: A. Hermann (ed.): "Deutsche Nobelpreisträger" [German winners of the Nobel prize] Heinz Moos Verlag, München (1969) p. 131–132.

[15] Ibidem, p. 132.

This is the modern version of the law and it is amply confirmed in experiments by comparing the entropies calculated from measurements of specific heats with the known value of entropy in the ideal gas phase of a substance, see below.

Planck's form of the third law goes far beyond Nernst's, because it is not restricted to chemical reactions, or phase transitions. It allows us to calculate the *absolute value* of the entropy of any *single* body. The handbooks used by physicists and chemists provide these values as parts of their tables of constitutive properties.

Note that this is more than the chemists need, because in their formulae it is only the entropy of reaction that is needed, that is to say a *combination* of the entropy constants of the reactants and resultants, see Chap. 5.

This is just like with energy: Chemists need only the heat of reaction, but Einstein's formula $E = mc^2$ furnishes the absolute value of energy for all reacting constituents in terms of their mass. This, however, is not useful knowledge for the chemist. Indeed, the mass defect of chemical compounds is too small to be measured by weighing (say). Yet, in summary it may be said that the first decade of the 20th century furnished both: the theoretical possibility for the determination of the absolute values of energy and entropy.

Liquefying Gases

It is not easy to lower temperatures and the creation of lower and lower temperatures is in itself a fascinating chapter in the history of thermodynamics which we shall now proceed to consider. The chapter is not closed, because low-temperature physics is at present an active field of research. Currently the world record for the lowest temperature in the universe[16] stands at 1.5 μK, which was reached at the University of Bayreuth in the early 1990's. Naturally the cold spot was maintained only for some hours. Such a value was, of course, far below the scope of the pioneers in the 19th century who set themselves the task of liquefying the gases available to them and then, perhaps, reach the solid phase.

The easiest manner to cool a gas is by bringing it in contact with a cold body and let a heat exchange take place. But that requires the cold body to begin with, and such a body may not be available. No gas – apart from water vapour – could be liquefied in this manner in the temperate zones of Europe where most of the research was done.

Since liquids occupy only a small portion of the volume of gases at the same pressure, it stands to reason that a high pressure may be conducive to liquefaction, just as a low temperature is. Both together should be even

---

[16] The universe, through its *background radiation*, imparts a temperature of 3K to bodies that are not otherwise heated or cooled.

better. That idea occurred to Michael Faraday – a pioneer of both electromagnetism and cryogenics, the physics of low-temperature-generation – in 1823. He combined high pressure and low temperature in an ingenious manner by using a glass tube formed like a boomerang, cf. Fig. 6.3. Some manganese di-oxide with hydrochloric acid was placed at one end. The tube was then sealed and gentle heating liberated the gas chlorine which mixed with the air of the tube and, of course, raised the pressure. The other end was put into ice water and it turned out that chlorine condensed at that end and formed a puddle at 0°C and high pressure.

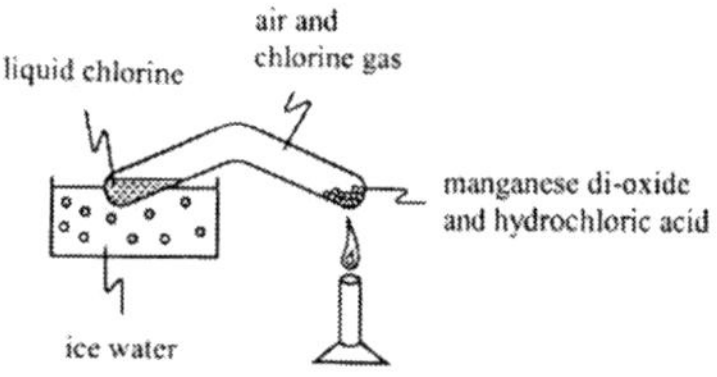

Fig. 6.3. Michael Faraday (1791–1867) Liquefaction of chlorine

When the pressure is slowly released, some of the liquid chlorine evaporates and, if this is done adiabatically, the heat of evaporation comes in part from the liquid, which therefore cools. In this manner Faraday was able to determine the boiling point of chlorine at 1atm as being –34.5°C. A further decrease of pressure will cool the liquid chlorine beyond that point, provided of course, that any is left.

Other scientists joined the campaign for low temperatures, notably Charles Saint Ange Thilorier (1771–1833), a chemist, who liquefied carbon dioxide in a strong metallic *boomerang* under high pressures and then lowered the pressure – hence, by evaporation, the temperature – far enough to make it solid. When enough solid was accumulated to experiment with, it turned out that carbon dioxide at 1atm goes immediately from the solid phase into vapour and vice-versa – at –78.5°C – in a process called sublimation, or de-sublimation respectively. That makes solid carbon dioxide popular as *dry ice*. It cools an article without soaking it upon melting; after all, it does not melt, it sublimates.

Thilorier invented another trick as well. He mixed the strongly volatile ether[17] with solid carbon di-oxide. The ether evaporated and thus produced temperatures as low as −110°C, or 163 K. Having enough of this cold mixture available, Faraday and Thilorier could now liquefy other gases by simple heat exchange, although for some of them they needed high pressure to help in the process.

And yet, there are eight gases which cannot be liquefied at 163 K even under high pressure. They are oxygen, argon, fluorine, carbon monoxide, nitrogen, neon, hydrogen and helium of which Faraday knew five; he did not know the noble gases. So he called those five gases *permanent.* And that is where the further development was stuck for a while. Until Thomas Andrews (1813–1885) found out about the critical point or, in particular, the critical temperature.

Andrews worked with carbon dioxide CO_2, a gas that can be liquefied at room temperature under pressure. He took a sample of liquid CO_2 under high pressure – 60–70atm (say) – and watched the liquid evaporate at some fixed temperature upon heating. Then he raised the pressure and started again, and again. He observed that the phase separation became less pronounced for higher pressure and vanished altogether at p = 73atm and T = 31°C. That point was called the *critical point* by Andrews. For higher pressures the liquid did not evaporate upon heating nor did the vapour liquefy upon cooling; the vapour just became ever denser without any evidence of a separation between liquid and vapour.

Andrews conjectured that all substances have critical points and that these points had escaped the attention of thermodynamicists only, because they were far out of the usual and easily accessible ranges of pressure and temperature. Therefore he concluded that the *permanent* gases can also be liquefied, if only we start raising the pressure on a sample that is colder, or even considerably colder than 163 K, which at that time was the record minimum.

Eventually this proved to be the case. But there was the problem of reaching lower temperatures. This problem was solved by Louis Paul Cailletet (1832–1913) in 1877. He compressed oxygen to a pressure of p_H = 66atm (say) in a compressor and then cooled the compressed gas back to room temperature T_H = 298K. Afterwards he subjected the gas to an adiabatic expansion to p_L= 1atm through a turbine, regaining some of the compressor work. For the expansion the *adiabatic equation of state* may be used in the form $\frac{p_H}{p_L} = \left(\frac{T_H}{T_L}\right)^{z+1}$, and for $z = {}^5/_2$ – appropriate for a two-atomic ideal gas – it follows that the oxygen leaves the turbine with $T_L \approx 90\,\mathrm{K}$, very close to the condensation point and far below the previous record minimum of 163 K. Actually Cailletet observed a fog of liquid droplets

[17] Diethyl ether, not the luminiferous variety of Chap. 2, of course; that would have been something!

behind the turbine. Thus he had successfully liquefied oxygen although, of course, the droplets quickly evaporated. The same could be done for fluorine, carbon monoxide and nitrogen and – after the noble gases had been isolated – for argon and neon.[18]

Effective isolation eventually produced liquids of the *permanent* gases in quantities sizable enough to study their properties, e.g. the boiling points. Even hydrogen was eventually liquefied in 1898 by James Dewar (1842–1923) and its boiling point turned out to be 20.3 K; solidification happens at 14K. For isolation Dewar invented the *Dewar flask*, a kind of thermos bottle, in which cold liquids could be stored for a long time, because the flasks had a *vacuum-filled* double wall, whose surface was silvered, so that even radiation losses were kept at a minimum.

Dewar was a man of many interests and talents: He erred, however, when he saw a connection between the blue of the sky and the blue colour of liquid oxygen. He invented cordite, a smokeless gun powder, and that brought him into a bitter fight about an alleged patent infringement with Alfred Bernhard Nobel (1833–1896). So, understandably, there was no Nobel prize for Dewar, although the road to absolute zero was otherwise paved with those prizes. However, Dewar was knighted and became Sir James. After his work only helium remained a gas. It deserves its own section, see below.

Despite effective isolation, until 1895 the cold liquids remained a laboratory curiosity. But then Carl Ritter von Linde (1842–1934) invented a continuous process of successive adiabatic throttling which produced liquids of oxygen and nitrogen in quantity, to be filled into high-pressure bottles and put to industrial use.[19] Throttling occurs when a vapour or a liquid are pushed or sucked through a narrow opening so that the pressure decreases and so does the temperature in most substances. The cooling effect is known as the Joule-Thomson effect – or Joule-Kelvin effect. We have learned about this before, cf. Chap. 2. In an ideal gas the effect is nil, or very tiny indeed – to the extent that the gas is not really ideal. This means that before throttling can be applied efficiently, the gas has to undergo Cailletet's adiabatic expansion, which converts it into a vapour

[18] The reader has surely noticed the author's special liking for the science essays of Isaac Asimov. Actually the present treatment of gas liquefaction also makes use of two such essays, namely I. Asimov: "Liquefying gases" and "Toward absolute zero" both in "Exploring the earth and the cosmos." Penguin Books, London (1990). These essays, however, see Asimov wrong, because he confuses Cailletet's adiabatic expansion and the adiabatic Joule-Thomson effect. The former is an essentially reversible process at constant entropy, while the latter is an inherently irreversible process with an unchanged enthalpy between beginning and end.

[19] Oxygen, nitrogen and hydrogen come in blue, green and red bottles, respectively, under a pressure of 150 bar.

close to liquefaction. Linde used several steps of throttling and *regeneration*, i.e. he pre-cooled the incoming flow of vapour by making it exchange heat with the already throttled one. The Linde process is still used now. And Linde's firm – founded in 1879 – thrives on selling liquefied gases, although it is mostly putting out the ubiquitous compression refrigerators, another invention of Linde's.

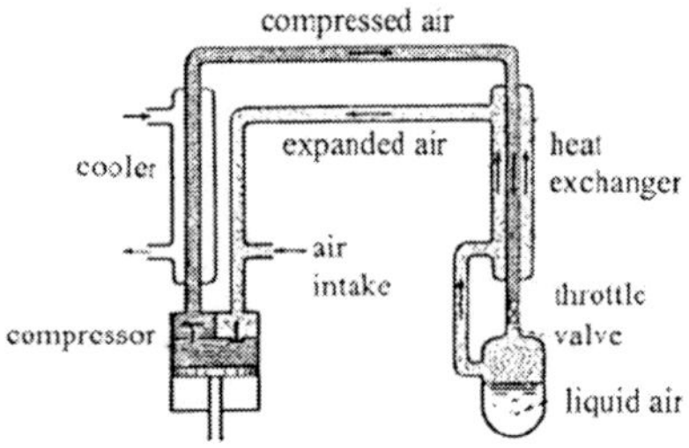

Fig. 6.4 . Carl Ritter von Linde (1842–1934). Schematic view of his air liquefying apparatus

Johannes Diderik van der Waals (1837 – 923)

Van der Waals was the person who made sense out of the concept of the critical point and who corroborated Andrew's conjecture that all gases should have such a point. He considered that the ideal gas law $p = \frac{1}{v}\frac{k}{\mu}T$ is an idealization which ignores inter-atomic forces. Van der Waals reasoned that the interaction force – now called *van der Waals force* – is mildly attractive at large distances and strongly repulsive when the atoms are close. Thus the potential $\varphi(r)$ of the force between two atoms in the distance r has the form shown qualitatively in Fig. 6.5.[20] On the grounds of this assumption van der Waals was able to derive a modified form of the ideal gas law, namely, cf. Insert 6.1

$$p = \frac{\frac{k}{\mu}T}{v-b} - \frac{a}{v^2} \text{ [21]} .$$

[20] Van der Waals could not know the nature of the attractive force. It is an electric dipole-dipole interaction, and the dipoles are due to a mutually induced differential shift of the electron shells and the nuclei of adjacent atoms.

[21] J.D. van der Waals: "Over de continuiteit van den gas- en vloeistoftoestand." [On the continuity of the gaseous and the liquid state]. Dissertation, Leiden (1873).

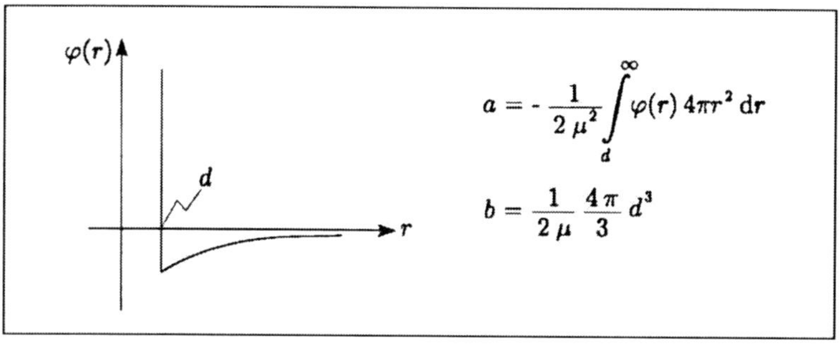

$$a = -\frac{1}{2\mu^2}\int_d^\infty \varphi(r)\,4\pi r^2\,dr$$

$$b = \frac{1}{2\mu}\,\frac{4\pi}{3}\,d^3$$

Fig. 6.5. Schematic form of the interatomic interaction potential as a function of the distance of two atoms. Also: van der Waals coefficients

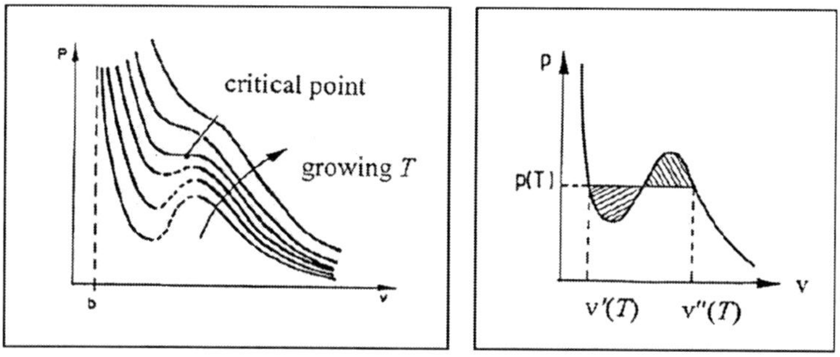

Fig. 6.6. Isotherms of a van der Waals gas. Also: Maxwell construction

This has become known as the *van der Waals equation* for a *real gas.* Obviously the modification lies in the positive coefficients a and b. The coefficient b represents the volume of an atom which clearly must detract from the total available volume. And the coefficient a represents the range and size of the attractive interaction which reduces the pressure exerted on the wall.

In a certain range of temperatures the van der Waals equation describes isotherms in a (p,v)-diagram that are non-monotone, as shown qualitatively in Fig. 6.6. Thus there is the possibility to have two – actually three – specific volumes for one pressure and one temperature. Ignoring the middle one, van der Waals interpreted the two remaining volumes as those of the liquid and the vapour, and came up with the surprising conclusion that his theory, intended for real gases – as opposed to ideal gases –, could perhaps describe a liquid-vapour transition. This is what the title of his work suggests. Accordingly *the* temperature, whose isotherm develops a horizontal point of inflection, has to be interpreted as the critical isotherm and the inflection point itself as the critical point. By the van der Waals equation that point has the coordinates

$$v_C = 3b, \qquad p_C = \frac{1}{27}\frac{a}{b^2} \qquad \frac{k}{\mu}T_C = \frac{8}{27}\frac{a}{b}.$$

Although van der Waals's work was presented as a doctoral thesis, – rather than in a scientific journal – it became quickly known. Boltzmann recognized it as a masterpiece, and he was so enthusiastic about the derivation that he called van der Waals the *Newton of real gases.*[22] And Maxwell discovered a graphical method for the determination of the saturated vapour pressure $p(T)$ for the van der Waals gas, see. Fig. 6.6. He wrote the phase-equilibrium condition of Insert 3.7 for the free energy $F = U\text{-}TS$ in the form

$$F'' - F' = -p(T)(V'' - V') \qquad \text{or with} \qquad p = -\left(\frac{\partial F}{\partial V}\right)_T$$

$$\int_{V'}^{V''} p(V,T)dV = p(T)(V'' - V'),$$

where the integration must be taken along the isotherm. Thus $p(T)$ is *the* isobar that makes the two shaded areas in Fig. 6.6 equal in size, This graphical method to determine $p(T)$, and $v'(T)$, $v''(T)$ has become known as the construction of the *Maxwell line.*

An interesting corollary of the van der Waals equation emerges when one introduces dimensionless variables

$$\pi = \frac{p}{p_C}, \qquad \nu = \frac{v}{v_C} \qquad \tau = \frac{T}{T_C},$$

because in that case the equation becomes *universal*, i.e independent of parameters relating to the particular fluid

$$\pi = \frac{8\tau}{3\nu - 1} - \frac{3}{\nu^2}.$$

Van der Waals called this relation the *law of corresponding states*: States with equal non-dimensional variables correspond (sic) to each other irrespective of the material properties. This implies that the liquid-vapour properties of all substances are alike:

- convex, monotonically increasing vapour pressure curves,
- similar wet steam regions and, of course
- critical points.

The underlying reason for such conformity is the fairly plain (φ, r)-relation, cf. Fig. 6.5, which is common – qualitatively – to all gases.

---

[22] In: Encyclopädie der mathematischen Wissenschaften, Bd. V.1. p. 550.

From a practical point of view, and with regard to liquefying gases, the most important conclusion from the van der Waals equation concerns the Joule-Thomson effect in a throttling experiment. It turned out that throttling did not necessarily lead to cooling. One thing was well-known, however: The energy flux before and behind an adiabatic throttle must be equal; therefore the first law requires that the specific enthalpy h is unchanged, provided that the kinetic energy of the flow can be neglected. That condition could be used for the calculation of the temperature change ΔT for a given pressure drop Δp, cf. Insert 6.2. One obtains the criteria

$$\frac{1}{v}\left(\frac{\partial v}{\partial T}\right)_p - \frac{1}{T} \begin{array}{c} > \\ = \\ < \end{array} 0 \text{ for } \begin{array}{c} \text{cooling} \\ \text{no change} \\ \text{heating} \end{array} .$$

Rather obviously the equality holds for ideal gases, so that ideal gases do not change their temperature upon adiabatic throttling. And for a van der Waals gas the criteria imply that the initial state must lie below the graphs which define the *inversion curve* in the (ν,τ)- , the (π,τ)-, or the (π,ν)-diagram, viz.

$$\nu = \frac{1}{3 - 2\sqrt{\frac{1}{3}\tau}}, \qquad \pi = 24\sqrt{3\tau} - 12\tau - 27, \qquad \pi = -\frac{9}{\nu^2} + \frac{18}{\nu}.$$

Obviously we have used here the dimensionless variables of the law of corresponding states. If a state lies on the inversion curves, it does not change temperature upon throttling; if it lies above the curves, the gas heats up.

Figure 6.7 shows the inversion curves in the (π,τ)-diagram and in the (π,ν)-diagram along with – for better orientation the critical isochor and the critical isotherm, respectively. Inspection of the (π,τ)-diagram – and of the mini-table in Fig. 6.7 with critical data for oxygen and hydrogen – shows that hydrogen of 1atm *heats up*, if throttled above T = 140 K. Therefore the Linde process for the liquefaction of hydrogen must start at a lower temperature. For oxygen, on the other hand, the process may start at room temperature. To be sure, it is not very efficient there; the cooling effect at room temperature was barely big enough to have been noticed by Joule and Kelvin.

The van der Waals equation with its two parameters a and b is quantitatively not good for any actual gas no matter how a and b are chosen. It does, however, have great heuristic value, because it is based on molecular considerations, cf. Insert 6.1, and it represents a fairly simple analytic thermal equation of state. It is therefore revisited over and over again. Fairly recently I have come across an instructive article entitled

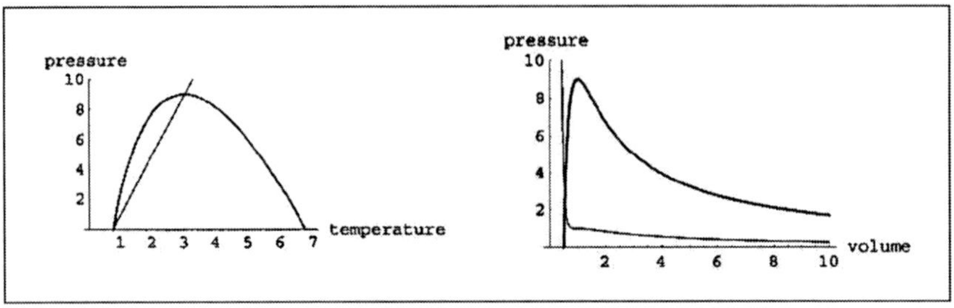

	T_c [°C]	p_c [atm]
Oxygen	-182.97	49.7
Hydrogen	-252.78	12.8

Fig. 6.7. Inversion curves and critical isochor and isotherm Also: Mini-table of critical data

"Thirteen ways of looking at the van der Waals equation".[23] And I believe that in a recent book [24] I have presented a fourteenth way.

Students of thermodynamics are often mystified by the non-monotone isotherms exhibited in Fig. 6.6 and, in particular, by the branch with a positive slope, which suggests instability. These features are reflections of the non-convex character of the function $\varphi(r)$, but we shall not go into that, although at present – while I write this – there is great interest in similar phenomena occurring in phase transitions in solids, like shape memory alloys. An instructive mechanical model for non-monotone stress-strain curves has been proposed and investigated by the author.[25]

Van der Waals equation

All N atoms of a monatomic gas in a volume V with the surface ∂V and outer normal $\boldsymbol{n}$ move according to Newton's law of motion

$$\ddot{x}_\alpha = K_\alpha \qquad (\alpha = 1,2,\ldots).$$

If that equation is multiplied by $\boldsymbol{x}_\alpha$, and then averaged over a long time τ, and summed over all α, one obtains

$$-\left\langle \sum_{\alpha=1}^{N} \dot{x}_\alpha^{\,2} \right\rangle = \left\langle \sum_{\alpha=1}^{N} K_\alpha x_\alpha \right\rangle \qquad \text{(angular brackets denote averages).}$$

[23] M.M. Abbott: Chemical Engineering Progress, February (1989).
[24] I. Müller, W. Weiss: "Entropy and energy,..." loc.cit. (2001).
[25] I. Müller, P. Villaggio: "A model for an elastic-plastic body" Archive for Rational Mechanics and Analysis. 65 (1977).

The left hand side is equal to $-3NkT$, since each atom has an average value $^1/_2\, kT$ of kinetic energy. The right hand side was called *virial* by Clausius. The virial has two parts W_s and W_i due to forces on atoms from the surface and from other atoms respectively. Therefore we write

$$-3\, NkT = W_s + W_i\,.$$

Assuming that only atoms in the immediate neighbourhood of the surface element dA of ∂V feel the effect of the surface, and that the sum of forces from the surface on those atoms is equal to $-p\boldsymbol{n}dA$ on average, we obtain $W_s = -3\, pV$. Hence follows

$$pV = N\,k\,T + {}^1/_3\, W_i.$$

Without the inner virial W_i we thus have regained the ideal gas law.

The force on atom α from atom β may be written $K_{\alpha\beta} = -K\left(\left|x_\alpha - x_\beta\right|\right)\dfrac{x_\beta - x_\alpha}{\left|x_\beta - x_\alpha\right|}$. It follows for W_i

$$W_i = -\sum_{\alpha=1}^{N}\sum_{\beta=1}^{N}\left\langle K\left(\left|x_\alpha - x_\beta\right|\right)\frac{x_\beta - x_\alpha}{\left|x_\alpha - x_\beta\right|}\, x_\alpha\right\rangle$$

$$W_i = \frac{1}{2}\sum_{\alpha=1}^{N}\sum_{\beta=1}^{N}\left\langle K\left(\left|x_\alpha - x_\beta\right|\right)\left|x_\alpha - x_\beta\right|\right\rangle$$

$$= \frac{N}{2}\sum_{\alpha=1}^{N}\left\langle K\left(\left|x_\alpha - x_\beta\right|\right)\left|x_\alpha - x_\beta\right|\right\rangle \quad \text{for any } \beta.$$

The last step requires that on average each atom is surrounded by others in the same manner. We set $\left|x_\beta - x_\alpha\right| - r$ and convert the sum in an integral by defining the particle density $n(r)$.

$$W_i = \frac{N}{2}\int_V K(r)\, r\, n(r)\,\mathrm{d}V \qquad \text{or, by isotropy}$$

$$= 2\pi N\int_0^\infty K(r) n(r) r^3 \mathrm{d}r.$$

The force $K(r)$ and the potential $\varphi(r)$ of Fig. 6.5 are related by $K(r) = -\frac{\mathrm{d}\varphi}{\mathrm{d}r}$ and the particle density $n(r)$ may approximately be given by $\frac{N}{V}\exp(-\frac{\varphi}{kT})$, so that an atom on average is surrounded by a cloud of other atoms which is densest, where $\varphi(r)$ has its minimum. Insertion provides

$$W_i = 2\pi \tfrac{N^2}{V} kT\, 3\int_0^\infty (1 - \exp(-\tfrac{\varphi}{kT}))r^2 \mathrm{d}r.$$

We set $\varphi = \infty$ for $r < d$ and $\frac{\varphi}{kT} << 1$ for $r > d$ as indicated in Fig. 6.5 and obtain

$$W_i = 3\left[NkT\frac{\frac{1}{2}N\frac{4\pi}{3}d^3}{V} + \frac{N}{2}\int_d^\infty \varphi(r)\frac{N}{V}4\pi r^2 \mathrm{d}r\right]$$

or, with a and b from Fig. 6.5,

$$W_i = 3\left[NkT\frac{b}{v} - V\frac{a}{v^2}\right].$$

Elimination of W_i between this and the equation for pV provides the van der Waals equation, provided we assume that $b << v$ holds which is reasonable.

Insert 6.1

Throttling

Adiabatic throttling is an isenthalpic process of lowering the pressure by $\Delta p < 0$. The temperature changes accordingly so that $\Delta h = 0$ can be satisfied. Therefore we have

$$\frac{\Delta T}{\Delta p} = -\frac{\left(\frac{\partial h}{\partial p}\right)_T}{\left(\frac{\partial h}{\partial T}\right)_p}.$$

The denominator is the specific heat c_p, and the numerator may be rewritten – by use of the Gibbs equation – in the form $\left(\frac{\partial h}{\partial p}\right)_T = v - T\left(\frac{\partial v}{\partial T}\right)_p$. Hence follows

$$\frac{\Delta T}{\Delta p} = \frac{vT}{c_p}\left(\frac{1}{v}\left(\frac{\partial v}{\partial T}\right)_p - \frac{1}{T}\right)$$

and we conclude that cooling occurs, if the thermal expansion coefficient $\alpha = \frac{1}{v}\left(\frac{\partial v}{\partial T}\right)_p$ is greater than $^1/_T$. For ideal gases we have $\alpha = {}^1/_T$.

Insert 6.2

Helium

Helium deserves its own section, although it was liquefied in the same manner as hydrogen, by adiabatic expansion and throttling. It just took more time, because the boiling point was lower: 4.2K. It was Heike Kammerlingh-Onnes (1853–1926) who succeeded in 1908 and who eventually reached 0.8K by adiabatic evaporation of the liquid. Kammerlingh-Onnes received the Nobel prize for his efforts in 1912.

He did not succeed, however, to *freeze* helium, and later it turned out that it cannot be done, no matter how far the temperature is lowered, at least not under ordinary pressures. It took pressures of 20 atm, or so, to make helium solid.

The reason for the persistence of the liquid phase is supposed to be quantum mechanical. According to quantum mechanics a particle with momentum p and energy $\frac{p^2}{2\mu}$ may be considered as a *de Broglie wave* with a wave length $\lambda = \frac{h}{p}$ and a frequency $\nu = \frac{1}{h}\frac{p^2}{2\mu}$.[26] Such a particle has an equal probability to be anywhere in space, so that it cannot be localized. A particle, however, which we know to be boxed in, in a range of linear dimension Δx, is represented in quantum mechanics by a *packet* of de Broglie waves, i.e. a superposition of such waves with momenta in the range Δp. Between Δx and Δp there is the relation $\Delta x \, \Delta p = h$, which is called Heisenberg's *uncertainty relation.* Thus either x or p may be fixed, but not both.

The above is a subject of single-particle quantum mechanics, governed by the Schrödinger equation. The uncertainty relation is extrapolated to thermodynamics by the assumption that Δp may be interpreted as the momentum of a particle of a liquid (say) during its thermal motion. Thus we may write

$$\Delta p = \sqrt{2\mu E} \quad \text{or with} \quad E = \tfrac{1}{2}kT: \quad \Delta p = \sqrt{\mu kT}\,,\ \Delta x = \frac{h}{\sqrt{\mu kT}}.$$

Δx is therefore a typical *de Broglie wave length* of a particle of a body of temperature T, such that Δx^3 represents the smallest volume element in which such a particle can be localized. For an atom of liquid helium at T = 1K the uncertainty Δx of position comes out as $\Delta x = 2\cdot 10^{-9}$m. This is considerably more than if the particle were confined to an elementary cell in a solid lattice structure. Therefore the solid lattice cannot form, and that is why helium remains liquid, or so they say.

Once liquid helium was available, it could be used to cool other substances down to the neighbourhood of absolute zero. And it turned out that some metals, like mercury and lead develop a very strange behaviour indeed. They lose their electrical resistance at some characteristic temperature. We say that they become *super-conductors,* materials with zero resistivity in which a current, once induced, moves round and round forever.

Actually helium itself, below 2.19K – the so-called λ-point – exhibits a somewhat similar unique phenomenon of its own: It behaves like a mixture of a normal fluid with a small viscosity and a *super-fluid,* which has no viscosity at all. That liquid *mixture* is called He II as opposed to He I, liquid helium *above* the λ-point. The lower the temperature is, the higher is the proportion of the super-fluid.

Strange phenomena occur in He II, or they should occur and do not. The substance dumbfounded eminent scientists like Lev Davidovich Landau (1908–1968) and Evgenii Michailovich Lifshitz (1915–1985), Landau's colleague, collaborator and frequent co-author. It is, perhaps, worthwhile to describe two of the more illustrative snares in which those scientists found

---

[26] $h = 6.626\cdot 10^{-34}$Js is the Planck constant.

themselves entangled for years. If nothing else, that will be a consolation for those of us – less eminent than Landau and Lifshitz – who find it difficult to adjust their minds to the evidence of the new and unusual. Let us consider:

Sound first: Sound in air – essentially a two-constituent mixture of nitrogen and oxygen – permits two wave modes, both longitudinal. One consists of the joint oscillation of both constituents with no relative velocity, while for the other one the two constituents move relative to each other with no motion of the mixture as a whole. Those modes may be called the *first and second sound* respectively. Both propagate with different speeds and both are usually coupled so that, if the first sound is stimulated, the second sound follows, and vice versa. We never actually hear the second sound in air, because it is damped away within the distance of less than 1 mm from our vibrating vocal cords; this may be a good thing, because it saves us from hearing everything twice. Also temperature oscillations are associated with both sounds, although in air they cannot be detected, at least not by our coarse human senses.

Sound in helium below the λ-point is qualitatively similar, since it behaves like a mixture. But quantitatively it differs, chiefly on account of the fact that one constituent, the super-fluid, is free of friction so that damping is absent. The theory of first and second sound was first worked out by Lazlo Tisza (1907-).[27] A little later Landau developed essentially the same theory [28] and therefore the governing equations are very often called the *Landau equations.* According to those equations the second sound should be detectable, but it was not, or not for years. The first sound came through helium loud and clear at one side when it was excited by a vibrating membrane at the other side, but no second sound could be detected. At the end, after many vain attempts, a frustrated Lifshitz sat down and did a simple calculation, a calculation that should have been done beforehand: He calculated the amplitudes.[29] Then it turned out that, according to the Landau equations, the first and second sound were now uncoupled, so that the second sound could not be stimulated by a vibrating membrane, and that the first sound was not accompanied by a temperature oscillation, but the second sound was. So Lifshitz suggested to use an electric coil with an alternating current instead of a membrane. The Joule heating of the coil produced temperature oscillations, and there was the second sound immediately, – as a thermal wave, in a manner of speaking. It propagated with the speed predicted by the Landau equations.

---

[27] L. Tisza: "Transport phenomena in He II." Nature 141 (1938).

[28] L.D. Landau: "The theory of superfluidity of Helium II." Journal of Physics (USSR) 5 (1941).

[29] E.M. Lifshitz: "Radiation of sound in Helium II." Journal of Physics (USSR) 8 (1944).

For the Landau equations – and other achievements – Landau obtained the Nobel prize in 1962. From January of that year, and for the whole year, he lay in a coma after an automobile accident. They say that he passed away several times but was brought back to life by drastic methods. Lifshitz presented the award to him in the hospital. Landau survived, but not as an active physicist.

Another peculiarity of He II – apart from second sound – occurs under rotation: Since the super-fluid has no viscosity, it should be impossible to impart a rotation to it. Accordingly Landau – faithful to the Landau equations – predicted that the surface of super-fluid helium should remain flat, even if its container sits on a rotating turntable. That, of course, presented a challenge for experimentalists and it was not long before D.V. Osborne[30] came up with a rotating container of liquid helium. The surface turned out to be a perfect paraboloid, just like for any other incompressible liquid in rigid rotation, – in contradiction to Landau's expectations.

In that case it was Lars Onsager (1903–1976) who proposed an ingenious solution of the dilemma during a panel discussion of the Osborne phenomenon. Onsager knew that a homogeneous distribution of potential vortices [31] *mimics* a rigid rotation, i.e. has the same velocity field. Therefore he suggested that Osborne's rotating helium was a superposition of such potential vortices. In this way he saved Landau's theory and yet explained Osborne's experiment. Moreover, sceptics were quickly convinced, because the vortices could in fact be made visible when it turned out that an electron beam could pass through the cores of the vortices and nowhere else.

It remains to understand the physical reason for super-fluidity of helium. The usual assumption seems to be that this phenomenon is a case of Bose-Einstein condensation, which we shall come to know later in this chapter.

Adiabatic Demagnetisation

The wish to study super-conductivity of metals and super-fluidity of helium has motivated a drive for lower and lower temperatures. New methods were needed to get below 0.5K and they were found. Peter Joseph Wilhelm Debye (1884–1966) and William Francis Giauque (1895–1982) came up independently with the idea of *adiabatic de-magnetization.* A magnetic salt – gadolinium sulfate in Giauque's case – was put under a strong magnetic field so that the magnetic dipoles of the salt lined up in the direction of the field, because energetically that is the most favourable position. That material – still under the magnetic field – was cooled with

[30] D.V. Osborne: "The rotation of liquid Helium II" Proceedings Physical Society A 63 (1950).

[31] Potential vortices are like the vortex in an emptying bathtub, or like a tornado. Ideally they are free of dissipation and thus should be able to exist in super-fluid helium.

liquid helium and then adiabatically isolated. Afterwards the field was slowly switched off, so that the thermal motion of the dipoles could randomise their orientation by sending the dipoles *uphill*, as it were, in the energetic landscape, against the direction of the remaining field. This means that the salt was cooled and the salt in turn cooled the surrounding helium. Giauque reached 0.25K with gadolinium sulfate and later, with other salts, temperatures as low as 0.02K. The technique was refined and eventually produced temperatures as low as 3mK. Further cooling proved to be impossible in this way, because the dipoles of the electron shells start to align *themselves*, so that the magnetic field had no effect, nor does randomisation take place.

He^3 – He^4 Cryostats

However, there are also *nuclear* dipoles, of copper (say). In order to align *them,* very strong magnetic fields and sustained small temperatures are needed and those can be provided by a He^3-He^4 cryostat. The method for *maintaining* low temperatures by evaporation of He^3 was first conceived by Heinz London (1907–1970) in 1962. Let us consider this.

Helium comes in two isotopes He^3 and He^4. Under natural conditions there are about a million times more He^4 atoms than He^3 atoms. But the mixture can be enriched and, when this is done, it turns out that below 0.87K – in the liquid phase – a miscibility gap opens up, cf. Fig. 6.8, because the now sluggish thermal motion cannot supply the energy needed to form (He^3-He^4)-neighbours. Roughly speaking that gap is bell-shaped in the (T,X)-diagram, see Fig. 6.8.[32] Since He^3 is lighter, it floats on top, where it may be made to evaporate. As always for adiabatic evaporation the temperature drops – by ΔT – and, since the light constituent is more volatile, the system loses He^3, even though the He^3-rich solution on top becomes even more enriched, cf. Fig. 6.8. Thus more He^3 evaporates and so it happens that a low temperature of 10μK can be reached and maintained for days. The copper is eventually just as cold and the magnetic field keeps its nuclear dipoles aligned. Afterwards, when demagnetisation occurs, the dipoles randomise and the copper cools to 1.5 μK, the lowest temperature reached so far.

Physicists have so much faith in the third law that a jargon has developed among them according to which *the third law forces miscibility gaps to appear* in alloys and mixed isotopes because, after all, the mixture must shed its entropy of mixing, if the entropy is to go to zero.

---

[32] I am told that the bell is not quite symmetric and that it does not seems to cover the whole range $0 < X < 1$ when T tends to zero. For the present consideration this is not important.

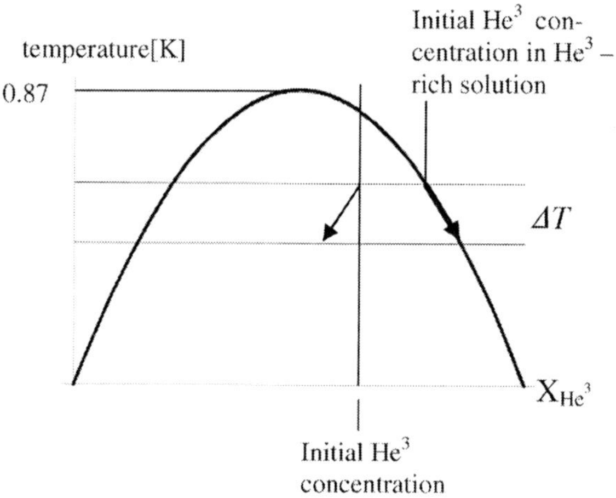

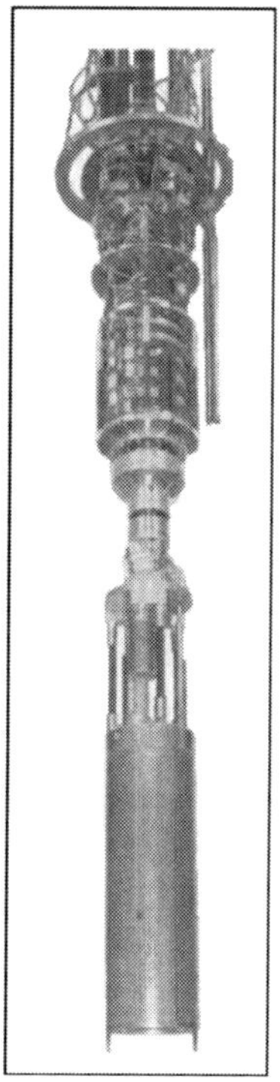

Fig. 6.8. Miscibility gap in He^3-He^4 phase diagram (schematic). Enriching He^3 in the He^3-rich phase by evaporation. He^3-He^4 cryostat of the Physikalisch Technische Bundesanstalt in Berlin.[33]

Entropy of Ideal Gases

Although in this chapter we are dealing with low and lowest temperatures, we have to consider ideal gases for several reasons, but primarily because we wish to have further confirmation of the third law. Also we wish to understand super-fluidity, perhaps.

We recall that Boltzmann's extrapolation

$$S = k \ln W \quad \text{with} \quad W = \frac{N!}{\prod\limits_{xc} N_{xc}!}$$

---

[33] The photograph is taken from an article by P. Strehlow: "Die Kapitulation der Entropie – 100 Jahre III. Hauptsatz der Thermodynamik." [The capitulation of entropy – 100 years of 3rd law of thermodynamics] Physik Journal 4 (12) 2005.

was seriously flawed, cf. Chap. 4. The basic reason is the *way of counting* realizations of the distribution $\{N_{xc}\}$, because Boltzmann believed – as everybody did in his time – that an interchange between identical particles in different elements of the (***x***,***c***)-space leads to a new realization. According to quantum mechanics of many particles this is *not* the case. Also Boltzmann could not know about bosons and fermions and de Broglie waves. So, if we wish to repair Boltzmann's reasoning, we have to take two observations from quantum mechanics into account:

- **There is no way to distinguish between identical particles**
  The classical idea is that we may mark particles, e.g. *paint* them in different colours. But this is not only impractical, it is incompatible with quantum mechanics, where the particles are de Broglie waves, as it were.
- **There are two types of particles, fermions and bosons**
  No two fermions may occupy the same state, but there is no such restriction on bosons; they may all pile up in one state.

For a unified treatment of fermions and bosons we assume here that each state may be occupied by up to d particles. Of course, $d = 1$ holds for fermions and $d = \infty$ for bosons and these seem to be the only two cases that occur in nature.[34]

The new argument was prompted by Satyendra Nath Bose (1893–1974), who made two important contributions when he improved the derivation of Planck's radiation formula in 1924:

- Bose was the first person to take seriously Boltzmann's *cells* in ***x***,c-space and to give them a definite *volume*.[35] We recall that Boltzmann himself had considered those cells as a calculational trick without physical significance, cf. Chap. 4. Not so Bose; he quantized the phase space – spanned by coordinates and momenta – into cells of size h^3. He needed that value in order to arrive at Planck's formula.[36]
- Also Bose introduced a new way of characterizing realizations and distributions. He does that without any fanfare as a matter of course, and without commenting on the move, and without showing a sign that he was aware of revolutionizing statistical mechanics. Bose sent his

[34] The idea of having an occupancy of an arbitrary number d was introduced by G. Gentile: "Osservazioni sopra le statistiche intermedie." [Observations on intermediate statistics] Nuovo Cimento 17, p. 493–497.

[35] We shall review Bose's contribution in detail in the next chapter which deals with radiation. Let it be said here – in anticipation – that Planck's radiation formula had resulted from an interpolation between two empirical functions. This was not satisfactory, at least not for Bose. Einstein had already improved Planck's derivation by introducing stimulated emission; but he, too, relied on classical thinking when he adopted the Boltzmann factor, cf. Chap. 4, for the relative frequency of atoms in different energetic states. Again Bose found this unsatisfactory.

[36] Accordingly the measure factor Y which I have used heretofore will henceforth be chosen as $Y = \mu^3/h^3$.

4 page-paper to Einstein who translated it into German and had it published in the Zeitschrift für Physik.[37]

Einstein added a *note of the translator* saying that *Bose's derivation of the Planck formula represents ... an important step forward. The method used* [by Bose] *furnishes also a quantum theory of the ideal gas as I shall explain elsewhere.* And indeed, Einstein let himself be inspired by Bose's paper. He followed it up with two papers of his own which he read in July 1924 and January 1925 to the Preußische Akademie der Wissenschaften.[38] In these papers Einstein develops the novel theory of degenerate gases, i.e. ideal gases at low temperature and large density, which I proceed to describe.

Of course, $S = k\ln W$ had to be retained, because of its inherent plausibility, and neither Bose nor Einstein touched that relation. But the realization of a distribution, and the distribution itself, were modified, and so was W. As before, cf. Insert 4.6, we concentrate on the infinitesimal element $\mathrm{d}\boldsymbol{x}\mathrm{d}\boldsymbol{c}$ at $(\boldsymbol{x},\boldsymbol{c})$ in $(\boldsymbol{x},\boldsymbol{c})$-space where we have

$$P_{\mathrm{d}x\mathrm{d}c} = Y\,\mathrm{d}x\mathrm{d}c \qquad - \quad \text{No. of cells in } \mathrm{d}x\mathrm{d}c$$
$$N_{\mathrm{d}x\mathrm{d}c} = \sum_{P_{\mathrm{d}x\mathrm{d}c}} N_{xc} = f(x,c)\mathrm{d}x\mathrm{d}c \qquad - \quad \text{No. of atoms in } \mathrm{d}x\mathrm{d}c.$$

The new *distribution* in $\mathrm{d}\boldsymbol{x}\mathrm{d}\boldsymbol{c}$ is given by the set

$$\{p_l^{xc}\} = \{p_0^{xc}, p_1^{xc}, \ldots p_d^{xc}\}$$

which represents the number of cells which are occupied by $0,1,\ldots d$ atoms. Obviously the values p_l^{xc} must satisfy the constraints

$$\sum_{l=0}^{d} p_l^{xc} = P_{\mathrm{d}x\mathrm{d}c} \qquad \text{and} \qquad \sum_{l=0}^{d} l p_l^{xc} = N_{\mathrm{d}x\mathrm{d}c}\,.$$

A *realization* of this distribution is given by $\{N_{xc}\}$, the number of atoms sitting in the individual cells $(\boldsymbol{x},\boldsymbol{c})$ in $\mathrm{d}\boldsymbol{x}\mathrm{d}\boldsymbol{c}$. Thus by the rules of combinatorics the number of realizations of the distribution $\{p_l^{xc}\}$ is equal to

$$W_{xc} = \frac{P_{\mathrm{d}x\mathrm{d}c}!}{\prod_{l=0}^{d} p_l^{xc}!}. \quad \text{Hence} \quad S_{xc}\mathrm{d}x\mathrm{d}c = k\ln\frac{P_{\mathrm{d}x\mathrm{d}c}!}{\prod_{l=0}^{d} p_l^{xc}!}$$

---

[37] S.N. Bose: "Planck's Gesetz und Lichtquantenhypothese." [Planck's law and the hypothesis of light quanta] Zeitschrift für Physik 26 (1924).

[38] A. Einstein: "Quantentheorie des einatomigen idealen Gases." [Quantum theory of a monatomic ideal gas] Sitzungsberichte physikalisch mathematische Klasse, September 1924 pp. 261–267.

A. Einstein: "Quantentheorie des einatomigen idealen Gases II" [Quantum theory of a monatomic ideal gas. II] Sitzungsberichte physikalisch mathematische Klasse, February 1925 pp. 3–14.

is the entropy of the atoms in the element d$\boldsymbol{x}$d$\boldsymbol{c}$, and

$$S = k \ln \prod_{xc} \frac{P_{dxdc}!}{\prod\limits_{l=0}^{d} p_l^{xc}!}$$

is the total entropy of the gas, where $\prod\limits_{xc}$ is the product over all elements d$\boldsymbol{x}$d$\boldsymbol{c}$ of the space $(\boldsymbol{x},\boldsymbol{c})$.

This new form of entropy lacks the inherent perspicuity of Boltzmann's entropy, because the relation to $N_{\mathrm{d}x\mathrm{d}c}$, or to the distribution function $f(\boldsymbol{x},\boldsymbol{c})$ is not explicit. However, for fermions such an explicit relation does exist, and for bosons it does exist in *local equilibrium*, where there is no knowledge about N_{xc} in d$\boldsymbol{x}$d$\boldsymbol{c}$ except about the average value which is

$$N_{xc} = \frac{N_{\mathrm{d}x\mathrm{d}c}}{P_{\mathrm{d}x\mathrm{d}c}}.$$

In those cases the entropy may be written in the form

$$S = -k\int \left[\ln\frac{f}{Y} \pm \frac{Y}{f}\left(1 \mp \frac{f}{Y}\right)\ln\left(1 \mp \frac{f}{Y}\right)\right] f\mathrm{d}c\mathrm{d}x \quad \begin{matrix}\text{fermions}\\ \text{bosons}\end{matrix} .$$

This is the proper form of the entropy in a monatomic gas; the expression generalizes Boltzmann's relation

$$S = -k\int \ln\frac{f}{b} f\mathrm{d}c\mathrm{d}x \quad \text{with} \quad b = eY ,$$

found – by accident or luck, as it were – in the kinetic theory of gases, cf. Chap. 4. And it coincides with Boltzmann's form, if the difference between fermions and bosons, i.e. the ± - alternative, becomes unimportant. This happens for $f/Y \ll 1$ or

$$N_{xc} = \frac{N_{\mathrm{d}x\mathrm{d}c}}{P_{\mathrm{d}x\mathrm{d}c}} \ll 1 \quad \text{i.e. for sparse occupancy of each element d}x\text{d}c.$$ [39]

This observation is eminently plausible because, if there is much less than one atom per element d$\boldsymbol{x}$d$\boldsymbol{c}$ on average, it makes no difference whether the atom is a fermion or a boson, since even a double occupancy of a cell practically does not occur, let alone higher occupancies.

Obviously S in terms of the distribution function f is a non-equilibrium entropy in general. In a closed adiabatic gas, i.e. for a fixed number N of atoms and for a fixed energy U, we expect S to tend to a maximum S_{equ} in equilibrium. The calculation provides

[39] Recall that for Boltzmann it was a matter of course, that N_{xc} was greater than 1. In fact, it had to be big enough that the Stirling formula could be applied.

$$f_{equ} = \frac{Y}{\exp[-\frac{\mu g}{kT} + \frac{\mu c^2}{2kT}] \pm 1} \qquad \begin{matrix}\text{fermions} \\ \text{bosons}\end{matrix} ,$$

where g is the specific Gibbs free energy, and T the temperature, of course. This expression replaces the Maxwellian distribution function in a *degenerate gas* , i.e. a gas for which the quantum effects – evidenced by the ± – alternative – make themselves felt. The thermal and caloric equations of state

$$p = \frac{2}{3}\frac{U}{V} = p\left(\tfrac{N}{V}, T\right) \text{ and } g = g\left(\tfrac{N}{V}, T\right)$$

are given implicitly

$$\frac{N}{V} = 4\pi Y \sqrt{\tfrac{2}{\mu}}^{3} \int_{o}^{\infty} \frac{x^2 dx}{\exp[-\frac{\mu g}{kT} + \frac{x^2}{kT}] \pm 1}, \quad \frac{U}{V} = 4\pi Y \sqrt{\tfrac{2}{\mu}}^{3} \int_{0}^{\infty} \frac{x^4 dx}{\exp[-\frac{\mu g}{kT} + \frac{x^2}{kT}] \pm 1},$$

and the equilibrium entropy S_{equ} reads

$$T S_{equ} = -N\mu g + {}^{5}/_{3}\, U.$$

Classical Limit

The Boltzmann limit occurs – just like in the non-equilibrium case – when the ±-alternative for the fermions and bosons does not matter, i.e. for $\frac{\mu g}{kT} << -1$. In that case we have

$$p = \frac{N}{V} kT \text{ and with } \frac{\mu^3}{Y} = h^3 : \frac{\mu g}{kT} = \ln\left[\frac{N}{V}\left(\frac{h}{\sqrt{2\pi\mu kT}}\right)^3\right].$$

It follows that the classical limit is the one, in which an element of phase space of the dimension of a typical *thermal* de Broglie wavelength, see above, contains practically no particle. In contrast, degeneration therefore appears as *the* state, where the particles are so dense, or the temperature is so low, that the de Broglie wavelengths overlap.

Note that for particles with a small mass the de Broglie wavelength is big. It is for that reason that even at room temperature – and even for a few thousand K – the *electron gas* in a metal is strongly degenerate, – also of course, because the electron density ${}^{N}/_{V}$ is large.

For the non-degenerate state the equilibrium entropy has the form

$$S_{equ} = Nk\left\{\frac{5}{2} - \ln\left[\frac{N}{V}\left(\frac{h}{\sqrt{2\pi\mu kT}}\right)^3\right]\right\}.$$

That value is entirely explicit! Thanks to Bose's choice $Y = \mu^3/h^3$ there is no unknown constant. The expression provides the absolute value of the entropy for a rarefied ideal gas. Hence, by integration over $c_p(T,p)/T$ – and summation of latent heats divided by the temperatures of their occurrence – downward to lower temperatures, one may obtain the absolute value of entropy of liquids and solids at absolute zero, or as close as we can get there.

If one proceeds with that integration – after having made all those caloric measurements – one obtains the value zero for entropy *in most cases* and thus confirms Planck's extension of the third law of thermodynamics. Sometimes, however, the value zero is *not* obtained. That seems to happen only when the solid phase is amorphous, – rather than crystalline – so that the third law must be qualified: the entropy at absolute zero for amorphous solids is *not* zero. Handbooks record the value as the *zero point entropy.*

Full Degeneration and Bose-Einstein Condensation

The opposite of the classical limit – the limit of full degeneracy – is different for fermions and bosons.

Fermions

For fermions the limit is characterized by $\frac{\mu g}{kT} >> 1$ so that

$$f_{equ} = \begin{cases} Y \\ 0 \end{cases} \text{for} \quad \begin{matrix} \frac{1}{2}c^2 < g \\ \frac{1}{2}c^2 > g \end{matrix}.$$

At low temperature all atoms tend to assemble at zero kinetic energy, but that desirable state cannot be achieved, since each velocity can only be assumed by just one atom.[40] Therefore the atoms do the next best thing and fill all states with the lowest velocities. N and U are given by

[40] Actually, two atoms may assume the same velocity, if they have different spins.

ELGIN COMM COLL LIB
1700 SPARTAN DRIVE
ELGIN IL 60123

BAKE

INV

FED TAX ID: 56-1761729
SHIPPED FROM: MOMENCE
CUSTOMER SERVICE:
CREDIT: 800.340.5370/INTL 704.998.3399

GST/TAX ID#: 00 000000000000000
PO#: 5866ZZ

ALL CLAIMS MUST BE MADE WITHIN 45 DAYS OF INVOIC

QTY	TITLE		AUTHOR	TYPE
	BT ORDER #	CUSTOMER PO #	FUND #	CUST REF
1	HISTORY OF THERMODYNAMICS THE DOCTRINE O 92339807	5866ZZ	MULLER, INGO	HRD 00513715
1				SUB TO USD CU

TERMS:

R&TAYLOR

the future delivered

OICE

INVOICE #: 2024763228
INVOICE DATE: 06/17/10
ACCOUNT #: 203409 U074245 2 B00000
ATS #: MOM9119141

PAGE: 001

BILL TO: ACCOUNT #: 203409 U074245 2 B00000
SAN #:
NAME: ELGIN COMMUNITY COLLEGE
ADDRESS: JUNIOR COLLEGE DIST 509
1700 SPARTAN DR
ELGIN IL 60123

SHIP TO: ACCOUNT #: 203409 U074245 2 000000
SAN #: 3041530 0001
NAME: ELGIN COMM COLL LIB
ADDRESS: 1700 SPARTAN DRIVE
ELGIN IL 60123

:. RETURN AUTHORIZATION REQUIRED. NOT RESPONSIBLE FOR GOODS SENT UNINSURED.

	ISBN	PUB.	PRICE	DISC.	NET PRICE	EXTENDED PRICE
#	ISBN-10			VAS		
'7	9783540462262 3540462260	SPRIV	119.00	5.0%	113.05	113.05
AL :RENCY						113.05

FREIGHT SURCHARGE 0.28

TOTAL AMOUNT DUE 113.33

REMIT TO: BAKER & TAYLOR
P.O. BOX 277930
ATLANTA, GA 30384-7930
NEW REMITTANCE ADDRESS

PLEASE INDICATE INVOICE # ON YOUR REMITTANCE

00 NET 30 DAYS
AMOUNTS BILLED IN USD

$$\frac{N}{V} = 4\pi Y \sqrt{\frac{2}{\mu}}^3 \frac{1}{3} g^{3/2} \quad \text{and} \quad \frac{U}{V} = 4\pi Y \sqrt{\frac{2}{\mu}}^3 \frac{1}{5} g^{5/2},$$

so that the energy is large, but the entropy vanishes.

Bosons

For bosons – with the lower sign – we must realize that the biggest value of g must be $g = 0$, lest negative values of the distribution function appear. Therefore $g = 0$ and

$$f_{equ} = \frac{Y}{\exp[\frac{\mu c^2}{2kT}] - 1}$$

characterize the Bose case of full degeneracy. The properties of the distribution are much as expected, because it implies that there are less particles with larger speeds. However, there is a problem, since f_{equ} is singular for $c = 0$: To be sure, the values of $^N/_V$ and $p = {^2/_3}\, {^U/_V}$ are finite, namely[41]

$$\frac{N}{V} = Y \sqrt{\frac{2}{\mu}}^3 \sqrt{2\pi \frac{k}{\mu} T}^3 \zeta\left(\frac{3}{2}\right) \quad \text{and} \quad p = Y \sqrt{2\pi \frac{k}{\mu} T}^5 \zeta\left(\frac{5}{2}\right),$$

but there is something strange. Indeed $^N/_V$ and p are functions of T only, a circumstance that we have come to expect as an equilibrium condition for saturated vapour coexisting with a boiling condensate.

That observation may serve as a hint that the equation for the number N of atoms is incorrect, because N cannot possible depend on T. And indeed, the equation holds only for the number of particles with $\boldsymbol{c} \neq 0$, while N_0, the number of particles with $\boldsymbol{c} = 0$, has somehow slipped through the (Riemann)-integration, although its density is singular. Therefore the $^N/_V$–equation must be rewritten as

$$N = N_0 + YV \sqrt{\frac{2}{\mu}}^3 \sqrt{2\pi \frac{k}{\mu} T}^3 \zeta\left(\frac{3}{2}\right).$$

And, if $YV\sqrt{\frac{2}{\mu}}^3 \sqrt{2\pi \frac{k}{\mu} T}^3 \zeta(\frac{3}{2})$ is the number of particles in the *vapour*, N_0 is the number of particles in the *condensate*. One says: The N_0 particles with

41 $\zeta(^3/_2)$ and $\zeta(^5/_2)$ are values of the Riemann zeta function which occurs in the integration of the distribution function for $g = 0$.

$c = 0$ form the *Bose-Einstein condensate.*[42] For $T \rightarrow 0$ there will be more and more condensate, whose entropy is zero. The entropy of a Bose gas for full degeneracy vanishes therefore for $T \rightarrow 0$.

The observed decomposition of liquid helium into a normal fluid and a super-fluid is often seen to be a reflection of the Bose-Einstein condensation. The idea is appealing, although, of course, the reflection – if that is what it is – must be distorted, since helium is not a gas when the decomposition occurs at 2.19K. The whole argument about degeneracy ignores the van der Waals forces which enforce liquefaction of helium at the comparatively high temperature of 4.2K.

Erwin Schrödinger (1887–1961), the pioneer of quantum mechanics, has published a thoughtful and well-written small book on statistical ermodynamics,[43] in which he discusses quantum effects in gases of fermions and bosons in some detail. He calls the theory of degeneracy of gases *satisfactory, disappointing* and *astonishing*. He finds the theory *satisfactory*, because for high temperature and small density it tends to the classical theory of ideal gases. At the same time the theory is *disappointing*, because all its fascinating peculiarities occur at temperatures that are so low, that van der Waals forces have overwhelmed the gases – and made them liquid – long before the effects of degeneracy can be expected to appear.[44] The most *astonishing* feature of the theory occurs, because in the classical limit we have $N_{xc} \ll 1$, while the classical theory itself has $N_{xc} \gg 1$, in fact, N_{xc} must be big enough in the classical case that the Stirling formula can be applied.

The fact that the entropies of gases of both bosons and fermions vanish in the state of full degeneracy is often quoted as collateral support of the third law. The support is somewhat precarious, however, since no gas exists close to absolute zero.

Satyendra Nath Bose (1893–1974)

As a student Bose had been a member of a small and isolated, but dedicated group of scholars in Calcutta, and then for long years he was an underpaid lecturer at a measly salary of 100 rupees. In the opinion of Dutta,[45] his obsequious biographer, Bose was thus being punished for his *outspokenness*. Dutta gives no examples for this characteristic, but he does not forget to praise the youthful Bose as a person who – in his college days – prepared

---

[42] We have seen that, if velocity and momentum of a particle are zero, it cannot be localized because of the uncertainty relation. That effect seems to be secondary in the present context and we have ignored it in the preceding argument.

[43] E. Schrödinger: "Statistical thermodynamics." Cambridge at the University Press (1948).

[44] This is not true for the electron gas in a metal as I have explained and, perhaps, liquid helium shows vestiges of gas-degeneracy in the phenomenon of super-fluidity.

[45] M. Dutta: "Satyendra Nath Bose – life and work." Journal of Physics Education. 2 (1975).

bombs. Presumably those were to be used for patriotic – terroristic (?) – deeds against the colonial power.

Bose had treated a photon gas, then called a gas of light quanta.[46] As I have mentioned before, Einstein translated his paper and it inspired him to develop the statistical mechanics of degenerate gases, in which he discovered the condensation-like phenomenon which is now called the Bose-Einstein condensation, see above. Fritz Wolfgang London (1900–1954) and his brother Heinz London (1907–1970) were first to suggest – in 1937 – that the super-fluidity of Helium II might be due to the Bose-Einstein condensation.

Soon after the Bose-Einstein statistics Enrico Fermi (1901–1954) formulated a statistics for particles which satisfy the Pauli exclusion principle. In his honour we call those particles fermions. It seems that Fermi's work was independent of Bose's and Einstein's; at least that is what Belloni implies in a somewhat diffuse article.[47] Paul Adrien Maurice Dirac (1902–1984) showed that quantum mechanics of many particles permits two *types of statistics,* i.e. *ways of counting*: Bose-Einstein for bosons and Fermi-Dirac for fermions.[48]

Still as a young man, but after the publication of his salient paper with the help of Einstein, Bose spent two years in Europe; in France and Germany. Then he returned to India and became an influential physics teacher and administrator. He finished his career as an honoured elder scientist; except when, after his retirement, he tried to continue his activity. According to Dutta this attempt violated the maxims laid down by the poet Rabindranath Tagore (1861–1941), and there was some public debate and severe criticism of Bose.

Bosons and Fermions. Transition Probabilities

The equilibrium distributions f_{equ} for fermions and bosons acquire a certain interpretability by the following argument which concerns the transition probabilities in a collision between atoms with velocities $\boldsymbol{c}$ and $\boldsymbol{c}^1$ which,

---

[46] An account of Bose's arguments is given in Insert 7.4 below.

[47] L. Belloni: "On Fermi's route to Fermi-Dirac statistics." European Journal of Physics 15 (1994).
Belloni informs us that *Fermi's detailed and definitive theory for the quantization of the ideal gas was published in German.* He does not say when and where, and merely cites someone else's opinion about the paper. Thus he provides a good example for modern writing in the history of science, where historians of science cite other historians of science rather than the original authors.

[48] Actually Fermi's article appeared in: E. Fermi: Zeitschrift für Physik 86 (1926) p. 902. Diracs contribution may be found in: P.A.M. Dirac: Proceedings of the Royal Society (A) 41 (1927) p. 24.

after the collision, have velocities $\boldsymbol{c}'$ and $\boldsymbol{c}^{1\prime}$. We assume that the transition probability is of the form.

$$P_{cc^1 \to c'c_1'} = cN_{xc}N_{xc^1}(1 \mp N_{xc'})(1 \mp N_{xc'^1}) \quad \begin{matrix}\text{fermions}\\ \text{bosons}\end{matrix}$$

so that it depends not only on the occupation numbers N_{xc} of the elements dxdc before the collision, but also on those numbers after the collision. c is a factor of proportionality. Thus the transition of fermions is less probable, if the target elements are well-occupied, – maximally with $N_{xc} = 1$ – while the transitions of bosons into such target elements become more probable when they are already well-occupied.

For the reverse transition we assume an analogous expression for the transition probabilities, viz.

$$P_{c'c'^1 \to cc_1} = c\, N_{xc'}N_{xc'^1}(1 \mp N_{xc})(1 \mp N_{xc^1}) \quad \begin{matrix}\text{fermions}\\ \text{bosons}\end{matrix}.$$

In equilibrium, where both transition probabilities are equal, we conclude that

$$\ln\frac{N_{xc}}{1 \mp N_{xc}} \quad \text{is a collisional invariant.}$$

Therefore this expression must be a linear combination of the collisional invariants mass and energy of the atoms and we may write

$$\ln\frac{N_{xc}}{1 \mp N_{xc}} = \alpha + \beta\frac{\mu}{2}c^2 \ \text{ hence } \ N_{xc} = \frac{1}{\exp(\alpha + \beta\frac{\mu}{2}c^2) \pm 1}. \quad \begin{matrix}\text{fermions}\\ \text{bosons}\end{matrix}$$

This agrees with the equilibrium distribution calculated before by a maximization of entropy. Thus the ansatz for the transition probabilities acquires some credibility. Comparison of the whole argument with analogous arguments by Maxwell and Boltzmann for the classical case, cf. Chap. 4, highlights the modification made necessary by quantum mechanics. Classically an effect of the target element on the transition probability is unthinkable. Of course the classical formula is recovered for the special case $N_{xc} \ll 1$.

7 Radiation Thermodynamics

All energy available on earth – except nuclear and volcanic energy – comes from the sun through empty space by radiation, – or it came in previous geological eras and was stored as coal, mineral oil, or natural gas.

- Animals on the surface of the earth have evolved so as to see with their eyes those frequencies, – from red to violet – where the sunlight has its maximal intensity.
- Plants utilize the red and yellow part of the visible spectrum for the thermodynamically precarious process of photosynthesis that has evolved for the production of glucose and cellulose, the biomass of plants.
- And all creatures take advantage of the heating-part of the solar radiation which lies in the range of frequencies $3{\cdot}10^{12}\,\mathrm{Hz} < \nu < 3{\cdot}10^{14}\,\mathrm{Hz}$ or in the range of wavelengths $10^{-6}\,\mathrm{m} < \lambda < 10^{-4}\,\mathrm{m}$.

Despite the appearance of the numbers, these are small frequencies and long wavelengths. That is to say that the wavelengths (say) are long compared to the dimensions of atoms and molecules. However, the solar radiation does contain shorter wavelengths which are of the dimension of atoms and smaller. It stands to reason that the interaction of such high-frequency-radiation with matter is strongly influenced by the atomic structure, which in turn is governed by the laws of quantum mechanics.

Therefore the scientific research into radiation led to the discovery and development of quantum mechanics. This, of course, is no longer thermodynamics, but the pioneers of radiation physics, Stefan, Boltzmann, Planck, and Einstein were either thermodynamicists themselves or they were trained to think thermodynamically. Therefore we follow their arguments in this chapter up to the point where they turn into quantum mechanics proper.

Not only does radiation carry solar energy to the earth, the *radiation pressure* inside the sun serves to maintain the star in a stable mechanical equilibrium. Stellar physics is a paradigmatic application of the thermo-mechanical laws, and the consideration of radiation enriches the field in a non-trivial manner.

Black Bodies and Cavity Radiation

The history of the scientific study of light begins with Newton, of course, who concluded from his experiments with prisms that white light was a mixture of colours, from red to violet. Goethe, who occasionally dabbled in science – and usually drew the wrong conclusions – ridiculed the idea of white light as a mixture as *clerical*, because it reminded him of the Trinity, the hypostatic union of the Father, the Son, and the Holy Spirit in one godhead. Newton carried the day, although his prisms were not good enough to see more than just colours.

Actually those colours were a nuisance for the users of microscopes, field-glasses and telescopes; they inevitably appeared at the rim of the field of vision and spoiled the view. Joseph von Fraunhofer (1787–1826) addressed those difficulties. He was an optician with strong scientific interests and he became an expert in making *achromatic* lenses. Also the quality of his prisms allowed him to discover *lacking frequencies*, i.e. dark lines in the spectra of the sun and of stars, – several hundred of them. Fraunhofer's optical instruments served Bessel to discover the parallax of some stars, and therefore his gravestone carries the euphemistic engraving in Latin: *Approximavit sidera* – he brought the stars closer. Well, at least he did help to make astronomers appreciate how far away the stars really were. However, the significance of the dark lines was not recognized by Fraunhofer, or anybody else in Fraunhofer's time.

The study of hot gases and the light which they emit became a popular and important field of research in the mid 19th century and Gustaf Robert Kirchhoff (1824–1887) was the most conspicuous researcher in that field. He worked with Robert Wilhelm Bunsen (1811–1899), the inventor of the Bunsen burner, which burns with the emission of so little light that everything burning in it can be clearly distinguished. Kirchhoff discovered that each element, when heated to incandescence, sends out light of frequencies that are characteristic for the element. Thus with his spectroscope he discovered several new elements, e.g. cesium and rubidium, both named – in Latin – for the colour of their spectral lines: blue and red respectively.

Moreover, Kirchhoff found that when light passes through a thin layer of an element – or through its vapour – it would lose exactly those frequencies which the hot element emits. That observation is sometimes called *Kirchhoff's law*, enunciated in 1860. So, since the sunlight lacks the frequencies that heated sodium (say) emits, Kirchhoff concluded that sodium vapour must be present at the solar surface. This was considered a great feat, since it gave evidence of the composition of the sun, something which had been deemed impossible before. Asimov writes[1]

---

[1] I. Asimov: "Biographies ..." loc.cit. p. 377.

> Thus was blasted the categorical statement of the French philosopher Auguste Comte who, in 1835, had declared the composition of the stars to be an example of the kind of information science would be eternally incapable of obtaining. Comte died (insane) two years too soon to see spectroscopy developed.

Kirchhoff conceived of a *black body*, a hypothetical body that sends out radiation of all frequencies and that should therefore – by Kirchhoff's law – also absorb all radiation, and reflect none, so that it appears black. Such black bodies came to play an important role in radiation research, although in the early days no real good black body existed to serve as a reliable object of study. Therefore Kirchhoff suggested an ingenious surrogate in the form of a cavity with blackened, e.g. soot-covered interior walls, which could be heated. Any radiation that enters the cavity by a small hole is absorbed or reflected when it hits a wall. If reflected, the light will most likely travel to another spot of the wall, being absorbed or reflected there, etc. etc. In this way virtually no reflected light comes out through the hole so that the hole itself absorbs radiation *as if it were* a black body. The radiation emitted through the hole is called cavity radiation and it can be studied at leisure for any temperature of the walls.

Kirchhoff himself found that the energy flux density $J_\nu \mathrm{d}\nu$ emitted by a black body, or a cavity between frequencies ν and $\nu + \mathrm{d}\nu$ depends on the temperature of the body *universally,* i.e. it is independent of the mechanical, or electrical, or magnetic properties of the body.[2] Thus Kirchhoff focused the interest of physicists on the universal function $J_\nu(\nu,T)$, the *spectral energy flux density.*

Of course, at that time it was already well-known that there is more to radiation than can be seen. As early as 1800 the eminent astronomer Friedrich Wilhelm Herschel, – Sir William since 1816, the discoverer of the planet Uranus – had placed a thermometer below the red end of the solar spectrum and noticed that it registered a fast increasing temperature. Thus he discovered *heat radiation* which came to be called *infrared radiation.* And then Johann Wilhelm Ritter (1776–1810), an apothecary, discovered in 1801 that silver chloride, which was known to break down under light – changing colour from white to black, the key to photography – continued to do so, if placed beyond the blue and violet end of the spectrum. In this manner he detected *ultraviolet radiation.*

---

[2] It is always difficult to prove experimentally that some property of bodies is universal, because one would have to test all existing bodies. However, in Kirchhoff's time progressive scientists knew the then new second law very well and its universal prohibition that heat pass from cold to hot. So Kirchhoff used a cumbersome thought experiment to prove that, if $J_\nu(\nu,T)$ were dependent on material, the second law could be contradicted. The argument is convincing enough, but somewhat boring; therefore I skip it. The same is true for some arguments by Wien, see below.

In 1879 Josef Stefan (1835–1893), Boltzmann's mentor in Vienna found by careful experimentation that the radiant energy flux density $J = \int_0^\infty J_\nu d\nu$ emanating from a black body – as black as possible – was proportional to the fourth power of its absolute temperature. Thus a body of 600K emits sixteen times more energy than at 300K. Stefan's experiments also provided a rough value for the factor of proportionality which, of course, is universal, since $J_\nu(\nu, T)$ is universal.

Kirchhoff's cavity-model was much more than a means of obtaining good-quality black body radiation. It proved to be an important heuristic tool for theoretical studies. One feature that attracted physicists to the radiation-filled cavity was its similarity to a cylinder filled with a gas. The similarity becomes even more pronounced when one wall of the cavity is considered a movable piston, thus making it possible to apply work to the radiation, or to extract work from it – at least in imagination. Moreover, the energy density e of the cavity radiation can easily be measured, because $e = \frac{4}{c} J$ holds, where J – as before – is the measurable energy flux density emitted by the hole in the cavity wall.

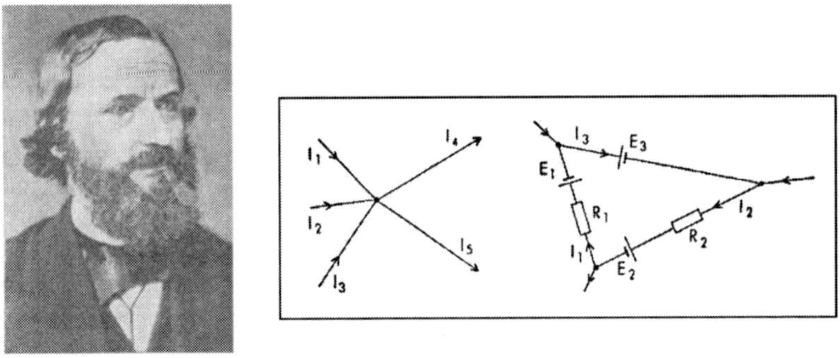

Fig. 7.1. Gustav Robert Kirchhoff (1824–1887) a pioneer of electrical engineering and of radiation thermodyanmics. Kirchhoff is best known for the *Kirchhoff rules* about currents and voltage drops in electric circuits

Boltzmann utilized the cavity model in 1884 to corroborate Stefan's T^4-law: With considerable courage – or deep insight – he wrote a Gibbs equation for the radiation in the cavity in the form

$$dS = \frac{1}{T}[d(eV) + pdV].$$

Now, Boltzmann was also an eager student of Maxwell's electromagnetism and so he knew that the radiation pressure p and the energy density e of radiation are related so that $p = \frac{1}{3}e$ holds, see Chap. 2. Therefore the integrability condition implied by the Gibbs equation reads

$\mathrm{dln}e = 4\cdot\mathrm{dln}T$ so that e must be proportional to T^4 just as Stefan had found it to be. The T^4-law has been called the *Stefan-Boltzmann law* ever since.

And this was just the beginning of the scientific return – experimental or conceptual – from the cavities. Experimentalists used them to *measure* the graph $J_\nu(\nu,T)$, cf. Fig. 7.2 and theoreticians used them to *derive* the function that fitted the graph.

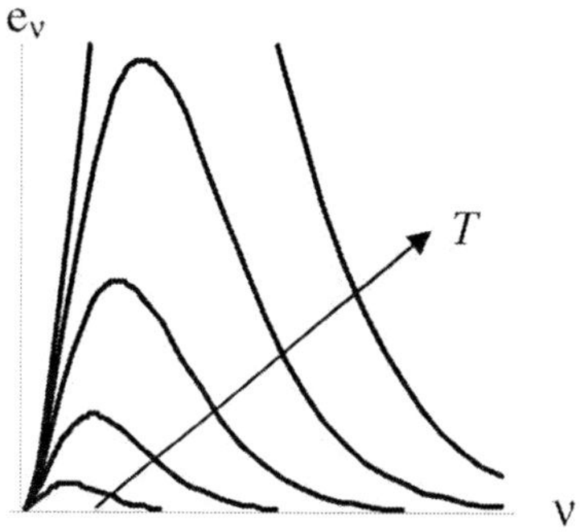

Fig. 7.2. Wilhelm Wien (1864–1928). Spectral energy density of black body radiation as observed (not the Wien ansatz!). For small values of ν the graphs are parabolic

One of the experimentalists was Wilhelm Wien (1864–1928): He found that the peak of the graph shifts to larger frequencies in a manner proportional to T,[3] and he fitted a function of the type[4]

$$J_\nu(\nu,T) = B\nu^3 e^{-\frac{h\nu}{kT}} \qquad \text{(Wien´s ansatz)}$$

to the descending branch of $J_\nu(\nu,T)$ for large frequencies.[5] B and h are constants, universal ones of course, since the whole function is universal.

The opposite limit for small frequencies deserves its own section, since its explanation baffled the scientists in the 1890's.

Violet Catastrophe

While actual cavities had soot-blackened walls for practical purposes, theoreticians did not see why the walls should not be perfectly reflecting in most parts, as long as they contained a tiny black spot of temperature T. The

---

[3] This observation became known as *Wien's displacement law.*

[4] Of course Wien did nor write h, he combined h/k into a universal constant α. Wien's ansatz is not altogether too bad: It satisfies the T^4-law and Wien's own displacement law. However, the ν^3-dependence for small frequencies was contradicted by experiments. The curves should start with ν^2.

[5] W. Wien: Wiedemann's Annalen 58 (1896) p. 662.

effect on the cavity radiation should be the same, at least if the hole was small enough; after all, the radiation is universal, independent of the nature of the wall. As long as there is *something* somewhere to absorb the radiation and reemit it, the intermediate reflections are irrelevant. In fact, a single charge e with mass m connected to the wall by a linearly elastic spring capable of motion in the x-direction (say) should be sufficient. The spring must only be in thermal contact with the wall so that the oscillating mass has the mean energy $\varepsilon = kT$, cf. Insert 7.1. And there must be one spring of eigen-frequency ν for every frequency of radiation.

Now, if physicists know anything very well, it is the harmonic oscillator; so they were on home ground with the one-oscillator model of a cavity. It is true that in the present case the oscillating mass m has a charge e so that there is radiation damping, but that was no difficulty for the top scientists in the field. Actually, as early as 1895, Planck had written a long article[6] in which he showed that the equation of motion of a one-dimensional oscillator with mass m, charge e, and eigen-frequency ν in an electric field $E(t)$ reads approximately, i.e. for weak damping[7]

$$\ddot{x} + \frac{8\pi^2 e^2}{3mc^3}\nu^2\,\dot{x} + 4\pi^2\nu^2\,x = \frac{e}{m}E(t)\,.$$

It is true that $E(t)$ is a strongly and irregularly varying function in the cavity, but only *the* Fourier component will appreciably interact with the oscillator which has its eigen-frequency ν. Let the energy density residing in that component be $^1/_2\varepsilon_0 E_\nu^2$, see Chap. 2. This represents $^1/_6$ of the spectral energy density e_ν of the cavity radiation, because the y- and z-components of the electric field also contribute to the energy density, and so do the components of the magnetic field; all of them contribute equal amounts. Thus it turns out – from the solution of the equation of motion – that the mean kinetic and potential energy ε of the oscillator is related to the radiative energy density e_ν, or the energy flux density $J_\nu - {}^c/_4\, e_\nu$ by

$$J_\nu = \frac{c}{4}\frac{8\pi\nu^2}{c^3}\varepsilon\,.$$

[6] M. Planck: "Über elektrische Schwingungen, welche durch Resonanz erregt und durch Strahlung gedämpft werden." [On electrical oscillations excited by resonance and damped by radiation] Sitzungsberichte der königlichen Akademie der Wissenschaften in Berlin, mathematisch-physikalische Klasse, 21.3.1895. Wiedemann's Annalen 57 (1896) p. 1.
Planck was much interested in radiation; primarily because he believed for a long time that radiation damping is the essential *mechanism* of irreversibility. Boltzmann opposed the idea and eventually Planck disabused himself of it.

[7] This equation and the following argument are too complex to be derived here, even as an Insert. However, they are replayed in all good books on electrodynamics. I found a particularly clear presentation in R. Becker, F. Sauter: "Theorie der Elektrizität." [Theory of electricity.] Vol. 2 Teubner Verlag, Stuttgart (1959).

Therefore, all that John William Strutt (1842–1919) – Lord Rayleigh since 1873 – had to do was to insert the mean energy ε of the oscillator in order to come up with $J_\nu(\nu,T)$, the spectral energy flux density of the black body radiation. According to the best of Rayleigh's – or anybody else's – knowledge at the time, that mean energy is kT, cf. Insert 7.1, so that Rayleigh obtained[8]

$$J_\nu(\nu,T) = \frac{c}{4}\frac{8\pi\nu^2}{c^3}kT \qquad \text{(Rayleigh - Jeans formula)}.$$ [9]

The formula fits the observed curve well for small frequencies, but it is a disaster for large ones: To begin with, the expression is not even integrable and, besides, it increases monotonically. These circumstances became known as the *violet catastrophe,* – or *ultraviolet catastrophe*[10] – because the high frequencies, beyond the violet in the visible spectrum, were very badly represented by the formula indeed.

Obviously, in order to agree with observations, cf. Fig. 7.2, oscillators with high eigen-frequencies ν must get less than their classical share $\varepsilon = kT$ of energy. And the share must depend on the value of the eigen-frequency and decrease with it. Planck asked the question: *How much* do the oscillators get? *How much* in Latin is *quantum* – with plural *quanta*– and so Planck's answer to the question, and all it entailed, became eventually known as *quantum mechanics.*[11]

The violet catastrophe of cavity radiation heralded the fall of classical physics which amounted to a scientific revolution. It started in 1900 with Planck's paper: "Zur Theorie des Gesetzes zur Energieverteilung im Normalspektrum."[12] Ironically nobody at the time noticed the full signify-cance of what had begun, certainly not Planck himself, – and not for many years. We proceed to consider this.

---

[8] Lord Rayleigh: Philosophical Magazine 49 (1900) p. 539.

[9] We shall discuss Jeans's contribution below.

[10] So named by Paul Ehrenfest in 1910, – *posthumously* says S.G. Brush: "The kind of motion we call heat ..." loc. cit. p. 306. And indeed, by that time, to all intents and purposes, the Raleigh-Jeans theory was dead.

[11] Of course, Planck did not write Latin, but the Latin word *Quantum* is routinely used in the German language meaning portion, or share, or ration.

[12] [On the theory of the law of energy distribution in the normal spectrum] M. Planck: Verhandlungen der deutschen physikalischen Gesellschaft 2 (1900) p. 202.
Normal spectrum is Planck's word for the black body spectrum.

Expectation value of the energy of a classical oscillator

We recall the Boltzmann factor, by which the probability of a body to have an energy ε_n ($n = 0,\dots\infty$) is proportional to $\exp\left(\frac{\varepsilon_n}{kT}\right)$. Therefore the expectation value ε of the energy is given by

$$\varepsilon = \frac{\sum_{n=0}^{\infty} \varepsilon_n e^{-\frac{\varepsilon_n}{kT}}}{\sum_{n=0}^{\infty} e^{-\frac{\varepsilon_n}{kT}}} = kT^2 \frac{\partial}{\partial T}\left(\ln \sum_{n=0}^{\infty} e^{-\frac{\varepsilon_n}{kT}}\right).$$

If ε_n is the energy $\frac{m}{2}(\dot{x}^2 + \nu^2 x^2)$ of an oscillator of mass m, and eigen-frequency ν, the index n is a double index $(x, \dot{x})$ and we may write

$$\ln\left(\sum_{n=0}^{\infty} e^{-\frac{\varepsilon_n}{kT}}\right) = \ln\left[\sum_{x,\dot{x}} \exp\left(-\frac{m}{2kT}(\dot{x}^2+\nu^2x^2)\right)\right]$$

$$= \ln\left[Y^{1/3} \int_{-\infty}^{+\infty} \exp\left(-\frac{m}{2kT}(\dot{x}^2+\nu^2x^2)\right)\mathrm{d}x\mathrm{d}\dot{x}\right] .$$

$$= \ln\left(Y^{1/3}\frac{2\pi kT}{m\nu}\right)$$

Hence follows $\varepsilon=kT$ by insertion. [The summation over $(x, \dot{x})$ was converted here into an integration by virtue of the measure factor Y used before, cf. Chaps. 4 and 6. Since that factor does not influence the result, the conversion – from sum to integral – might be considered as an auxiliary mathematical tool. Certainly Boltzmann considered it so, as we have discussed in Chap. 4.]

Insert 7.1

Planck Distribution

The revolution started as an interpolation project between the Wien ansatz and the Rayleigh-Jeans formula which were good for high and low frequencies respectively. Actually given the task, a student can do the interpolation, – and identify the coefficient B of the Wien ansatz –, simply by studying the two relations given above like the pieces of a puzzle. He obtains the following formula after a little time which, admittedly, may be shortened by hindsight.

$$J_\nu(\nu,T) = \frac{c}{4}\frac{8\pi\nu^2}{c^3}\frac{h\nu}{e^{\frac{h\nu}{kT}}-1}.$$

This is *Planck's radiation formula*, or the *Planck distribution*. Planck apparently could not see how easy it was to get. Therefore he proceeded along a cumbersome route which I replay in Insert 7.2, for historical correctness, as it were.

The value of h may be determined by fitting the function to the observed curves. Thus h turns out to be equal to $6.55 \cdot 10^{-34}$Js. This is sometimes called the *action quantum*, because it has the dimension of an action. More often it is called the *Planck constant.*

I believe that the true history of the interpolation that led to the Planck radiation formula will never be known. Planck himself gave slightly conflicting accounts. To be sure, textbook folklore has it that there was an interpolation between Wien's ansatz and the Rayleigh-Jeans formula. I have so argued myself above. However, in the relevant papers by Planck in 1900/01 [13] there is no mention of Rayleigh, let alone Jeans. So maybe Planck did not know Rayleigh's work which, after all, had appeared only in the same year 1900. Planck says that he was convinced of the deficiency in Wien's formula by the results of low-frequency experiments made known to him by the experimentalists F. Kurlbaum and H. Rubens who confirmed earlier measurements by O. Lummer and E. Pringsheim.[14] And then he says, referring to the arguments reported in Insert 7.2

> Pursuing this idea I came to construct arbitrary expressions for the entropy which were more complicated than those of Wien … but acceptable. Among those expressions my attention was caught by
>
> $$\frac{\partial^2 s_\nu}{\partial e_\nu^2} = \frac{\alpha}{e_\nu(\beta + e_\nu)}$$
>
> which comes closest to Wien's in simplicity and … deserves to be further investigated.

On the other hand, in his Nobel lecture of 1920[15] Planck says that the measurements of Kurlbaum, Lummer et al. convinced him that for low frequencies the expression should read $\frac{\partial^2 s_\nu}{\partial e_\nu^2} \sim \frac{1}{e_\nu^2}$:

---

[13] There were three such papers. Apart from the one cited above they are
M. Planck: "Über eine Verbesserung der Wien'schen Spektralgleichung." [On an improvement of Wien's spectral equation] Verhandlungen der deutschen physikalischen Gesellschaft 2 (1900) pp. 202–204.
M. Planck: "Über das Gesetz der Energieverteilung im Normalspektrum." [On the law of energy distribution in the normal spectrum.] Annalen der Physik (4) 4 (1901) pp. 553–563.

[14] These researches were published in 1901:
H. Rubens, F. Kurlbaum: Annalen der Physik 4 (1901) p. 649.
O. Lummer, E. Pringsheim: Annalen der Physik 6 (1901) p. 210.

[15] M. Planck: "Die Entstehung und bisherige Entwicklung der Quantentheorie." [The origin and subsequent development of quantum theory] Nobel lecture to the Royal Swedish Academy of Sciences in Stockholm, held on June 2nd, 1920.

> Nothing was then more plausible than to set [the reciprocal of] this expression equal to the sum of a term with the first power and a term with the second power of the energy.

Of course, it was trial and error both ways, but a little less so in the second manner. Obviously Planck did not quite remember his arguments after 20 years. Maybe this is the place to quote a thoughtful remark by Einstein:[16] *Every reminiscence is coloured by today's being what it is, and therefore by a deceptive point of view.*

Planck's derivation of the radiation formula

Planck, steeped in thermodynamics, as he was, replaced $^1/_T$ in Rayleigh-Jeans's and Wien's laws by $\frac{\partial s_\nu}{\partial e_\nu}$ using the Gibbs equation for the spectral entropy density s_ν. Thus he obtained respectively

$$\frac{\partial s_\nu}{\partial e_\nu} = -\frac{8\pi\nu^2 k}{c^3}\frac{1}{e_\nu} \quad \text{and} \quad \frac{\partial s_\nu}{\partial e_\nu} = -\frac{k}{h\nu}\ln\frac{e_\nu}{\frac{4}{c}B\nu^3} .$$

Differentiation with respect to e_ν provides

$$\frac{\partial^2 s_\nu}{\partial e_\nu^2} = -\frac{8\pi\nu^2 k}{c^3}\frac{1}{e_\nu^2} \quad \text{and} \quad \frac{\partial^2 s_\nu}{\partial e_\nu^2} = -\frac{k}{h\nu}\frac{1}{e_\nu} ,$$

and *it was between those two algebraic functions that Planck interpolated* to obtain

$$\frac{\partial^2 s_\nu}{\partial e_\nu^2} = -\frac{k}{h\nu}\frac{1}{e_\nu + \frac{c^3}{8\pi h\nu^3}e_\nu^2} .$$

Integration provides $^1/_T$ again on the left hand side and thus

$$\frac{1}{T} = -\frac{k}{h\nu}\ln\frac{\frac{c^3}{8\pi h\nu^3}e_\nu}{1+\frac{c^3}{8\pi h\nu^3}e_\nu} ,$$

if one fixes the constant of integration by requiring that $e_\nu \to \infty$ for $T \to \infty$. Solving for e_ν one obtains the Planck distribution.

Insert 7.2

[16] P.A. Schilpp (ed.): "Albert Einstein: Philosopher – Scientist." Library of living philosophers, New York (1949).

However, Sir James Hopwood Jeans (1877–1946), a mathematician much interested in astronomy, was not convinced that the Rayleigh formula was wrong for high frequencies. He kept a campaign going till the end of the first decade of the 20th century in which he criticizes the cavity model and maintains that no stationary state can prevail in such a cavity.[17] His arguments faded away with the growing confidence in the Planck distribution. But the battle leaves its traces in the textbooks, because the violet catastrophe is a handy tool for the illumination of the scientific terrain of classical physics before quantum physics prevailed. As late as 1910 Planck was moved to refute Jeans's arguments.[18] He says:

> The radiation theory of J.H. Jeans is the most satisfactory one according to the present state of physics; however, it must be rejected, because it leads to a contradiction with observations.

Note that Planck, even in 1910, ten years after his radiation formula, does not consider his own contribution as belonging to the *present state of physics.*

Note also that the low-frequency limit of the Planck distribution – the Raleigh-Jeans formula – provides a possibility to determine the Boltzmann constant k. We may recall here Loschmidt's complicated and inaccurate argument for the calculation of k, in order to determine the molecular mass μ, cf. Chap. 4. This argument can now be considered obsolete and indeed Einstein in his reminiscences speaks of *...Planck's determination of the true size of the atom from the law of radiation.*[19] On the other hand, in his work on Brownian motion in 1905 Einstein proposes to measure k by observation of a Brownian particle, see Chap. 9; that would be a cumbersome method in comparison.

Energy Quanta

From the above we conclude that according to Planck's interpolation the mean energy ε of the oscillator must be equal to

$$\varepsilon = \frac{h\nu}{e^{\frac{h\nu}{kT}} - 1} = kT^2 \frac{\partial}{\partial T}\left(\ln \frac{1}{1 - e^{-\frac{h\nu}{kT}}}\right).$$

If that is compared with the generic expression for ε derived from the Boltzmann factor, cf. Insert. 7.1, namely

---

[17] J.H. Jeans. Philosophical Magazine, February 1909 p. 229.
J.H. Jeans: Ibidem, July 1909 p. 209

[18] M. Planck: "Zur Theorie der Wärmestrahlung." [On the theory of heat radiation] Annalen der Physik (4) 31 (1910) pp. 758–768.

[19] P.A. Schilpp (ed.): In: "Albert Einstein: Philosopher – Scientist." "Autobiographical notes." loc. cit.

$$\varepsilon = \frac{\sum_{n=0}^{\infty} \varepsilon_n \exp[-\frac{\varepsilon_n}{kT}]}{\sum_{n=0}^{\infty} \exp[-\frac{\varepsilon_n}{kT}]} = kT^2 \frac{\partial}{\partial T} \ln\left(\sum_{n=0}^{\infty} \exp[-\frac{\varepsilon_n}{kT}]\right),$$

we obtain

$$\sum_{n=0}^{\infty} \exp[-\frac{\varepsilon_n}{kT}] = \frac{1}{1-e^{\frac{h\nu}{kT}}}.$$

Obviously the equation represents the summation of an infinite geometric series provided that $\varepsilon_n = nh\nu$ holds.

Thus one may conclude – or must conclude – that the oscillator is not able to accommodate all energies, but only equidistant energies 0, $h\nu$, $2h\nu$,... The oscillator can absorb – and emit – only *energy quanta* of size $h\nu$ and, if the eigen-frequency grows, those quanta become ever bigger. For large eigen-frequencies the quanta are so big that the thermal motion of the particles of the wall of the cavity cannot provide them. Therefore high frequency oscillators are inactive, i.e. they remain at rest, – at least that was the idea at first. It is because of that, that the spectral energy density e_ν of the radiation is concentrated at relatively low frequencies. However, when the temperature grows, the range of accessible frequencies becomes bigger and the bulk of the area below $e_\nu(\nu,T)$ shifts to the right, as observed, cf. Fig. 7.2, and as expressed by Wien's displacement law.

It is this – formally, and in retrospect – fairly straightforward argument by which Planck has introduced the concept of *quantized* energy levels of an oscillator.[20] Of course, the argument was totally at odds with classical thinking. Therefore physicists – foremost Planck himself – *suspected that the whole thing might be a piece of mathematical jugglery without any correspondence to anything real in nature.* [Planck] *struggled for years to find a way around his own discovery.*[21]

At some time during this struggle Planck came up with the idea that maybe the *emission* of radiation from the oscillator indeed happened in steps of size $h\nu$, *but that absorption was continuous.*[22] According to the new hypothesis the oscillator was supposed to accumulate absorbed radiation between two steps so that on average it would be found half-way between $nh\nu$ and $(n + 1)h\nu$. This led Planck to an alternative equation for the expectation value ε, namely

---

[20] Since molecules usually represent high frequency oscillators, their vibrational *degrees of freedom* do not contribute to the specific heat at normal temperatures. The same is true for the *rotation* of a two-atomic molecule about the axis that links the atoms. Thus quantum mechanics finally explained that puzzling observation about specific heats.

[21] According to I. Asimov: "Biographies" loc.cit. p. 506.

[22] M. Planck: "Eine neue Strahlungshypothese." [A new hypothesis about radiation] Verhandlungen der deutschen physikalischen Gesellschaft, February 3, 1911.

$$\varepsilon = \frac{h\nu}{2} + \frac{h\nu}{e^{\frac{h\nu}{kT}} - 1}.$$

Accordingly, in effect the oscillator had to have energy levels $\varepsilon_n = (n+{}^1/_2)h\nu$ – instead of $\varepsilon_n = nh\nu$ – so that it could never be quite without energy; even for $T = 0$ there had to be a *zero point energy.*

Miraculously this equation – and the concept of zero point energy – was later confirmed by proper quantum mechanics, based on the Schrödinger equation, although *continuous absorption* was never taken seriously, – or not to my knowledge. The zero point energy is nowadays taken to be a reflection of Heisenberg's uncertainty relation applied to the oscillator.

Max Karl Ernst Ludwig Planck (1858–1947)

Max Planck was 42 years old when he derived the radiation formula. He had studied under Helmholtz, Kirchhoff and Weierstraß. His doctoral thesis[23] is a rehash of Clausius's ideas which Planck admired greatly. He claimed that Helmholtz had not read his work. Kirchhoff read it and disapproved, while Clausius was not interested.

Planck's great achievement is the formulation of the correct radiation formula and – in consequence – the realization that the formula required quantized energy levels of an oscillator. Of course, Planck sent the paper around. Boltzmann received a copy and, according to Planck,[24] *he expressed his interest and basic agreement with my reasoning.* As there is no reflection of this reaction in Boltzmann's work, it was probably no more than politeness. Indeed, according to Lindley[25] *Boltzmann had never had much time for Planck.* The two scientists had been in contact over Planck's idea that the explanation of irreversibility required electro-magnetic radiation damping and could not be explained by the kinetic theory. Boltzmann won this argument hands down. And then there was the Zermelo controversy, see Chap. 4, which must have soured relations.

Planck himself remained sceptical for many years of his own discovery, calling it *an act of desperation.*[26] When Einstein went ahead and took quanta seriously, Planck did not wish to follow. Instead he continued to search for a way to reconcile the new concept with classical physics. He says: *My vain efforts to incorporate the quantum of action somehow into the classical theory took several years and much work. Some of my colleagues*

---

[23] M. Planck: "Über den zweiten Hauptsatz der mechanischen Wärmetheorie." [On the second law of the mechanical theory of heat] Dissertation, Universität München (1879).

[24] Planck: Nobel lecture. loc. cit.

[25] D. Lindley: "Boltzmann's atom." loc.cit. p. 212.

[26] A. Hermann (ed.): "Deutsche Nobelpreisträger" loc.cit. p. 91.

have seen this as tragic. But I disagree...[27] Ironically Planck's well-known and oft-quoted dictum about the non-acceptance of new ideas, cf. Fig. 7.3, is therefore primarily applicable to himself.

Planck's own achievement, along with his partisanship of the works of his colleagues Nernst and Einstein, and his soft-spoken but steadfast rectitude in politically turbulent times made Planck one of the most renowned physicist of his time, second only to Einstein. Thus it happened at the end of the second world war, – when Planck was fleeing the rampaging Russian army, and was picked up at the roadside by an American passport-checking patrol – that his name was recognized and he was given VIP-transport to Göttingen in a jeep. There at the age of nearly ninety years, he became acting head of the Kaiser Wilhelm Institute, – the last one, because, when a worthy younger director was appointed, the institute was renamed Max Planck Institute.

Planck's head was used on early 2 deutsch-mark coins, – not for long though, because soon a more deserving politician was found to replace him.

The only way to get revolutionary advances in science accepted is to wait for all old scientists to die.[28]

Fig. 7.3. Max Planck (1858–1947)

[27] Ibidem.

[28] This is the somewhat shortened quotation from M. Planck: "A Scientific autobiography and other papers." Williams and Norgate, London (1950).
Brush writes: *I suppose that most people who read (or repeat) this quotation think Planck is referring to his quantum theory, but in fact he was talking about his struggle to convince scientists in the* 1880*'s and* 1890*'s that the second law of thermodynamics involves a principle of irreversibility, and that the flow of energy from hot to cold is not analogous to the flow of water from a high level to a low one, as Ostwald and the energeticists claimed.* Cf. S.G. Brush: "The kind of motion we call heat, ..." loc. cit. p. 640.

Photoelectric Effect and Light Quanta

Heinrich Hertz had noticed that light falling upon metals stimulates the emission of electrons. This became known as the photoelectric effect, or simply the photo-effect. Philipp Eduard Anton von Lenard (1862–1947) investigated the effect systematically in 1902 and he found that the energy of the emitted electrons does not depend on the intensity of the incident light. A brighter light just produces more electrons, not more energetic ones. Instead, light of a higher frequency creates more energetic electrons. There was no explanation until Einstein stepped forward with an extrapolation of Planck's energy quanta.[29]

Einstein argued that, if an oscillator could only exchange quanta of energy $h\nu$ with the surrounding radiation field, the emitted radiation itself should appear as quanta; they came to be called *light quanta* at first and could, perhaps, be considered as little particles of light with the energy $h\nu$. If such a light quantum hits an electron, bound to a metal with less energy than $h\nu$, the light may kick the electron loose and make it move off with the surplus. The higher the frequency, the higher the surplus and the quicker the electron moves. On the other hand, if the light quantum – for low frequency– carries less than the binding energy of the electron to the metal, there is no emission of electrons. The threshold frequency, when emission started, was found to be a characteristic property of the metal.

This is all simple enough except that one has to accept the idea of light quanta. Since the idea was based on Planck's theory of energy quanta, its success was a first confirmation of that theory other than radiation itself. Einstein's hypothesis of the photo-effect *went a long way, perhaps even all the way toward establishing the new quantum theory.*[30] Einstein received the Nobel prize for this in 1921. However, among the scientists who remained sceptical, was Planck.[31]

Simple as the explanation of the photo-effect may be, it had a truly far-reaching consequence on natural philosophy. Indeed, Einstein thus *cancelled out the luminiferous ether as unnecessary by assuming that light travelled in quanta and therefore had particle-like properties and was not merely a wave that required some material* [the ether] *to do the waving.*[32] So the question of absolute space, in which the ether was at rest was finally done away with.

---

[29] A. Einstein: "Über einen die Erzeugung und Verwandlung des Lichtes betreffenden heuristischen Standpunkt." [On a heuristic point of view concerning the creation and reaction of light.] Annalen der Physik (4) 17 (1905).

[30] I. Asimov: "Biographies ..." loc.cit. p. 517.

[31] A. Hermann (ed): "Deutsche Nobelpreisträger." loc.cit. p. 91.

[32] Asimov: "Biographies ..." loc.cit. p. 589.

Radiation and Atoms

Time went on and Planck's concept of energy quanta of hypothetical oscillators in cavity walls found its way into the atom. Niels Henrik David Bohr (1885–1962) constructed a model of the atom in 1913, whose essential feature is quantized energy levels for electrons in the electric field of the nucleus. That model prevailed with slight modifications to this day and by now it is taught in elementary schools.

Thus it became possible to think about atoms in equilibrium with a radiation field and – not surprisingly – Einstein was first and foremost to develop the idea.[33] He introduced the novel concept of stimulated emission and derived Planck's radiation formula without Planck's interpolation. The matter is simple enough so that we can replay it here in an understandable form on less than one page.

We are interested in radiation with frequency ν and spectral energy density $e_\nu(\nu,T)$. If the frequency is such that $h\nu = \varepsilon_n - \varepsilon_m$ holds, the radiation may be emitted and absorbed when the electron moves between the levels with ε_n and ε_m. The emission and absorption probabilities are respectively

$$p_{n\to m} = A + Be_\nu(\nu,T) \qquad \text{and} \qquad p_{m\to n} = Ce_\nu(\nu,T)\,.$$

Two of the three terms – those with A and C – represent *spontaneous emission* and absorption. They are eminently plausible. But the third term – the one with B – is not. It represents what Einstein called *induced* or *stimulated emission* and at the end, upon reflection, we shall recognize that that concept was introduced *ad hoc* so that the argument leads to the Planck distribution. Einstein expresses this by saying:

> In order for the desired result to come out we need to extend our hypotheses.

The probabilities of finding atoms with energies ε_n and ε_m are proportional to the Boltzmann factors $\exp(-\varepsilon_n/kT)$ and $\exp(-\varepsilon_m/kT)$. Therefore the expectation values for emission and absorption are

$$(A + Be_\nu(\nu,T))\frac{e^{-\frac{\varepsilon_n}{kT}}}{\sum e^{-\frac{\varepsilon_i}{kT}}} \quad \text{and} \quad Ce_\nu(\nu,T)\frac{e^{-\frac{\varepsilon_m}{kT}}}{\sum e^{-\frac{\varepsilon_i}{kT}}}\,.$$

In equilibrium both expressions must be equal so that the equilibrium spectral energy density has the form

---

[33] A. Einstein: "Strahlungsemission und –absorption nach der Quantentheorie." Deutsche physikalische Gesellschaft, Verhandlungen 18 pp. 318–323 (1916).
A. Einstein: "Quantentheorie der Strahlung." Physikalische Gesellschaft Zürich, Mitteilungen 16 pp. 47–62 (1916).
A. Einstein: "Quantentheorie der Strahlung." [Quantum theory of radiation] Physikalische Zeitschrift 18 pp. 121–128 (1917).

$$e_\nu(\nu,T) = \frac{A}{C}\frac{1}{e^{\frac{h\nu}{kT}} - \frac{B}{C}}.$$

Since e_ν (ν,∞) may be expected to be infinite, B must be equal to C and, since for small ν the Raleigh-Jeans formula ought to hold, we may determine A/C and obtain

$$e_\nu(\nu,T) = \frac{8\pi\nu^2}{c^3}\frac{h\nu}{e^{\frac{h\nu}{kT}} - 1},$$

which is the Planck distribution.

The new and original feature in Einstein's argument is *stimulated emission.* Thus he envisages a process by which the radiation energy e_ν amplifies itself by shaking a quantum $h\nu$ loose from the atom and the probability for this amplification is proportional to the *extant value* of e_ν, so that a run-away amplification is conceivable.

In the 1917-paper there is a thoughtful but inconclusive discussion about the *momentum exchange* between matter and radiation, and about the *recoil* of size $\frac{h\nu}{c}$, or actually $\frac{h\nu}{c^2}c$ of an atom that emits a light quantum $h\nu$. Although momentum is much on his mind, Einstein seems to shy away from definitely assigning the momentum $\frac{h\nu}{c}\boldsymbol{n}$ to a light quantum moving in the direction $\boldsymbol{n}$.

Thus, although he came close, Einstein missed the full import of stimulated emission, which amplifies the energy of the emission-stimulating ray of radiation by a light quantum that moves *in the direction of the ray*. This fact was later – in the 1920's and 1930's – recognized and incorporated into the treatment of the *photon gas* by astrophysicists, see below. But then Einstein did not look back and so he and everybody else failed to recognize the potential applicability of the phenomenon for the creation of coherent, unidirectional, and monochromatic light. The result lay dormant for 50 years, before some clever electrical engineers used it in the 1960's to construct an amplifier that became known by the acronym *maser* = ***m****icrowave* ***a****mplifier by* ***s****timulated* ***e****mission of* ***r****adiation*. Shortly afterwards the same was done for light in the *laser*.

Still, Einstein's improved derivation of the Planck formula was eagerly accepted. Bose[34] comments on the argument and calls it *a remarkably elegant derivation.*[35] And yet, Bose had some reservations, essentially based on the fact that Einstein's final result needs to refer to the Rayleigh-Jeans formula which is purely classical. Bose's own argument avoids this. Bose

---

[34] S.N. Bose: "Plancks Gesetz ..." loc. cit.

[35] Actually it is Einstein who calls Einstein's argument *bemerkenswert elegant* [remarkably elegant], because he translated Bose's paper. However, we may assume that Bose's unpublished original English version used words to that extent.

was the first to take the *cells* of phase space seriously. We recall that Boltzmann had previously introduced cells as the smallest elements that can accommodate a *point* (***x,c***), or (***x,p***); Boltzmann had considered this – cf. Chap. 4 – as a conceptual artifact introduced for mathematical convenience, and he did not need to speak about the cell-size, because it dropped out of his final results. For Bose that size had to be equal to h^3, if he wished to obtain the Planck distribution. Also Bose introduced the new way of characterizing a distribution of light quanta and counting the number of realizations. We review Bose's paper in the briefest possible manner in Insert 7.4.

Photons, A New Name for Light Quanta

Einstein's hypothetical light quanta had the energy $h\nu$, but they could not really be considered particles until they were firmly endowed with a momentum. Einstein had come close to doing that in his paper on stimulated emission, see above. His expression $\frac{h\nu}{c}$ for the recoil of an emitting atom is in fact the magnitude of the momentum. This can easily be confirmed, since light – being electro-magnetic radiation – exerts a pressure $p = {}^1/_3\, e$ on a wall, where e is the energy density, cf. Chap. 2. From this result it follows that the momentum $\boldsymbol{p}$ of the light quanta is in fact equal to $\frac{h\nu}{c}\boldsymbol{n}$, where $\boldsymbol{n}$ is the direction of their motion, see Insert 7.3.

Arthur Holly Compton (1892–1962) proved this expression for the momentum *directly* when he observed collisions of light quanta with electrons, in which – naturally – momentum and energy had to be conserved. The observed *Compton effect* settled the matter. Thus the light quantum now had energy *and* momentum and could be considered a particle Compton proposed the name *photon* and that was generally accepted after some time.

Radiation pressure and momentum of light quanta

As in Insert 4.1 we consider that ${}^1/_6$ of the photons with the energy $h\nu$ and the (unknown) momentum p_ν move in the six spatial directions perpendicular to the sides of a cube. The walls reflect them elastically. In this manner the photons with momentum p_ν exert a pressure $2p_\nu c\frac{n_\nu}{6}$ on a wall, where n_ν is the number density. The energy density is obviously $h\nu{\cdot}n_\nu$ and since – by Maxwell's equations – the energy density equals three times the pressure, the momentum p_ν of a quantum equals $\frac{h\nu}{c}$ in magnitude.

Insert 7.3

Bose's derivation of the Planck distribution

Let V be a volume, homogeneously filled with N_ν photons with frequencies between ν and $\nu + d\nu$. Accordingly the spectral energy is $E_\nu - N_\nu h\nu$. The photons occupy a spherical shell of volume

$$V 4\pi \left(\frac{h\nu}{c} \right)^2 \frac{h d\nu}{c}$$

in the phase space spanned by space and momentum coordinates. The phase space has cells of size h^3 which can accommodate only two photons, – one each for the two possible polarizations. Therefore there are $A_\nu = V 4\pi \frac{\nu^2}{c^3} d\nu$ cells in the spherical shell.

Bose introduced the idea that the distribution of photons is characterized by p_r^ν, the number of cells occupied by r photons in the range $d\nu$. Their spectral entropy is therefore

$$S_\nu = k \ln W_\nu \qquad \text{where} \qquad W_\nu = \frac{A_\nu!}{\prod_{r=0}^{\infty} p_r^\nu!}.$$

Maximizing this under the constraints $A_\nu = \sum_r p_r^\nu$ and $N_\nu = \sum_r r p_r^\nu$ we obtain

$$S_\nu = -k \left(\ln \frac{N_\nu}{A_\nu} - \left(1 + \frac{A_\nu}{N_\nu} \right) \ln \left(1 + \frac{A_\nu}{N_\nu} \right) \right) N_\nu .$$

With $N_\nu - E_\nu / h\nu$ we get

$$\frac{\partial S_\nu}{\partial E_\nu} = \frac{1}{T} = -\frac{k}{h\nu} \ln \frac{\frac{N_\nu}{A_\nu}}{1 + \frac{N_\nu}{A_\nu}} \quad \text{hence} \quad N_\nu = \frac{A_\nu}{\exp\left(\frac{h\nu}{kT}\right) - 1}$$

and with the above value for A_ν and $E_\nu - e_\nu(\nu, T) V d\nu$

$$e_\nu(\nu, T) = 8\pi \frac{\nu^2}{c^3} \frac{h\nu}{\exp\left(\frac{h\nu}{kT}\right) - 1}$$

which is the Planck distribution once again, but now derived without any reference to classical thinking and classical formulae and, of course, without any interpolation between empirical functions.

Insert. 7.4

Photon Gas

Now that photons may be considered as particles, endowed with momentum and energy, we may write an equation of transport for a *photon gas*. Let $f(\boldsymbol{x},\boldsymbol{p},t)\mathrm{d}\boldsymbol{p}$ be the number density of photons with momenta between $\boldsymbol{p} = \frac{h\nu}{c}\boldsymbol{n}$ and $\boldsymbol{p} + \mathrm{d}\boldsymbol{p}$. Since all photons have the speed c, the density function satisfies the *photon transport equation*

$$\frac{\partial f}{\partial t} + cn_k \frac{\partial f}{\partial x_k} = S(f),$$

which represents an equation of balance for the number of photons with $\boldsymbol{x}$ and t and with momentum $\boldsymbol{p}$. The equation is a little like the Boltzmann equation, cf. Chap. 4, except that the right hand side, which represents the source density of photons, is not specific yet. Since the photons do not interact among themselves – at least not normally – the right hand side is due exclusively to interaction of the photons with matter. $S(f)$ is zero, when the radiation is in equilibrium with matter, and, of course, when there is no matter, there is no production either.

Multiplication of the photon transport equation by a generic function $\psi(\boldsymbol{x},\boldsymbol{p},t)$ and integration leads to the equation of transport for radiative quantities

$$\frac{\partial \int \psi f \mathrm{d}\boldsymbol{p}}{\partial t} + \frac{\partial \int \psi cn_k f \mathrm{d}\boldsymbol{p}}{\partial x_k} = \int \left(\frac{\partial \psi}{\partial t} - cn_k \frac{\partial \psi}{\partial x_k} \right) f \mathrm{d}\boldsymbol{p} + \int \psi S(f) \mathrm{d}\boldsymbol{p}.$$

The right hand side represents the production density of photons. $^1/_y$ is the volume of a cell of $(\boldsymbol{x},\boldsymbol{p})$-space, and it is equal to h^3 according to Bose.

For $\psi = 1$, $p_j = \frac{h\nu}{c} n_j$, $cp = h\nu$, and $-k(\ln\frac{f}{y} - \frac{y}{f}(1+\frac{f}{y})\ln(1+\frac{f}{y}))$ we obtain equations of balance for the number of photons and for momentum, energy and entropy with densities, fluxes and source densities as indicated in Table 7.1.

The entropic terms in the table are those appropriate for a Bose gas, for which the photon gas is the prototype, see above and Chap. 6. For equilibrium the entropy has to have a maximum and that occurs for the density function

$$f_{equ}(p,T) = \frac{y}{e^{\frac{h\nu}{kT}} - 1},$$

where T is the temperature of the matter with which the radiation is in equilibrium. The equilibrium density function is the Planck distribution, of course; it is homogeneous and isotropic. Insertion of f_{equ} into the table provides the entries of Table 7.2, most of which are zero.

Of some interest are beams emanating from a spherical source S into empty space. Inside the source the radiation is supposed to be in equilibrium and the temperature is T_S. Therefore in a point outside the source the density function is given by, cf. Fig. 7.4

$$f_{equ}(\boldsymbol{x},t,\boldsymbol{p}) = \begin{cases} f_{equ}(\nu, T_S) & \text{for} \quad 0 \le \varphi \le 2\pi, \quad 0 \le \beta \le \beta_0 = \arcsin\frac{r}{R} \\ 0 & \text{else.} \end{cases}$$

The distribution is strongly non-homogeneous and non-isotropic and therefore it is a *non-equilibrium distribution*, although within the spherical cone of angle β_0 it is a Planck distribution appropriate for the temperature T_S. We may calculate the entries of Table 7.1 for this distribution and obtain the results of Table 7.3.

Table 7.1 Thermodynamic fields of radiation. [*] stands for $-k\,(\ln\frac{f}{y} - \frac{y}{f}(1+\frac{f}{y})\ln(1+\frac{f}{y}))$

	Density	Flux	Source Density
number	$n = \int f\mathrm{d}\boldsymbol{p}$	$\int cn_k f\mathrm{d}\boldsymbol{p}$	$\int S\mathrm{d}\boldsymbol{p}$
Momentum	$P_j = \int \frac{h\nu}{c} n_j \, f\mathrm{d}\boldsymbol{p}$	$P_{jk} = \int h\nu n_j n_k f\mathrm{d}\boldsymbol{p}$	$\int \frac{h\nu}{c} n_j \, S\mathrm{d}\boldsymbol{p}$
energy	$e = \int h\nu \quad f\mathrm{d}\boldsymbol{p}$	$J_k = \int h\nu c n_k \quad f\mathrm{d}\boldsymbol{p}$	$\int h\nu S\mathrm{d}\boldsymbol{p}$
entropy	$h = \int [*] \quad f\mathrm{d}\boldsymbol{p}$	$\varphi_k = \int [*] c n_k \quad f\mathrm{d}\boldsymbol{p}$	$\int k\ln(1+\frac{y}{f}) S\mathrm{d}\boldsymbol{p}$

Table 7.2 Equilibrium values of radiative fields. $a = \frac{8\pi^5}{15}\frac{k^4}{h^3c^3} = 7.8\cdot 10^{-16}\,\frac{\mathrm{J}}{\mathrm{m^3K^4}}$, $\zeta(3) = 1.202$

	Density	Flux	Source Density
number	$\frac{15\zeta(3)}{\pi^4}\frac{a}{k}T^3$	0	0
Momentum	0	$\frac{1}{3}aT^4\delta_{ij}$	0
energy	aT^4	0	0
entropy	$\frac{4}{3}aT^3$	0	0

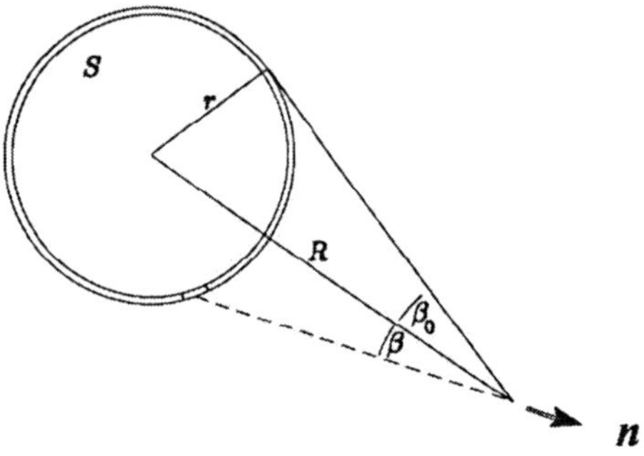

Fig. 7. 4. Radiation from a spherical source

Table 7.3. Radiative thermodynamic quantities of rays emanating from a spherical source S. $\sqrt{\ }$ stands for $\sqrt{1-\frac{r^2}{R^2}}$

	Density	Flux	Source
number	$n=\frac{1}{2}\frac{a}{k}\frac{15\zeta(3)}{\pi^4}T_S^{\,3}(2-\sqrt{\ })$	$\frac{c}{4}\frac{a}{k}\frac{15\zeta(3)}{\pi^4}T_S^{\,3}\frac{r^2}{R^2}\begin{bmatrix}0\\0\\1\end{bmatrix}$	0
Momentum	$P_j=\frac{1}{4c}aT_S^{\,4}\frac{r^2}{R^2}\begin{bmatrix}0\\0\\1\end{bmatrix}$	$P_{jk}=\frac{1}{2}aT_S^{\,4}\cdot\begin{bmatrix}\frac{1}{6}\sqrt{\ }^3-\frac{1}{2}\sqrt{\ }+\frac{1}{3} & 0 & 0\\0 & \frac{1}{6}\sqrt{\ }^3-\frac{1}{2}\sqrt{\ }+\frac{1}{3} & 0\\0 & 0 & \frac{1}{3}-\frac{1}{3}\sqrt{\ }\end{bmatrix}$	0
energy	$e=\frac{1}{2}aT_S^{\,4}(1-\sqrt{\ })$	$J_k=\frac{c}{4}a\,T_S^{\,4}\frac{r^2}{R^2}\begin{bmatrix}0\\0\\1\end{bmatrix}$	0
entropy	$h=\frac{1}{2}a\frac{4}{3}T_S^{\,3}(1-\sqrt{\ })$	$\varphi_k=\frac{c}{4}a\frac{4}{3}T_S^{\,3}\frac{r^2}{R^2}\begin{bmatrix}0\\0\\1\end{bmatrix}$	0

Some of the entries in the table permit simple calculations for the temperatures of the sun and the planets as follows. The sun has the radius $r_\odot = 0.7 \cdot 10^9$m and it is at the distance $R_E = 150 \cdot 10^9$m from the earth. Also we know from measurements that the energy flux density reaching the earth from the sun equals 1341 W/m^2, – the so-called *solar constant.* Therefore we have

$$\frac{c}{4} a T_\odot^{\,4} \frac{r_\odot^{\,2}}{R_E^{\,2}} = 1341 \frac{\mathrm{W}}{\mathrm{m}^2} .$$

From this relation the surface temperature of the sun may be calculated and it comes out as $T_\odot = 5700$ K.

If r_P is the radius of a planet with the distance R_P from the sun, the temperature T_P of the planet follows from the equation

$$\frac{c}{4} a T_\odot^{\,4} \frac{r_\odot^{\,2}}{R_P^{\,2}} \pi r_P^{\,2} = \frac{c}{4} a T_P^{\,4} 4\pi r_P^{\,2} ,$$

since it absorbs solar radiation on the circle $\pi r_P^{\,2}$ exposed to the sun and emits radiation on its whole surface $4\pi r_P^{\,2}$. Since we know the distances R_P of all planets, we may prepare a table of planetary temperatures as shown in Table 7.4. The value for the earth is a trifle low – the mean temperature of the earth is 288K – but this is due to the fact that all kinds of secondary effects have been ignored by the calculation, e.g. the albedo, or coefficient of reflection, and the cloud cover. The same is true for the values of other planets.

Let us also be interested in incoming and outgoing entropy fluxes of a body under solar radiation. The entropy flux density from the sun to the earth reads, according to Table 7.3

$$\varphi_\downarrow = \frac{c}{4} a \frac{4}{3} T_\odot^{\,3} \frac{r_\odot^{\,2}}{R_E^{\,2}} = 0.30 \frac{\mathrm{W}}{\mathrm{m}^2\mathrm{K}} .$$

Table 7.4 Planetary temperatures

	Mercury	Earth	Mars	Jupiter
R_P [m]	$50 \cdot 10^9$	$150 \cdot 10^9$	$230 \cdot 10^9$	$770 \cdot 10^9$
T_P [K]	475	275	222	122

On the other hand, a body with the temperature T= 298K, the leaf of a plant (say), emits entropy at the rate

$$\varphi_{\uparrow} = \frac{c}{4} a \frac{4}{3} T^3 = 2.00 \frac{\text{W}}{\text{m}^2\text{K}} .$$

Thus between absorption and emission the leaf has produced radiative entropy, because it emits more than it absorbs.

A more detailed investigation of this phenomenon was recently presented by Wolf Weiss as part of a memoir on the entropy sources of the earth's atmosphere.[36] As a preliminary exercise Weiss considers radiative and material entropy sources in a black stone plate exposed to the sun. This exercise shows what *can* be done without using explicit expressions for the source terms, if only conditions are stationary, cf. Insert 7.5. In that case the sources may be calculated from the balance of in- and effluxes of entropy and energy, and it turns out that the scattering of radiation provides the biggest contribution to the entropy production; far bigger than the dissipation of matter.

Therefore it is conceivable – at least from the entropic point of view – that the entropy source of matter is negative, if only it is accompanied by radiative scattering. Schrödinger seems to advocate that possibility when he declares[37] radiation to be the cause, when plants decrease their entropy during growth in the process of photosynthesis of glucose. We shall review that proposition in Chap. 11.

Dissipative and radiative entropy sources

We consider a black stone plate of thickness L = 0.1m exposed to solar radiation perpendicular to the plate. The plate absorbs the radiation in a thin surface layer of temperature T_1. That layer reemits part of the absorbed energy and the rest is transmitted through the plate by heat conduction. On the dark side – away from the sun – the plate emits radiation according to its temperature T_2 and according to the Stefan-Boltzmann law. The emitted radiation on the dark side again comes from a thin surface layer. We look at stationary conditions. The heat flux is governed by Fourier's law, cf. Chap. 8 so that we have

$$q = -\kappa \frac{T_1 - T_2}{L} \quad \text{and} \quad T(x) = T_2 + \frac{T_1 - T_2}{L} x \quad (0 \le x \le L) .$$

First we determine T_1 and T_2. We balance the in-and effluxes in the whole plate and in the surface layer on the dark side and obtain respectively

[36] W. Weiss: "The balance of entropy on earth." Thermodynamics and Continuum Mechanics 8, (1996).

[37] In his booklet: E. Schrödinger: "What is Life ?" Cambridge: At the University Press. New York: The Macmillan Company (1945).

$$Q_{\odot} - \frac{c}{4} a T_1^4 = \frac{c}{4} a T_2^4 \quad \text{and} \quad \frac{c}{4} a T_2^4 = \kappa \frac{T_1 - T_2}{L}.$$

With $\kappa = 0.74 \frac{\text{W}}{\text{mK}}$, appropriate for stone, and $Q_{\odot} = 1341 \frac{\text{W}}{\text{m}^2}$, the solar constant, the only relevant solution is

$$T_1 = 355\text{K} \quad \text{and} \quad T_2 = 296\text{K}.$$

The area density of entropy sources has four terms in principle which we denote by

Σ_{rr} – due to photon-photon interaction. Here absent.
Σ_{rm} – source of radiative entropy due to matter.
Σ_{mr} – source of material entropy due to radiation.
Σ_{mm} – dissipative entropy source due to heat conduction.

Σ_{rm} may be calculated from the entries of Table 7. 3 as the balance of in- and effluxes of radiative entropy

$$\Sigma_{rm} = -\frac{c}{4} a \frac{4}{3} T_{\odot}^3 \frac{r_{\odot}^2}{R_E^2} + \frac{c}{4} a \frac{4}{3} T_1^3 + \frac{c}{4} a \frac{4}{3} T_2^3 = 5.032 \frac{\text{W}}{\text{m}^2\,\text{K}}.$$

Σ_{mr} may be calculated according to Clausius, cf. Chap. 3, as $\frac{\dot{Q}}{T}$, i.e. as heat absorbed or emitted divided by the appropriate temperature. Thus

$$\Sigma_{mr} = \frac{1}{T_1}\left(Q_{\odot} - \frac{c}{4} a T_1^4\right) - \frac{1}{T_2} \frac{c}{4} a T_2^4 = -0.243 \frac{\text{W}}{\text{m}^2\,\text{K}}.$$

And $\Sigma_{mm}+\Sigma_{mr}$ must together be zero, because outside the plate there is no material entropy flux. Therefore we have

$$\Sigma_{mm} = 0.243 \frac{\text{W}}{\text{m}^2\,\text{K}}.$$

We conclude that, whatever entropy is produced by heat conduction is balanced by a decrease of the entropy of matter due to absorption and emission of radiation. We also see that the radiative entropy source is about 20 times bigger than the dissipative material one. Absorption, emission and scattering of radiation seems to be the prevalent *mechanism* of entropy production in the plate.

Insert 7.5

The most interesting – and most important – application of radiation thermodynamics is the physics of stars. And yet, the physicists of the 19th century, who raised their eyes to the stars, as it were, were unaware of the decisive role of radiation for stellar structure. They thought, perhaps, that the only role of radiation in a star was to carry the energy away from it.

Although they were mistaken in this assumption, their work laid a foundation and – by good luck – it could be used later as a basis for Eddington's more informed work. Let us review this preliminary work first, before we discuss the radiation thermodynamics of stars.

Convective Equilibrium

In the 19th century the only conceivable source of solar energy – or stellar energy – was the contraction of the stars under their gravitational pull as first envisaged by Helmholtz, cf. Chap. 2 and Insert 2.2. According to the contraction hypothesis, the heating occurs everywhere in the star while, of course, the cooling by radiation occurs near the surface. Thus it makes sense to think of a star as hot inside and cool – relatively cool – near the surface. And it was known that heat conduction could not account for the transfer of heat from the inner regions of a star to the surface, because the thermal conductivity is much too small. On the other hand, the important role of radiation inside the star was not recognized at the time. Therefore the transfer had to happen by *convection*, the same mechanism that distributes the heat from the stove throughout the living room. Let us consider this.

The situation of *hot below and cool above* is akin to the state of our atmosphere on a nice summer day, when the sun heats up the ground in the morning, and the ground heats up the air-layer next to it, which thus becomes warmer than the air on top, and lighter than it should be for equilibrium.[38] If that situation is only slightly disturbed, it causes *thermal convection*, i.e. a vertical rise of the warm air. Since the rising air enters zones of lower pressure, it expands and, since heat conduction is negligible, it cools adiabatically. When this goes on for some hours the air reaches a *convective equilibrium* by mid-day. In that equilibrium the pressure P, and the density ρ within the lower layers – as far up as the convection reaches – obey the *adiabatic equation of state*

$$P = \kappa \rho^{\gamma} .$$

The specific entropy is homogeneous in convective equilibrium. γ is the ratio of specific heats, equal to $^7/_5$ in air and accordingly the air temperature drops by 1K for every 100 meters of height. In the atmosphere the convection stops at night and convective equilibrium breaks down.

In a star there is no night, of course, and therefore the convection may be supposed to persist until the whole star is in convective equilibrium with

[38] Not lighter than the air on top, however, as scientific folklore sometimes has it.

$$\frac{\mathrm{d}P}{\mathrm{d}r} = -G\rho\frac{M_r}{r^2} \quad \text{as mechanical equilibrium condition or momentum balance and}$$

$$P = \frac{P_c}{\rho_c^\gamma}\rho^\gamma \quad \text{as (pressure, density) – relation .}$$

The index c refers to the centre of the star and M_r is the mass inside the sphere of radius r.

Of course, this cannot have been acceptable for all, because the adiabatic equation of state refers to ideal gases and the sun has a mean density of $1.4\frac{\mathrm{g}}{\mathrm{cm}^3}$, larger than the density of water and a thousand times denser than air. Could that matter possibly behave like an ideal gas? Well it does, at least approximately, but the physicists in the 19th century – without any knowledge of the atomic structure – could not begin to understand that. They put the problem on the shelf and proceeded anyway to calculate the potential energy of a gas sphere with radius R and mass M_r, cf. Insert 7.6:

$$E_{pot} = -\frac{3(\gamma-1)}{5\gamma-6}G\frac{M_R}{R^2}.$$

Potential energy of a star

According to Insert 2.2 the potential energy of a spherical mass is equal to

$$E_{pot} = -\frac{1}{2}G\frac{M_R^2}{R} - \frac{1}{2}G\int_0^R \frac{M_r^2}{r^2}\mathrm{d}r,$$

where the second term depends on the mass *distribution* in the star. That term may be rewritten, in terms of E_{pot} itself, for a star in convective equilibrium by the following string of equations involving partial integrations and the repeated use of the mechanical equilibrium condition, the adiabatic equation of state and the identity $\mathrm{d}M_r = \rho\, 4\pi r^2\, \mathrm{d}r$.

$$\frac{1}{2}G\int_0^R \frac{M_r^2}{r^2}\mathrm{d}r = -\frac{1}{2}\int_0^R M_r\frac{1}{\rho}\frac{\mathrm{d}P}{\mathrm{d}r}\mathrm{d}r = -\frac{1}{2}\frac{\gamma}{\gamma-1}\int_0^R M_r\frac{\mathrm{d}\frac{P}{\rho}}{\mathrm{d}r}\mathrm{d}r = \frac{1}{2}\frac{\gamma}{\gamma-1}\int_0^{M_r}\frac{P}{\rho}\mathrm{d}M_r$$

$$= \frac{1}{2}\frac{\gamma}{\gamma-1}4\pi\int_0^R P\, r^2\mathrm{d}r = -\frac{1}{6}\frac{\gamma}{\gamma-1}4\pi\int_0^R r^3\frac{\mathrm{d}P}{\mathrm{d}r}\mathrm{d}r = \frac{1}{6}\frac{\gamma}{\gamma-1}G\int_0^{M_R}\frac{M_r}{r}\mathrm{d}M_r$$

$$= -\frac{1}{6}\frac{\gamma}{\gamma-1}E_{pot}.$$

Insertion into the original equation for E_{pot} provides the equation given in the main text.

Insert 7.6

The pioneer of convective equilibrium was W. Thomson (Lord Kelvin) who conceived of the idea and suggested it for the atmosphere of the earth *and* for the sun.[39] He says:

> The essence of convective equilibrium is that the density and the temperature are so distributed throughout the whole fluid mass that the surfaces of equal density and equal temperature remain unchanged when currents are produced in it by any disturbing influence gentle enough that changes in pressure due to inertial motions are negligible.

And about stars he says that

> ...the natural stirring produced in a great free fluid mass like the Sun's by the cooling of the surface, must, I believe, maintain a somewhat close approximation to convective equilibrium throughout the whole mass.

J. Homer Lane investigated the problem thoroughly. The long title of his paper reveals his main assumption that the stellar material be considered as an ideal gas: "On the theoretical temperature of the sun under the hypothesis of a gaseous mass maintaining its volume by its internal energy and depending on the laws of gases known to terrestrial experiment."[40]

Lane obtained a fairly simple, albeit non-linear second order differential equations for $P(r)$, or $\rho(r)$, or $\frac{P(r)}{\rho(r)}$ by differentiating the momentum balance in convective equilibrium. That equation may be written in the form

$$\frac{\mathrm{d}^2 \frac{\gamma}{\gamma-1}\frac{P}{\rho}}{\mathrm{d}r^2} + \frac{2}{r}\frac{\mathrm{d}\frac{\gamma}{\gamma-1}\frac{P}{\rho}}{\mathrm{d}r} + \frac{4\pi G}{\left[\frac{\gamma}{\gamma-1}\frac{P_c}{\rho_c^{\gamma}}\right]^{\frac{1}{\gamma-1}}}\left(\frac{\gamma}{\gamma-1}\frac{P}{\rho}\right)^{\frac{1}{\gamma-1}} = 0\,,$$

and it is known as the *Lane-Emden equation*.[41] Lane set it up and Emden solved it by a laborious numerical scheme, and published reams and reams

---

[39] W. Thomson: "On the convective equilibrium of temperature in the atmosphere." Proceedings of the Literary and Philosophical Society of Manchester (3) II (1862) pp. 125–131.
See also: W. Thomson: Philosophical Magazine 22 (1887) p. 287 and
W. Thomson: "Mathematical and Physical Papers." 5 Cambridge (1911) p. 256.

[40] J. Homer Lane: American Journal of Science and Arts. Series 2 Vol. 4 (1870) p. 57.

[41] So called by S. Chandrasekhar: "An Introduction to the Study of Stellar Structure." University of Chicago Press (1939) p. 88. Reprinted by Dover Publications (1957). Page numbers refer to the Dover edition.
Chandrasekhar also gives much credit to A. Ritter who investigated the condition of convective equilibrium in a star independently of Lane's work. Ritter published 18 papers "Untersuchungen über die Höhe der Atmosphäre und die Constitution gasförmiger Weltkörper." [Investigations on the height of the atmosphere and the constitution of gaseous bodies] Wiedemann Annalen (1878–1883). Those papers says Chandrasekhar *...form a classic the value of which has never been adequately recognized...*and he is tempted to rename the Lane-Emden equation and call it the Lane-Ritter equation. However, Emden gave much credit to Ritter.

of tables for different values of γ.[42] Nowadays the solution is very easy with any one of the software systems for doing mathematics, e.g. Mathematica®. Thus Emden was able to calculate the central values ρ_c and P_c of density and pressure for the sun (say). He obtained, cf. Insert 7.7

$$\rho_c = 75.6 \cdot 10^3 \tfrac{\text{kg}}{\text{m}^3},\ P_c = 22.5 \cdot 10^{10}\,\text{bar, hence } T_c = \tfrac{\mu}{\mu_0} 4.03 \cdot 10^7\,\text{K},$$

where γ was chosen as $^4/_3$ [43] and where $\frac{\mu}{\mu_0}$ is the relative molecular mass of the particles of the sun. Such calculations suggested that the central temperatures of stars amount to several 10 million K, and that the central densities are many times higher than the density of the densest metal.

Solution of the Lane-Emden equation for $\gamma = {}^4/_3$

With dimensionless dependent and independent variables

$$u = \frac{P}{\rho}\frac{\rho_C}{P_C} \quad \text{and} \quad z = r\sqrt{\pi G}\,\frac{\rho_C}{\sqrt{P_C}}$$

the Lane-Emden equation reads for $\gamma = {}^4/_3$

$$\frac{d^2 u}{dt^2} + \frac{2}{z}\frac{du}{dz} + u^3 = 0 \quad \text{with} \quad u(0) = 1,\ \left.\frac{du}{dz}\right|_0 = 0.$$

The solution is shown in Fig. 7.5. The radius of the star occurs where $u(z)$ crosses the abscissa: the table in the figure shows that this happens for $z(R) = 6.90$. The table also shows values of $-z^2 \frac{du}{dz}$ and, in particular, its surface value $-z^2 \left.\frac{du}{dz}\right|_{z(R)} = 2.015$.

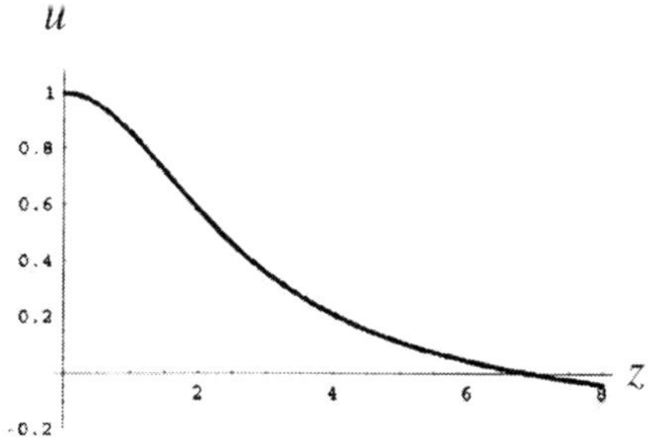

z	u	$-z^2 \frac{du}{dz}$
1	0.85505	0.2522
2	0.58282	1.0450
3	0.35921	1.6553
4	0.20942	1.9197
5	0.11110	2.0070
6	0.04411	2.0156
6.9011	0.00000	2.0150

Fig. 7.5. Solution of the Lane-Emden equation

[42] R. Emden: "Gaskugeln: Anwendungen der mechanischen Wärmetheorie." [Gas spheres: Applications of the mechanical theory of heat] Teubner, Leipzig and Berlin (1907).

[43] $\gamma = {}^4/_3$ proved later – in Eddington's work – to be the correct coefficient, although in that work it is not the ratio of specific heats.

These values are important for the calculation of P_c and ρ_c in terms of the radius R of the star and of its mass M_R. We have

$$z(R) = R\sqrt{\pi G}\,\frac{\rho_c}{\sqrt{P_c}} \quad \text{and} \quad -z^2 \left.\frac{du}{dz}\right|_{z(R)} = \frac{\sqrt{\pi G^3}}{4} M_R \frac{\rho_c^2}{\sqrt{P_c^3}},$$

the latter relation from the momentum balance. Hence follows

$$\rho_c = \frac{z^3(R)}{-z^2 \left.\frac{du}{dz}\right|_{z(R)}} \frac{1}{4\pi} \frac{M_R}{R^3} \quad \text{and} \quad P_c = \left(\frac{z^2(R)}{-z^2 \left.\frac{du}{dz}\right|_{z(R)}} \right)^2 \frac{G}{16\pi} \frac{M_R^2}{R^4}.$$

Insert 7.7

Neither γ nor μ were known to the physicists of the time. We recall that for ideal gases γ can have three values, namely $^5/_3$, $^7/_5$, or $^4/_3$ depending on whether the molecules have one, two or more atoms. Later, when the significance of the radiation pressure was recognized, it turned out that $\gamma = {}^4/_3$ is correct, see below, although there are no molecules in the sun. Indeed, the sun and other stars consist mostly of nuclei and free electrons, because the atoms are largely stripped of their electrons at the high temperatures which prevail.[44] The particles – nuclei, electrons and a few ions – fly freely through the space that is shielded by the electronic shells under normal conditions. It is for that reason that the matter of a star may be considered an ideal gas, even when the density in its center is a hundred times bigger than the density of the heaviest metal.

The strong ionisation is also responsible for a rather small value of μ, the mean mass of the particles, because the free electrons contribute a lot to the number density of particles, but only little to the mass density. Of course, μ depends on the composition of the star: The heavier the atoms are that compose a star, the more closely μ/μ_0 is equal to 2, because a heavy atom contributes approximately $^1/_2\,\mu/\mu_0$ electrons. If, on the other hand, the sun consists mostly of hydrogen – as we now think it does – μ/μ_0 is equal to ½, because a hydrogen atom provides two particles, a proton and an electron. Thus the above calculation suggest an interior solar temperature of 20.000.000K.[45]

---

[44] Of course, neither Lane nor Emden – nor anybody else at the time – knew anything about the atomic structure, or that an atom contains largely empty space, and that there are nuclei and electrons. This knowledge came with Rutherford in 1913, I believe, and it was primarily Eddington in the 1920's who made use of the new knowledge, see below.

[45] By good luck, perhaps, we have thus calculated an interior temperature of the sun that is currently considered to be right: According to the "Fischer Lexikon zur Astronomie" the temperature lies between 17 and 21 million K.

Arthur Stanley Eddington (1882–1944)

Although the pressure of radiation or, more generally, of electro-magnetic fields had been known since Maxwell's equations were known, cf. Chap. 2, its role in stellar physics was not recognized at first. The reason must have been that the radiation pressure p_{rad} exerts a minimal force in everyday life, so small that it cannot be felt when we hold out our hand to absorb sunlight (say). Note that, by Table 7.1, the momentum density P_j is related to the energy flux J_j by $P_j = \frac{1}{c^2} J_j$, so that the momentum density is a lot smaller than the energy flux. But then, p_{rad} equals $^1/_3\, aT^4$ according to Table 7.2. It grows with the fourth power of T and as it became clear, or at least probable, that the interior temperature of stars reaches millions of degrees, researchers decided that it might be worthwhile to look at the momentum balance equation of radiation rather than only at the energy balance.

It seems that Karl Schwarzschild (1873–1916) was first to take the radiation pressure into account in his investigations of the solar atmosphere.[46] Another influential astrophysicist was S. Rosseland.[47] Between them they worked out a plausible expression for the source density S in the photon transport equation. In obvious analogy to Einstein's ansatz [48] for absorption and emission – spontaneous and stimulated – of photons Rosseland sets

$$S = \rho(c_n\,(a + bf) - c_m cf)\,,$$ [49]

where c_n and c_m are the concentrations of atoms in the energetic state with ε_m, and $\varepsilon_n = h\nu + \varepsilon_m$.[50] ρ is the mass density. Following ideas of Schwarzschild [51] about thermodynamic equilibrium Rosseland gives this equation a suggestive form: In equilibrium the right hand side must vanish and, since we know $f_{equ.}$ we obtain $^a/_c = h^3$, $^b/_c = 1$, while $\frac{c_n}{c_m}$ must be equal

---

[46] K. Schwarzschild: "Über das Gleichgewicht der Sonnenatmosphäre." [On the equilibrium of the solar atmosphere] Göttinger Nachrichten 1906 p. 41.
K. Schwarzschild: "Über Diffusion und Absorption in der Sonnenatmosphäre." [On diffusion and absorption in the solar atmosphere] Berliner Sitzungsberichte 1914 p. 1183.

[47] S. Rosseland: "Note on the absorption of radiation within a star." Monthly Notices Vol. 84 (1924) p. 525.
S. Rosseland: "The theory of the stellar absorption coefficient." Astrophysical Journal Vol. 61 (1925) p. 424.

[48] A. Einstein: "Quantentheorie der Strahlung." (1917) loc.cit.

[49] Note that this expression goes a little beyond Einstein's ansatz in that the emission-stimulating ray produces a photon with its own ν *and in its own direction* $\boldsymbol{n}$. This is the phenomenon exploited in lasers.

[50] E.g. see: S. Rosseland: "Astrophysik auf atomtheoretischer Grundlage." [Astrophysics based on atomic theory] Springer, Berlin (1931).

[51] K. Schwarzschild: Göttinger Nachrichten (1906).

to $\exp(-\frac{h\nu}{kT})$ Assuming that the coefficients *a*, *b*, and *c* obey the same relations in non-equilibrium, or near-equilibrium, Rosseland could cast the source term into the plausible form

$$S = \underbrace{\rho c c_m (1-\exp[-\tfrac{h\nu}{kT}])}_{\mathsf{k}} (f_{equ} - f) \ .$$

Thus very plausibly the source is proportional to the difference between the photon density function and its equilibrium value with k – the ν- and T-dependent coefficient of absorption – as factor of proportionality. According to Tables 7.1 and 7.2 we may therefore write the momentum balance of radiation in the form

$$\frac{\partial \int \frac{h\nu}{c} n_j f \mathrm{d}p}{\partial t} + \frac{\partial \int h\nu n_j n_k f \mathrm{d}p}{\partial x_k} = -\rho \mathsf{k} \int \frac{h\nu}{c} n_j f \mathrm{d}p \, ,$$

where the ν-dependence of k has been neglected. The integral on the right hand side represents the momentum density or $\frac{1}{c^2} \cdot$ energy flux. An approximate value for the energy flux J results when we calculate the integrals on the left hand side in equilibrium. Thus we obtain

$$\frac{\partial p_{rad}}{\partial x_j} = -\rho \mathsf{k} \ \frac{1}{c^2} J_j \quad \text{or} \quad \frac{\partial \frac{1}{3} a T^4}{\partial x_j} = -\rho \mathsf{k} \frac{1}{c^2} J_j \, .$$

It was Arthur Stanley Eddington – Sir Arthur after 1930 – who used these equations decisively when he presented a consistent and complete *standard model* of a star.[52] He considered the pressure P inside the star as the sum of the gas pressure $p_{gas} = \rho \frac{k}{\mu} T$ and the radiation pressure $p_{rad} = \frac{1}{3} a T^4$. Eddington was lucky, because he found a (pressure, density)-relation of the form

$$P = \kappa \rho^{4/3}$$

which had been extensively studied already by Lane, Ritter and Emden, see above. It is true though that the exponent $^4/_3$ in Eddington's work had nothing to do with the ratio of specific heats of the stellar gas – it is a reflection of the T^4 law, see Insert 7.8. The coefficient κ, which is equal to $\frac{P_c}{\rho_c^{4/3}}$, turned out to be determined by the mass of the star. So, Eddington was able to transfer much of the mathematics from the earlier researches to

---

[52] A.S. Eddington: "The Internal Constitution of the Stars." Cambridge, University Press (1926).

his own. And in a few additional steps he could derive a relation between the luminosity L_R of a star – the total power emitted – and its mass M_R, cf. Insert 7.8. Using Eddington's data, one can find a rough analytical fit for the so-called *mass-luminosity relation* which reads

$$\frac{L_R}{L_\odot} = \left(\frac{M_R}{M_\odot}\right)^{3.5}$$

so that the luminosity of a star grows fairly steeply with its mass. This relation was confirmed for all stars whose mass was known, and that fact provided strong support for Eddington's model, e.g. for the ideal-gas-character of the stars, despite their large mean densities and their enormous central densities. After that structure was accepted for stars, the mass-luminosity relation allowed astronomers to determine the mass of a star from its brightness provided, of course, that the distance was known.

Mass-luminosity relation

The momentum balance equations for matter and radiation and for radiation alone read

$$\frac{\mathrm{d}P}{\mathrm{d}r} = -\rho G \frac{M_r}{r^2} \quad \text{and} \quad \frac{\mathrm{d}p_{rad}}{\mathrm{d}r} = -\frac{\kappa}{c^2}\rho J\,,$$

where $J = \dfrac{L_r}{4\pi r^2}$ is the radiative energy flux density. Elimination of ρ gives

$$\frac{\mathrm{d}p_{rad}}{\mathrm{d}r} = \frac{\kappa}{c^2}\frac{1}{4\pi G}\underbrace{\frac{L_r}{M_r}}_{\eta\frac{L_R}{M_R}}\frac{\mathrm{d}P}{\mathrm{d}r} \quad \text{and by integration} \quad p_{rad} = \frac{1}{4\pi c^2 G}\underbrace{\kappa\eta}_{opacity}\frac{L_R}{M_R}P,$$

where L_R is the *luminosity* of the star. In Eddington's *standard model* the *opacity* is considered homogeneous throughout the star and equal for all stars.

If βP and $(1-\beta)P$ are the partial pressures of matter and radiation respectively, we have

$$p_{rad} = (1-\beta)P = \tfrac{1}{3}aT^4$$

$$\text{hence} \quad T^3 = \frac{1-\beta}{\beta}\frac{3\,k/\mu}{a}\rho \quad \text{and} \quad P = \left[\frac{1-\beta}{\beta}\left(\frac{k}{\mu}\right)^4\frac{3}{a}\right]^{1/3}\rho^{4/3}\,.$$

$$p_{gas} - \beta\ P = \rho\frac{k}{\mu}T$$

Thus P is proportional to $\rho^{4/3}$ just like in the Lane-Emden theory for $\gamma = {}^4/_3$, where the factor of proportionality is $\frac{P_c}{\rho_c^{4/3}}$. Therefore comparison with the results of Insert 7.7 shows that we must set

$$\frac{1-\beta}{\beta}\left(\frac{k}{\mu}\right)^3 \frac{3}{a} = \frac{G^3\sqrt{\pi}^{-3}}{16}\left(\frac{M_R}{-z^2 \left.\frac{du}{dz}\right|_{z(R)}}\right)^2$$

so that β is only a function of M_R.

On the other hand, the formula for p_{rad} provides β as a function of $\frac{L_R}{M_R}$:

$$1-\beta = \frac{1}{4\pi c^2 G} k\eta \frac{L_R}{M_R}.$$

L_R is reliably measurable[53] for all stars, whose distance is known, and M_R is measurable for many binaries and, of course, both are known for the sun. Therefore $k\eta$ can be determined from solar data.

The mass-luminosity relation follows in an implicit form by elimination of β between the last two equations. Eddington solved that equation by numerical means, plotted it graphically, and compared the curve with astronomical data for many stars, finding good agreement.

Insert 7.8

His partisanship for relativity secured Eddington a place in 1919 on the expedition to Príncipe island in the gulf of Guinea, where the bending of light rays by the sun – predicted by Einstein's theory of general relativity – was first observed during a solar eclipse.

> Eddington was so busy changing photographic plates that he did not actually see the eclipse.[54]

Since we are dealing with radiation in this chapter, the ratio of radiation pressure and gas pressure to the total pressure is of interest. Eddington's calculations suggest, that that ratio depends only on the mass of the star and that it grows with the mass, cf. Insert 7.8. For the relatively small sun the radiation pressure amounts to only 5% of the total, but it runs up to 80% for a massive star of 60 times the solar mass. Since there are very few more massive stars than that, Eddington assumes that a high radiation pressure *is*

---

[53] Eddington remarks that…*it is said that the apparatus on Mount Wilson* [in California] *is able to register the heat radiation of a candle on the bank of the Mississippi river.* That was in 1926; I wonder what astronomers can do *now*.

[54] According to I. Asimov: "Biographies …" loc. cit. p.603.

dangerous for the stability of a star [55] ... *although one cannot, a priori, see a good reason why the radiation pressure acts more explosively than the gas pressure.*[56]

Eddington was an infant prodigy of the best type, – the type that grows into an adult prodigy. He was one of the first persons to appreciate Einstein's theory of relativity, and advertised it to British scientists.

At that time it was generally said that only three persons in the world understand the theory of relativity. When Eddington was asked about that by a journalist he answered: *Oh? And who is the third?* [57]

Fig. 7.6. Arthur Stanley Eddington

There is a group of fairly massive stars – between 5 and 50 solar masses– which exhibit a possible sign of instability by a regularly oscillating luminosity. These are the Cepheids, named after Delta Cephei for which that behaviour was first observed. Naturally Eddington's attention was drawn to the phenomenon, and he investigated it without, however, clearly relating it to the predominance of the radiation pressure. I suspect that *now* stellar physics can answer that question decisively; if so, I would not have heard about it.

The Cepheids play an important role in astronomy, because the astronomer Henrietta Swan Leavitt (1868–1921) has detected – in 1912 – a clear relation between the mean luminosity of those stars and the period of their oscillation: The more luminous stars oscillate more slowly. No reason was at first known, but nevertheless the observation led to the *Cepheid yardstick* for measuring the distance of galaxies. Since the brightness of equally luminous Cepheids depends on their distance, while the period of oscillation does not, of course, the relative distance of two Cepheids from the observer could be determined. Eddington's mass-luminosity relation provides a plausible explanation for Leavitt's observation: Indeed, more massive stars are more luminous and presumably more sluggish in their oscillations.

---

[55] A.S. Eddington: "The internal Constitution of the Stars." loc.cit p. 145.
[56] Ibidem, p. 21.
[57] Nowadays meetings on Relativity Theory are visited by up to 2000 participants. One must assume that, perhaps, all of them understand what the theory is about.

Eddington's book "The Internal Constitution of Stars" – written in 1924 and 1925 – is crystal clear in style and argument, and when assumptions occur, as they invariably must, they are made plausible either by reference to observations, or by convincing theoretical arguments. Some things he could only guess at, most notably the origin of the stellar energy. But he guessed well, albeit without being specific:

> ... after exhausting all other possibilities we find the conclusion forced upon us that the energy of a star can only result from subatomic sources.[58]

Eddington did not identify the *subatomic sources.* However, his insight into the enormous temperatures of stellar interiors made it feasible that nuclear fusion occurs which – basically – forms helium from hydrogen, at least to begin with. Hans Albrecht Bethe (1906–2005) is usually credited with having worked out the details of this nuclear reaction in 1938, although there were forerunners, most notably Jean Baptiste Perrin (1870–1924).

Strangely enough Eddington sticks to the obsolete ether waves when he speaks of radiation:

> Just as the pressure in a star must be considered partly as the pressure of ether waves and partly as pressure of material molecules, the heat content is also composed of ethereal and material components.[59]

It seems then, that despite his partisanship for Einstein's theory of relativity, Einstein's light quanta and Compton's photons did not impress Eddington – at least not at the time when he published the book.

Another peculiarity about Eddington is that he still believed in the element *coronium* – a hypothetical element of relative molecular mass of about 0.4 – which had been postulated by Dimitrij Iwanowitch Mendelejew (1934 – 1907)[60] in order to fill a perceived gap in the periodic table. Surely by 1926 atomic physicists did not give credence to this fictitious element, although Mendelejew's reputation was so great that many scientists clung to the coronium. So also the eminent geophysicist Alfred Lothar Wegener (1880 – 1930) – author of the continental drift theory – who says[61]

> ...because of Mendelejew's lucky shot with the prediction of germanium it seems to me that the hypothesis [about coronium] deserves our attention

[58] Ibidem, p. 31.

[59] Ibidem, p. 29.

[60] D.I. Mendelejew: Chemisches Centralblatt (1904) Vol. I p. 137.

[61] A.L. Wegener: "Thermodynamik der Atmosphäre" [Thermodynmics of the atmosphere] Verlag J.A. Barth, Leipzig (1911).

8 Thermodynamics of Irreversible Processes

Long before there was a thermodynamic theory of irreversible processes, there were *phenomenological equations,* i.e. equations governing the fluxes of momentum, energy and partial masses. They were read off from the observed phenomena of thermal conduction, internal friction and diffusion. Even the appropriate field equation for temperature was formulated correctly, – for special cases – before the first law of thermodynamics was pronounced and accepted. Thus it was that complex problems of heat conduction were being solved routinely in the 19th century before anybody knew what heat was.

It took more than a century after phenomenological equations had been formulated – and proved their reliability for engineering applications – before transport processes were incorporated into a consistent thermodynamic scheme. And the first theories of irreversible processes clung so closely to the laws of equilibrium – or near-equilibrium – that they achieved no more than confirmation of the 19th century formulae, and proof of their consistency with the doctrines of energy and entropy.

It is only most recently that non-equilibrium thermodynamics has been rephrased and given a formal mathematical structure with symmetric hyperbolic field equations. That structure is motivated by the classical laws, of course, but not in any obvious manner; no specific assumptions are carried over from equilibrium thermodynamics into the new theory of extended thermodynamics. It has thus been possible to modify the classical laws in an unprejudiced manner, and to extrapolate them into the range of rarefied gases and of non-Newtonian fluids. The kinetic theory of gases has provided a trustworthy heuristic tool for this extension of thermodynamics which, at this time, has only just begun.

Phenomenological Equations

Jean Baptiste Joseph Baron de Fourier (1768–1830)

Fourier came from poor parents and, besides, he became an orphan at the age of eight. So his ambitions to be a mathematician and artillery man seemed to be stymied and they would doubtless not have led him anywhere, were it not for the French revolution and Napoléon Bonaparte. As it was, the revolution happened in 1789 and Fourier could enter a military school – the later École Polytechnique of early 19th century fame, cf. Chap. 3 – and after graduation he stayed on as an instructor.

Napoléon took Fourier along on his disastrous Egyptian campaign and made him a baron in recognition of his great mathematical discoveries which were related to heat conduction and the calculation of temperature fields. Those discoveries were first published in the Bulletin des Sciences (Société Philomatique, année 1808). After that first work, Fourier continued a lively scientific production and eventually he summarized his life's work in the book "Théorie analytique de la chaleur" in 1824. This book is not available to me; therefore I refer to a German edition, published in 1884.[1] The translator claims that his work is identical to the original except that he *corrected numerous misprints*.

The work is essentially a book on analysis. It is completely unaffected by any speculations about the nature of heat, or whether heat is the weightless caloric or a form of motion. Fourier says:

> One can only form hypotheses on the inner nature of heat, but the knowledge of the mathematical laws that govern its effects is independent of all hypotheses.[2]

It is true that Fourier's pronouncements are couched in long and old-fashioned sentences like this one:

> If two corpuscles of a body lie infinitely close and have different temperatures, the warmer corpuscle transmits a certain amount of its heat to the other one; and this heat – given from the warmer corpuscle to the colder one at a given time and during a given moment – is proportional to the temperature difference, if that difference has a small value.[3]

However, Fourier also summarizes this cumbersome statement in the simple vectorial expression

$$q_i = -\kappa \frac{\partial T}{\partial x_i},$$

which is *Fourier's law* for the heat flux $\boldsymbol{q}$; κ is the thermal conductivity. Fourier calls it the *internal conductivity*. He proceeds from there by assuming that the rate of change of temperature of a *corpuscle* is proportional to the difference of the heat fluxes on opposite sides and thus he comes to formulate the *differential equation of heat conduction,* viz.

---

[1] M. Fourier: "Analytische Theorie der Wärme." Translated by Dr. B. Weinstein. Springer, Berlin (1884).

[2] Ibidem: Introduction, p. 11.

[3] Ibidem. p. 451/2.

$$\frac{\partial T}{\partial t} = \lambda \frac{\partial^2 T}{\partial x_i \partial x_i},$$

where λ is Fourier's *external conductivity,* in modern terms it is the ratio of κ and the density of the heat capacity. This equation is the prototype of all parabolic equations and Fourier presented solutions for a large variety of boundary and initial values in his book.

Among many other problems solved, there is the one – a particularly ingenious one – by which the yearly periodic change of temperature on the surface of the earth propagates as a damped wave into the interior, so that at certain depths the earth is colder in summer than in winter.

As a tool for the solution of heat conduction problems Fourier developed what we now call harmonic analysis – or Fourier analysis – by which any function can be decomposed into a series of harmonic functions, and he expresses his amazement about the discovery by saying:

> It is remarkable that the graphs of quite arbitrary lines and areas can be represented by convergent series [of harmonic functions] ... Thus there are functions which are represented by curves, ... which exhibit an osculation on finite intervals, while in other points they differ.[4]

The harmonic analysis has found numerous applications in mathematics, physics and engineering. It transcends the narrow field of heat conduction and proves its usefulness everywhere. Let me quote Fourier on the subject:

> The main property [of mathematical analysis] is clarity; [the theory] possesses no symbol for the expression of confused ideas. It combines the most diverse phenomena and discovers hidden analogies.[5]

His lifelong preoccupation with heat conduction had left Fourier with an *idée fixe*:

He believed heat to be essential to health so he always kept his dwelling place overheated and swathed himself in layer upon layer of clothes. He died of a fall down the stairs.[6]

Fig. 8.1. Jean Baptiste Joseph Baron de Fourier

---

[4] Ibidem. p. 160.
[5] Ibidem. Forword, p. XIV.
[6] I. Asimov: "Biographies..." loc.cit. p. 234.

Fourier's book has a distinctly modern appearance.[7] This is all the more surprising, if the book is compared with contemporary ones, like Carnot's, which appeared in he same year. Maybe that shows that physics is more difficult than mathematics, but the fact remains that every line of Fourier's book can be read and understood, while large parts of Carnot's book must be read, thought over and then discarded.

One of the eager readers of Fourier's book was the young W. Thomson (later Lord Kelvin). Fourier's results troubled him and in 1862 he wrote:

> For 18 years I have been worried by the thought that essential results of thermodynamics have been overlooked by geologists.[8]

Kelvin praises … *the admirable analysis which led Fourier to solutions* and he uses its results to determine the age of the *consistentior status* – the solid state – of the earth. That expression goes back to Leibniz. The prevailing idea was that, at some time in the past, the earth was liquid. Obviously it had to cool off to a solid of at most 7000°F before the geological history could begin. And Kelvin sets out to determine when that was.

Fourier had given the temperature field in two half spaces initially at temperatures $T_o \pm \Delta T$ as

$$T(x,t) = T_o + \frac{2\Delta T}{\sqrt{\pi}} \int_0^{\frac{x}{2\sqrt{\lambda t}}} e^{-z^2} \mathrm{d}z\,.$$

Kelvin took $\Delta T = 7000°\mathrm{F}$ and in effect fitted Fourier's solution to

- a constant surface temperature T_o of the earth,
- the known value of Fourier's *external conductivity*,
- the known value of the present temperature gradient near the earth's surface,

and calculated the corresponding value for t as 100 million years. Therefore the geological history of the earth had to be shorter than that.

That age was of the same order of magnitude as Helmholtz's result for the age of the earth, cf. Insert 2.2. So great was Kelvin's – and, perhaps, Helmholtz's – prestige that biologists started to revise their time tables for evolution. Geologists were at a loss, however. Fortunately for them it turned out in the end that both Kelvin and Helmholtz had made wrong assumptions. Indeed, the earth possesses within itself a source of heat by radioactive decay so that, whatever it loses by conduction is replaced by

---

[7] Well, that statement must be qualified. Let us say that the book has the appearance of a textbook on analysis written in the mid 20th century. Really modern books on the subject make even interested readers give up in frustration and bewilderment on the first half-page.

[8] W. Thomson: "On the secular cooling of the earth." Transactions of the Royal Society of Edinburgh (1862).

radioactivity. Thus the earth can maintain its present temperature for as long as needed to guarantee a geological – and biological – history of some billions of years. Yet Kelvin, who lived until 1907, would never accept radioactivity, he stuck to his old prediction till the end. Asimov says:

> In the 1880's Thomson settled down to immobility, ... and passed his last days bewildered by the new developments.[9]

Adolf Fick (1829–1901)

Fick was a competent physiologist who did much to increase our knowledge about the mechanical and physical processes in the human body. Later in life he became an influential professor in Zürich but at the time when he published his paper on diffusion[10] he was a *prosector,* i.e. the person who cut open dead bodies up to the point where the anatomy professor took over for his demonstrations to a class of medical students.

IV. *Ueber Diffusion; von Dr. Adolf Fick,*
Prosector in Zürich.

Fig. 8.2. Cut from the title page of Fick's paper

Fick was interested in diffusion of solutes in solvents and he adopted a molecular interpretation that sounds very peculiar indeed to modern readers, with regard to physics, grammar and style:[11]

> When one assumes that two types of atoms are distributed in empty space, of which some (the ponderable ones) obey Newton's law of attraction, while the others – the ether atoms – repel each other also in the combined ratio of masses, but proportional to a function $f(r)$ of the distance, which falls off more rapidly than the reciprocal value of the second power; when one assumes further that the ponderable atoms and ether atoms attract each other with a force, which again is proportional to the product of masses but also to another function $\varphi(r)$ of the distance which decreases even more rapidly than the previous one, when one – this is what I say – assumes all this, then one sees clearly, that each ponderable atom must be surrounded by a dense ether atmosphere, which if the ponderable atom may be thought of as spherical, will consist of concentric spherical shells, which all have the density of the ether, such that the ether density at some

---

[9] I. Asimov: "Biographies ..." loc. cit. p. 380.

[10] A. Fick: "Ueber Diffusion." [On diffusion] Annalen der Physik 94 (1855) pp. 59–86.

[11] Since all this was published, we must assume that it represented acceptable scientific reasoning at the time. And indeed, Navier and Poisson argued similarly when they derived their versions of the Navier-Stokes equations, see below.

> point at the distance r from the centre of an isolated ponderable atom may be expressed by $f_1(r)$, which must certainly for a large argument assume a value which equals the density of the general sea of ether.

Fick continues like that speculating about the form of the functions $f(r)$, $\varphi(r)$ and $f_1(r)$, and effectively weaving a Gordian knot of words and sentences until – on page 7(!) of his paper – he has the good sense of cutting the argument short with the words:

> Indeed, one will admit that nothing be more probable than this: The diffusion of a solute in a solvent ... follows the same rule which Fourier has pronounced for the distribution of heat in a conductor...[12]

This is a relief, because now he comes to what has become known as *Fick's law* for the diffusion flux J_i:

$$J_i = nv_i = -D\frac{\partial n}{\partial x_i}.$$

n is the number density of solute particles and v_i is their velocity, if one assumes that the solvent is at rest. D is the diffusion coefficient.

And again, in analogy to heat conduction, Fick assumes that the rate of change of n in a *corpuscle* is proportional to the balance of influx and efflux and thus obtains

$$\frac{\partial n}{\partial t} = D\frac{\partial^2 n}{\partial x^2}.$$

This is known as the *diffusion equation*; it is formally identical to the equation of heat conduction, so that Fourier's solutions can be carried over to boundary and initial value problems of diffusion.

In particular, for one-dimensional diffusion of a solute in an infinite solvent, if $n(x,t)$ is initially a constant n_0 in a small interval $X - {}^{\Delta}/_2 < x < X + {}^{\Delta}/_2$ and zero everywhere else, the solution reads[13]

$$n(x,t) = \frac{n_0\Delta}{\sqrt{4\pi Dt}}\exp\left(-\frac{(x-X)^2}{4Dt}\right).$$

It follows that a maximum of $n(x,t)$ passes through a given point x at the time

---

[12] I have taken the liberty to *prosect*, as it were, Fick's hemming and hawing from this sentence. He remarks that Georg Simon Ohm (1787–1854) has seen the same analogy for electric conduction.

[13] The solution refers to the limiting case $\Delta \to 0$ and $n_0 \to \infty$, but so that $n_0\,\Delta$ is equal to the total number of solvent particles.

$$t_{\max} = \frac{(x-X)^2}{2D} \qquad \text{hence} \qquad |x-X| = \sqrt{2Dt_{\max}} \,,$$

so that, in a manner of speaking, diffusion proceeds in time as $\sqrt{t}$. This is the hallmark of all random walk processes and we shall encounter it again in connection with Brownian motion, cf. Chap. 9. The maximum has the universal, i.e. D-independent value

$$n(x, t_{\max}) = \frac{n_o \Delta}{\sqrt{2\pi e (x-X)^2}} \,.$$

George Gabriel Stokes (1819–1903). Baronet Since 1889

At the age of thirty Stokes became Lucasian professor of mathematics at Cambridge; in 1854, secretary of the Royal Society; and in 1885, president of that institution. *No one had held all three offices since Isaac Newton.*[14] Stokes's mathematical and physical papers fill five volumes with a total of close to 2000 pages.[15] His main topic was fluid mechanics with an emphasis on viscous friction in liquids and gases and his name will always be connected with the *Navier-Stokes equations* which relate the viscous stress tensor $t_{ij} + p\delta_{ij}$ in a fluid to velocity gradients. In modern form they read[16]

$$t_{ij} + p\delta_{ij} = 2\eta \frac{\partial v_{\langle i}}{\partial x_{j\rangle}} + \lambda \frac{\partial v_l}{\partial x_l} \delta_{ij} \,.$$

To be sure, Stokes missed out on the second term with the bulk viscosity λ, but the other term is derived. η is now called the shear viscosity but Stokes does not seem to have named it. He derived the formula from the principle:

> That the difference between the pressure on a plane in a given direction passing through any point P of a fluid in motion and the pressure which would exist in all directions about P if the fluid in its neighbourhood were in a state of relative equilibrium depends only on the relative motion of the fluid immediately about P; and that the relative motion due to any motion

---

[14] I. Asimov: "Biographies..." loc. cit. p. 354.

[15] G.G. Stokes: "Mathematical and Physical Papers." Cambridge at the Universities Press (1880 – 1905).

[16] Angular brackets denote symmetric, trace-free tensors.

of rotation may be eliminated without affecting the differences of the pressure above-mentioned.[17]

Nowadays we would say concisely that the viscous stress is a linear isotropic function of the velocity gradient. But no matter, Stokes in his own way reached a result. After 13 pages of cumbersome, yet reproducible derivation Stokes came up with

$$\text{Stokes:} \qquad \frac{\partial p}{\partial x} - \eta\left(\frac{\partial^2 u}{\partial x^2} + \frac{\partial^2 u}{\partial y^2} + \frac{\partial^2 u}{\partial z^2}\right) - \frac{\eta}{3}\frac{\partial}{\partial x}\left(\frac{\partial u}{\partial x} + \frac{\partial v}{\partial y} + \frac{\partial w}{\partial z}\right).$$

This is the stress contribution to the x-component of the momentum balance.

Nobody at that time used vector and tensor notation, and (u, v, w) were the canonical letters for the velocity components in x, y, z direction.

As it was, Stokes had been anticipated by two scientists across the English Channel: Louis Navier[18] (1785–1836) and Siméon Denis Poisson[19] (1781–1840). Both had employed somewhat irrelevant molecular models – much in the manner of Fick whom I have cited at length – but they did come up with reasonable expressions, viz.

$$\text{Navier:} \qquad \frac{\partial p}{\partial x} - A\left(\frac{\partial^2 u}{\partial x^2} + \frac{\partial^2 u}{\partial y^2} + \frac{\partial^2 u}{\partial z^2}\right)$$

$$\text{Poisson:} \qquad \frac{\partial p}{\partial x} - A\left(\frac{\partial^2 u}{\partial x^2} + \frac{\partial^2 u}{\partial y^2} + \frac{\partial^2 u}{\partial z^2}\right) - B\frac{\partial}{\partial x}\left(\frac{\partial u}{\partial x} + \frac{\partial v}{\partial y} + \frac{\partial w}{\partial z}\right).$$

Thus we conclude that the credit should have gone to Poisson who, after all, had *two* coefficients which implies that he allowed for shear and bulk viscosity. However, Poisson is nowadays rarely mentioned in this context.

It is true though that Stokes did a lot more than set up the equations; he solved them in fairly complex situations. He was much interested in the motions of the pendulum and how this was affected by friction. In 1851 he wrote a long article on the question.[20] Section II of that article is entitled

> Solutions of the equations in the case of a sphere oscillating in a mass of fluid either unlimited, or confined by a spherical envelope concentric with the sphere in its position of equilibrium.

---

[17] G.G. Stokes: "On the theories of the internal friction of fluids in motion and of the equilibrium and motion of elastic solids." Transactions of the Cambridge Philosophical Society. III (1845) p. 287.

[18] L. Navier: Mémoires de l´Académie des Sciences VI (1822) p. 389.

[19] S.D. Poisson: Journal de l´´Ecole Polytechnique XIII cahier 20 p. 139.

[20] G.G Stokes: "On the effect of the internal friction of fluids on the motion of pendulums." Transactions of the Cambridge Philosophical Society IX (1851) p. 8.

The result could be specialized to the case of uniform motion of a sphere of radius r with the velocity v. The force to maintain the motion is given by

$$F = 6\pi\eta r v,$$

a formula that is universally called the *Stokes law of friction*. It is now derived as an exercise in all good books on fluid mechanics.

The solution of boundary value problems for the Navier-Stokes equation requires more than an able mathematician: A decision about the boundary values of the velocity components near the walls of a pipe or the surface of a sphere must be made. Stokes says:

> The most interesting questions connected with this subject require for their solution a knowledge of the conditions which must be satisfied at the surface of a solid in contact with the fluid[21]

SIR GEORGE GABRIEL STOKES, BART.,
SC.D., LL.D., D.C.L., PAST PRES. R.S.,
ET PRUSSIAN ORDER *POUR LE MÉRITE*, FOR. ASSOC. INSTITUTE OF FRANCE, ETC.
MASTER OF PEMBROKE COLLEGE AND LUCASIAN PROFESSOR OF MATHEMATICS
IN THE UNIVERSITY OF CAMBRIDGE.

Fig. 8.3. George Gabriel Stokes. His degrees and honours

Hesitantly he proposes the *no-slip-condition* which is now routinely applied for laminar flows:

> The condition which first occurred to me to assume … was, that the film of fluid immediately in contact with the solid did not move relatively to the surface of the solid.[22]

Stokes tends to consider this assumption as valid when the mean velocity of the flow is small. He is aware of the difficulties that turbulence might raise. But he is blissfully unaware, of course, of the problems that may arise in rarefied gases; these are problems that haunt the present-day researchers concerned with re-entering space vehicles.

---

[21] G.G. Stokes: "On the theories of the internal friction…." loc.cit. p. 312.
[22] Ibidem. p. 309.

Carl Eckart (1902–1973)

However convoluted the 19th century arguments of Fourier, Fick and Navier, and Stokes may have been, their works provided valid equations for the fluxes of mass, momentum and energy in terms of the basic fields of thermodynamics, viz. mass density, velocity and temperature. Yet, they did not provide a coherent picture of thermodynamics of processes, or non-equilibrium thermodynamics. The first such picture was created by Carl Eckart in 1940 in one stroke, or rather in two strokes, the first one concerning viscous, heat-conducting single fluids,[23] and the second one concerning mixtures.[24] Both papers form the basis of what came to be called TIP – short for **t**hermodynamics of **i**rreversible **p**rocesses. Let us review these papers in the shortest possible form:

One may say that the objective of non-equilibrium thermodynamics of viscous, heat-conducting single fluids is the determination of the five fields

$$\text{mass density } \rho(\boldsymbol{x},t), \quad \text{velocity } v_i(\boldsymbol{x},t), \quad \text{temperature } T(\boldsymbol{x},t)$$

in all points of the fluid and at all times.

For the purpose we need field equations and these are based upon the equations of balance of mechanics and thermodynamics, viz. the conservation laws of mass and momentum, and the equation of balance of internal energy, see Chap. 3

$$\dot{\rho} + \rho \frac{\partial v_j}{\partial x_j} = 0$$

$$\rho \dot{v}_i - \frac{\partial t_{ij}}{\partial x_j} = 0$$

$$\rho \dot{u} + \frac{\partial q_j}{\partial x_j} = t_{ij} \frac{\partial v_i}{\partial x_j}.$$

These equations are also known as the continuity equation, Newton's equation of motion and the first law of thermodynamics respectively.

While these are five equations – the proper number for five fields – they are not field equations for ρ, v_i, and T. The temperature does not even occur and, instead, the equations contain new quantities

[23] C. Eckart: "The thermodynamics of irreversible processes I: The simple fluid." Physical Review 58, (1940)

[24] C. Eckart: "The thermodynamics of irreversible processes II: Fluid mixtures." Physical Review 58, (1940).

- stress t_{ij}, • heat flux q_i, • specific internal energy u.

In order to close the system of equations, one must find relations between t_{ij}, q_i, and u and the fields ρ, v_i, T.

In TIP such relations are motivated in a heuristic manner from an entropy inequality that is based upon the Gibbs equation of equilibrium thermodynamics, cf. Chap. 3

$$\dot{s} = \frac{1}{T}\left(\dot{u} - \frac{p}{\rho^2}\dot{\rho}\right).$$

s is the specific entropy. u and p are considered to be functions of ρ and T as prescribed by the caloric and thermal equations of state, just as if the fluid were in equilibrium. This assumption is known as the *principle of local equilibrium.*

Elimination of $\dot{u}$ and $\dot{\rho}$ between the Gibbs equation and the equations of balance of mass and energy and some rearrangement lead to the equation[25]

$$\rho\dot{s} + \frac{\partial}{\partial x_i}\left(\frac{q_i}{T}\right) = -\frac{q_i}{T^2}\frac{\partial T}{\partial x_i} + \frac{1}{T}t_{\langle ij\rangle}\frac{\partial v_{\langle i}}{\partial x_{j\rangle}} + \frac{1}{T}\left(\tfrac{1}{3}t_{kk} + p\right)\frac{\partial v_n}{\partial x_n},$$

which may be interpreted as an equation of balance of entropy. That interpretation implies that

$\varphi_i = \dfrac{q_i}{T}$ is the entropy flux and

$\Sigma = -\dfrac{q_i}{T^2}\dfrac{\partial T}{\partial x_i} + \dfrac{1}{T}t_{\langle ij\rangle}\dfrac{\partial v_{\langle i}}{\partial x_{j\rangle}} + \dfrac{1}{T}\left(\dfrac{1}{3}t_{kk} + p\right)\dfrac{\partial v_n}{\partial x_n}$ is the dissipative source density of entropy.

Inspection shows that the entropy source is a sum of products of *thermodynamic fluxes* and *thermodynamic forces*, see Table 8.1

The dissipative entropy source must be non-negative. Thus results an entropy inequality – with $\varphi_i = q_i/T$ as entropy flux – which is often called the Clausius-Duhem inequality, because it represents Duhem's extrapolation of Clausius's second law to non-homogeneous temperature fields. Assuming only linear relations between forces and fluxes, TIP ensures the validity of the Clausius-Duhem inequality by constitutive relations – phenomenological equations in the jargon of TIP – of the type

[25] As before, angular brackets characterize symmetric traceless tensors.

Table 8.1. Fluxes and forces for a single fluid

Thermodynamic Fluxes	Thermodynamic Forces
heat flux q_i	temperature gradient $\frac{\partial T}{\partial x_i}$
deviatoric stress $t_{\langle ij\rangle}$	deviatoric velocity gradient $\frac{\partial v_{\langle i}}{\partial x_{j\rangle}}$
dynamic pressure $\pi = -\frac{1}{3}t_{ii} - p$	divergence of velocity $\frac{\partial v_n}{\partial x_n}$.

$$q_i = -\kappa \frac{\partial T}{\partial x_i} \qquad \kappa \geq 0 \qquad \text{Fourier}$$

$$\left.\begin{aligned} t_{\langle ij\rangle} &= 2\eta \frac{\partial v_{\langle i}}{\partial x_{\langle j}} && \eta \geq 0 \\ \pi &= -\lambda \frac{\partial v_n}{\partial x_n} && \lambda \geq 0 \end{aligned}\right\} \text{Navier - Stokes}$$

Together with the thermal and caloric equations of state $p = p(\rho,T)$ and $u = u(\rho,T)$ the phenomenological equations form the set of material properties characterizing a fluid. κ is the thermal conductivity, and η and λ are the shear- and bulk viscosities respectively; all three may be functions of ρ and T that must be found experimentally.

In this manner TIP incorporates Fourier's law and the law of Navier-Stokes into a consistent thermodynamic scheme. Neither Fourier, nor Navier, or Stokes had made use of thermodynamic arguments, or of the Gibbs equation, nor did they need them. They proposed their laws on the basis of plausible assumptions about the phenomena of heat conduction and internal friction.

The equations of state and the phenomenological equations combined with the equations of balance of mass, momentum and energy provide a set of field equations from which – given initial and boundary values – the fields $\rho(\boldsymbol{x},t)$, $v_i(\boldsymbol{x},t)$, and $T(\boldsymbol{x},t)$ may be calculated. And the solutions are satisfactory for nearly all *normal* cases. Indeed, it is no exaggeration to say that 99% of all flow problems in single fluids are solved by use of these field equations; and that begins with the calculation of pipe flow of a liquid and ends with the calculation of lift and drag on an airliner.[26] To be sure, both problems need numerical methods in general.

It is true that all this could have been done before Eckart – except for the numerical solutions, of course. After all Jaumann and Lohr did have the full set of equations.[27] Eckart's achievement is that he formulated a consistent and coherent theory with the phenomenological equations as part of it.

And Eckart did not stop with single fluids. He applied his scheme to mixtures of fluids as well. In that case he started with the Gibbs equation for a mixture, see Chap. 5 and identified thermodynamic fluxes and forces as shown in Table 8.2.

Table. 8.2. Fluxes and forces in a mixture of fluids

Thermodynamic Fluxes	Thermodynamic Forces
heat flux q_i	temperature gradient $\frac{\partial T}{\partial x_i}$
diffusion fluxes J_i^α	Chemical potential gradient $\frac{\partial \frac{1}{T}(g_\alpha - g_\nu)}{\partial x_i}$
deviatoric stress $t_{\langle ij \rangle}$	deviatoric velocity gradient $\frac{\partial v_{\langle i}}{\partial x_{j\rangle}}$
dynamic pressure $\pi = -\,^1/_3\, t_{ii} - p$	divergence of velocity $\frac{\partial v_n}{\partial x_n}$.
reaction rate densities λ^a	chemical affinities $\sum_{\alpha=1}^{\nu} g_\alpha \gamma_\alpha^a \mu_\alpha$

Obviously diffusion and chemical reactions are taken into account, and there are different chemical reactions $a = 1,2,\ldots n$. Vanishing of the chemical affinities implies the law of mass action, see Chap. 5. Phenomenological relations in the case of mixtures are more rich than for a single fluid; they read

[26] The exceptional 1%, that cannot be treated with the field equations described here, relate to rarefied gases, non-Newtonian fluids, ultra-low and ultra-high temperatures and exceptional cases like that.

[27] G. Jaumann: "Geschlossenes System ..." loc. Cit.
E. Lohr: "Entropie und geschlossenes Gleichungssystem,'' loc. cit.

$$\lambda^a = \sum_{b=1}^{n} l^{ab} \left(\sum_{\alpha=1}^{\nu} g_\alpha \gamma_\alpha{}^a \mu_\alpha \right) + l^a \frac{\partial v_i}{\partial x_i}$$

$$-\pi = \sum_{b=1}^{n} \tilde{l}^b \left(\sum_{\alpha=1}^{\nu} g_\alpha \gamma_\alpha{}^a \mu_\alpha \right) + \lambda \frac{\partial v_i}{\partial x_i}$$

$$q_i = L \frac{\partial \frac{1}{T}}{\partial x_i} + \sum_{\beta=1}^{\nu-1} L_\beta \frac{\partial \frac{1}{T}(g_\beta - g_\nu)}{\partial x_i}$$

$$J_i^\alpha = \tilde{L}_\alpha \frac{\partial \frac{1}{T}}{\partial x_i} + \sum_{\beta=1}^{\nu-1} L_{\alpha\beta} \frac{\partial \frac{1}{T}(g_\beta - g_\nu)}{\partial x_i}$$

$$t_{\langle ij \rangle} = 2\eta \frac{\partial v_{\langle i}}{\partial x_{j \rangle}}.$$

The entropy inequality is satisfied, if the matrices

$$\begin{bmatrix} l^{ab} & l^a \\ \tilde{l}^b & \lambda \end{bmatrix} \text{ and } \begin{bmatrix} L & L_\beta \\ \tilde{L}_\alpha & L_{\alpha\beta} \end{bmatrix} \text{ are positive semi - definite,}$$

and the viscosity η must be non-negative.

We note that the chemical potentials – functions of p, T, and the concentrations – play a central role in these laws, as they should. Clearly both Fourier's and Fick's laws are now made considerable more general than either Fourier or Fick had them. They allow for *cross effects* such that a temperature gradient may create diffusion and a concentration gradient may create heat conduction. Moreover, the concentration gradient of one constituent may cause the diffusion flux of another one. Analogous cross effects may occur between the reaction rates and the dynamic pressure, although I believe that they have never been observed.

Eckart never received much credit for his work, because shortly after his publications Josef Meixner (1908–1994) published a very similar theory,[28] and so did Ilya Prigogine (1917–).[29] In contrast to Eckart the latter authors stayed in the field and monopolized the subject, as it were. On somewhat uncertain grounds they added Onsager reciprocity relations for transport coefficients, see below. As a result it is not uncommon to hear

---

[28] J. Meixner: "Zur Thermodynamik der irreversiblen Prozesse in Gasen mit chemisch reagierenden, dissoziierenden and anregbaren Komponenten." [On thermodynamics of irreversible processes in gases with reacting, dissociating and excitable components] Annalen der Physik (5) 43 (1943) pp. 244-270.
J. Meixner: Zeitschrift der physikalischen Chemie B 53 (1943) p. 235.

[29] I. Prigogine: "Étude thermodynamique des phénomènes irréversibles." Desoer, Liège (1947).

Eckart's theory described as Onsager's theory. TIP became also known as the thermodynamics of the Dutch school, because many Dutch thermodynamicists contributed to it. The major monograph on the subject was written by de Groot and Mazur.[30] The book gives a fairly clear account of TIP; it puts some emphasis upon the so-called *Curie principle* by which thermodynamic forces and fluxes cannot be related linearly unless they have the same tensorial rank.

Clifford Ambrose Truesdell (1919–2000) recognized the Curie principle for what it is: a corollary of the representation theorems of isotropic functions. Truesdell was openly disdainful of TIP and in the 1950's and 1960's he waged war on *Onsagerism*[31,32] which, by reaction, made most thermodynamicists rally behind Onsager.

But Truesdell exempted Eckart to some degree from his criticism, because Eckart had been straightforward in his assumptions, not hiding them behind perceived principles. In fact Truesdell gives Eckart some faint praise when he says:

> ... C. Eckart, ... who attempted to split inequalities into parts without appeal to any non-existent theorem, ... – and who did not obfuscate the scene by any circular or inapplicable rule of symmetry.[33]

One must realize that Truesdell had his own axe to grind, because he felt called upon to advertise *rational thermodynamics,* see below, and in that endeavour he proved himself to be a master of subjectivity.

Before we leave Eckart, we must mention his third important paper[34] which appeared along with the two papers already cited. In that paper Eckart laid the foundation for relativistic irreversible thermodynamics of fluids, and he discovered the alternative form of Fourier's law which is appropriate for a relativistic gas. The thermodynamic force that drives heat conduction is no longer the temperature gradient alone, rather it is equal to

$$\frac{\partial T}{\partial x_i} + \frac{T}{c^2}\dot{v}_i,$$

where $\dot{v}_i$ is the acceleration, possibly the gravitational acceleration. Consequently, in equilibrium a gas in a gravitational field exhibits a temperature gradient. The reason is clear: higher temperature means higher energy, i.e. higher mass, i.e. higher weight and therefore the temperature field must be

---

[30] S.R. de Groot, P. Mazur : "Non-Equilibrium Thermodynamics" North Holland, Amsterdam (1963).

[31] C. Truesdell: "Six Lectures on Modern Natural Philosophy" Springer 1966.

[32] C. Truesdell: "Rational thermodynamics." McGraw-Hill series in modern applied mathematics (1969) Chap. 7.

[33] Ibidem, p. 141.

[34] C. Eckart: "The thermodynamics of irreversible processes III: Relativistic theory of the simple fluid." Physical Review 58 (1940).

barometrically stratified, just like the mass density. Of course the $^1/c^2$ in the denominator indicates that the effect is *relativistically small.*

Onsager Relations

Onsager relations in their proper form refer to some generic set of variables u_α ($\alpha = 1,2\ldots n$) which all vanish in equilibrium and which satisfy linear rate laws of the type

$$\frac{\mathrm{d}u_\alpha}{\mathrm{d}t} = -M_{\alpha\beta}u_\beta\,.$$

For obvious reasons we may call $\boldsymbol{M}$ a relaxation matrix.

The entropy S depends on the u's in such a manner that it has a maximum in equilibrium. Thus in second order approximation – which is sufficient for a linear theory – the entropy reads

$$S = S_{equ} + \frac{1}{2}\frac{\partial^2 S}{\partial u_\alpha \partial u_\beta}u_\alpha u_\beta = S_{equ} - \frac{1}{2}g_{\alpha\beta}u_\alpha u_\beta,$$

where $\boldsymbol{g}$ is symmetric and positive definite. In this case, where fluxes are absent, the entropy source is simply given by the rate of change of entropy

$$\dot{S} = \frac{\mathrm{d}u_\alpha}{\mathrm{d}t}\frac{\partial S}{\partial u_\alpha},$$

which may be considered as a sum of products of thermodynamic fluxes and forces as shown in Table 8.3.

Table 8.3. Generic fluxes and forces

Thermodynamic Fluxes	Thermodynamic Forces
$J_\alpha = \frac{\mathrm{d}u_\alpha}{\mathrm{d}t}$	$X_\alpha = \frac{\partial S}{\partial u_\alpha} = -g_{\alpha\beta}u_\beta$

Linear relations between fluxes and forces, namely

$$J_\alpha = L_{\alpha\beta}X_\beta \quad \text{with} \quad L_{\alpha\beta} - \text{positive semi-definite}$$

guarantee that the entropy source is non-negative. And Onsager relations [35] require that

$$L_{\alpha\beta} = M_{\alpha\beta}\, g^{-1}_{\alpha\beta} \quad \text{be symmetric.}$$

Onsager relations in this form – and *for these forces and fluxes* – can be proved on the basis of Onsager's hypothesis about the *mean regression of fluctuations*, cf. Chap. 9. A good presentation of the proof is contained in the popular monograph by de Groot and Mazur. The authors are remarkable candid when they call Onsager's hypothesis *not altogether unreasonable.*[36]

There are two qualifications of the Onsager relations, of which one is due to Onsager himself.[37] It concerns the presence of a magnetic flux density $\boldsymbol{B}$ and it refers to the well-known fact that the path of a charged particle in a magnetic field cannot be reversed by reversing the velocity, unless $\boldsymbol{B}$ is also reversed. The other qualification is due to Casimir[38] who distinguished even and odd variables among the u_α's with respect to time reversal. I shall not go into that and merely mention that the Onsager relations with Casimir's amendment are often cited under the acronym OCRR, for Onsager-Casimir-Reciprocity-Relations.

Meixner[39] has extrapolated the OCRR to transport phenomena in mixtures. To wit, he applied them to Eckart's phenomenological equations for mixtures, see above, where, according to Meixner, they read

$$l^{ab} = l^{ba} \quad l^{b} = -\tilde{l}^{\,b} \quad \text{and} \quad L_{\alpha\beta} = L_{\beta\alpha} \quad \tilde{L}_{\beta} = L_{\beta}\,.$$

A convincing proof in this more complicated case is not available.[40]

---

[35] L. Onsager: "Reciprocal relations in irreversible processes." Physical Review (2) 37 (1931) pp. 405-426 and 38 (1932) pp. 2265–2279.

[36] S.R. de Groot, P. Mazur : loc. cit. p. 102.
It is often said that *microscopic reversibility* is the key assumption in the proof of Onsager relations. And it is true that the proof makes use of the fact that atomistic trajectories are reversed when the velocities change sign. But this is so evident from the laws of microscopic physics that it barely needs to be mentioned. Certainly microscopic reversibility is infinitely more certain than the mean regression hypothesis.

[37] L. Onsager: (1932) loc.cit.

[38] H.B.G. Casimir: "On Onsager's principle of microscopic reversibility." Review of Modern Physics 17 (1945) pp. 343–350.

[39] J. Meixner, H.G. Reik: "Die Thermodynamik der irreversiblen Prozesse in kontinuierlichen Medien mit inneren Umwandlungen." [Thermodynamics of irreversible processes in continuous media with internal transformations] Handbuch der Physik III/2, Springer Heidelberg (1959).

[40] Again de Groot and Mazur, loc.cit. pp. 69–74 go farthest in the attempt to prove Onsager relations for transport processes, i.e. when the basic equations are partial differential equations rather than rate laws. They try to show that the tensor of thermal conductivity is symmetric, – Onsager's original problem. But they do not quite succeed: All they can show is, that the divergence of the anti-symmetric part vanishes.

However, there are some entirely macroscopic arguments which suffice to prove the symmetry of the matrix of diffusion coefficients $L_{\alpha\beta}$ on the basis of momentum conservation, and of the plausible assumption of *binary drag*, so that the interaction between two constituents is unaffected by the presence of a third constituent. This was shown by Truesdell,[41] and Müller[42] extrapolated that argument to show that in a mixture of Euler fluids we have $\tilde{L}_\beta = L_\beta$. The instances of valid Onsager relations often cited from the kinetic theory of gases are all of the type envisaged by Truesdell and Müller, so that there is not really confirmation for *general* Onsager relations to be found in the kinetic theory.

Also Meixner[43] has proved the symmetry of l^{ab} from the principle of detailed equilibrium of several chemical reactions, – again without reference to any hypothesis on the mean regression of fluctuations.

Rational Thermodynamics

If the truth were known and admitted, rational thermodynamics is not all that different from TIP. Both theories employ the Clausius-Duhem inequality and the Gibbs equation. It is true that the arguments are shuffled around some: The Curie principle of TIP is replaced by the principle of material frame indifference, and the Gibbs equation of rational thermodynamics is a *result,* whereas in TIP it is the basic *hypothesis.* With the Clausius-Duhem inequality it is the other way round. When applied to linear viscous, heat-conducting fluids, both theories lead to the same results. This is a good thing for both, because the field equations for such fluids were perfectly well known before either theory was formulated, and they were known to be reliable.

The difference between the theories lies in the claims of the protagonists: Whereas TIP was never intended to represent anything but a linear theory, and could not be extrapolated, there was no such *a priori* restriction in rational thermodynamics. Therefore the authors expected – and hoped for – more general validity. However, in that expectation they were eventually disappointed; they had overreached themselves, and the non-linear part of the theory crumbled. Let us consider this:

One new feature of the theory is the *principle of material frame indifference.*[44] This had been invented by Hanswalter Giesekus[45] in the

---

[41] C. Truesdell: "Mechanical Basis of diffusion." Journal of Chemical Physics 37 (1962).

[42] I. Müller: "A new approach to thermodynamics of simple mixtures." Zeitschrift für Naturforschung 28a (1973).

[43] J. Meixner: Annalen der Physik (1943) loc.cit.

[44] Also known as the *principle of material objectivity.*

[45] H. Giesekus: "Die rheologische Zustandsgleichung." [The rheological equation of state] Rheologica Acta 1 (1958) pp. 2–20.

context of non-Newtonian fluids and was formalized and extrapolated to continuum mechanics in general by Walter Noll (1925–) in 1958.[46] The principle refers to Euclidean transformations, i.e. time-dependent rotations and translations between frames such that, if x_i and x_i^* are the coordinates of a volume element in the frames S and S^*, we have

$$x_i^* = O_{ij}(t)\, x_j + b_i(t) \quad \leftrightarrow \quad x_i = O_{ji}\,(t)\,(x_j^* - b_j(t))\,.$$

The orthogonal matrix $\boldsymbol{O}(t)$ and the vector $\boldsymbol{b}(t)$ may be arbitrarily time-dependent and, if they are, at least one of the two frames is a non-inertial frame; in order to fix the ideas we take S as an inertial frame. The principle of material frame indifference states that the constitutive functions must not depend on the frame in which a body, or a volume element of a body, is at rest. This implies

- that only Euclidean vectors and tensors may occur as variables, and
- that the constitutive functions are isotropic functions.

The validity of hypotheses and postulates in continuum mechanics and thermodynamics – or at least their applicability to gases – can be checked by the kinetic theory of gases. And when such a check was made,[47] it turned out that the principle of material frame indifference was wrong, cf. Insert 8.1. To be sure, it was not *very wrong*, because the frame dependence is due to the curvature imparted to the mean free paths of the atoms by the Coriolis force. Therefore, in order to see an effect, one would have to use a very rapidly rotating frame indeed. In this sense the argument even *confirms* frame indifference as a *practical tool* and reconciles it with the idea – prevailing in non-relativistic physics – that the only true invariance of physical laws is Galilei invariance.[48]

But this was not the way, the protagonists of rational thermodynamics saw the matter. There were no *approximate principles* for them. Some were prepared to give up the kinetic theory in order to save material frame indifference. Noll suggested that the whole universe be turned to maintain the principle; in the meantime he changed the wording of the principle, thus excluding the influence of *external* forces which – in his

[46] W. Noll: "A mathematical theory of the mechanical behaviour of continuous media." Archive for Rational Mechanics and Analysis 2 (1958).

[47] I. Müller: "On the frame-dependence of stress and heat flux." Archive for Rational Mechanics and Analysis 45 (1972).

[48] Galilei transformations form a subgroup of Euclidean ones, where $\boldsymbol{O}$ is time-independent and $\boldsymbol{b}$ is a linear function of time. There are no inertial forces like the Coriolis force in Galilean frames.

understanding – include the inertial forces like the Coriolis force.[49] This is a somewhat strange idea, because frame indifference *can only* be violated by the effect of inertial forces; there is no other way![50] Truesdell,[51] referring to the argument of Insert 8.1, wondered caustically why the physics of a hollow cylinder should be different from the physics of a full cylinder[52] and afterwards ignored the objections. The subject was thus so successfully obfuscated that the discussion of material frame indifference never ended and is still going on in the years when I write this. However, nothing is said now that has not been said before.

Frame dependence of the heat flux

We consider a gas at rest between two concentric cylinders and focus the attention on a small volume element of the dimension of the mean free path of the atoms. There is a radial temperature gradient, see Fig. 8.4. The atoms at the bottom of the element have a greater mean kinetic energy than those on top, because the temperature is bigger. Therefore the atoms moving upwards through the plane H-H carry more energy through that plane than the downward moving atoms. Thus there is a net energy flux, a heat flux, in the upward-direction, opposite to the temperature gradient, just as predicted by Fourier's law. This is true, if the gas is at rest in an inertial frame. But then we take the cylinders and the gas and put them on a carousel with the axis of rotation coinciding with the axes of the cylinders. Then the paths of the atoms are curved by the Coriolis force so that there is a heat flux through the plane V-V as well as through the plane H-H, see figure. Therefore in the non-inertial frame of the carousel the heat flux has a component perpendicular to the temperature gradient and the size of that component is proportional to the angular velocity of the frame. The relation between the heat flux and the temperature gradient is therefore *frame-dependent.*

---

[49] W. Noll: "A new mathematical theory of simple materials." Archive for Rational Mechanics and Analysis 48 (1972).

[50] Logically the new principle of material frame indifference is at a par with Henry Ford's well-publicized advertisement of the customer service of his company: *The Model T may be had in all colours as long as they are black.*

[51] C. Truesdell: "Correction of two errors in the kinetic theory that have been used to cast unfounded doubt upon the principle of material frame indifference." Meccanica 11 (1976). One of the "errors" in Truesdell's opinion was supposed to occur in Müller's argument, cf. Insert 8.1. The other one was suspected by Truesdell to be contained in a paper by D.G.B. Edelen, T.A. McLennan: "Material Indifference: A Principle or a Convenience." International Journal of Engineering Science 11 (1973).

[52] The internal cylinder in the argument is needed for setting up a temperature gradient. In a full cylinder a radially symmetric, non-homogeneous temperature field cannot exist.

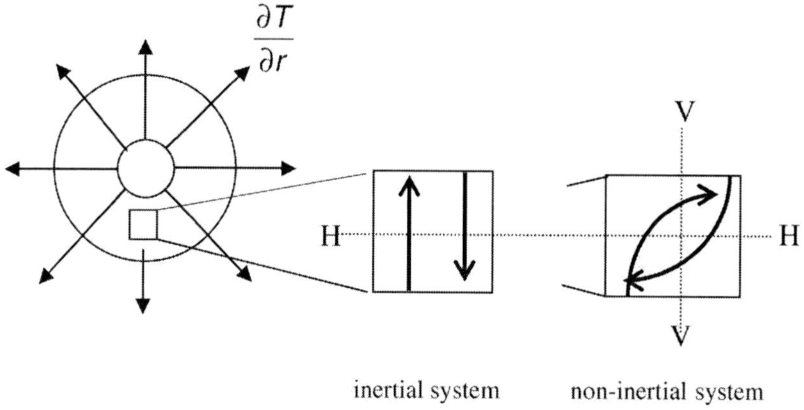

Fig. 8.4. On the frame dependence of the heat flux

A similar argument can be made for the relation between the stress and the velocity gradient. The kinetic theory of gases provides concrete equations for the suggestive argument presented in this insert.

Insert 8.1

More damage was suffered by rational thermodynamics when it was found that the theory could not be applied to non-Newtonian fluids. The early authors in the field were Bernard David Coleman (1930–) and Walter Noll, whose background was continuum mechanics and, in particular, continuum mechanics of visco-elastic solids and fluids.[53] Therefore from the outset rational thermodynamics has put a strong emphasis on constitutive *functionals*, by which the stress (say) depends on the *history* of the velocity gradient. This is fine as far as it goes. But for practical flow problems it has seemed appropriate to approximate the functional of the history by a *function* of a few time derivatives of the velocity gradient, say n of them. In this way one arrives at the theory of nth *grade fluids* whose stationary version was widely used to calculate solutions for viscometric flows.[54] However, then it turned out that rational thermodynamics predicts a *maximum* of free energy for a 2nd grade fluid in

---

[53] B.D. Coleman: "Thermodynamics of materials with memory." Archive for Rational Mechanics and Analysis 17 (1964).
B.D. Coleman, W. Noll: "An approximation theorem for functionals, with applications in continuum mechanics" Archive for Rational Mechanics and Analysis 6 (1960).

[54] E.g. see C. Truesdell: "The elements of continuum mechanics" Springer, New York (1966).
Also: B.D. Coleman, H. Markovitz, W. Noll: "Viscometric flows of non-Newtonian fluids." Springer Tract in Natural Philosophy 5 (1966).

equilibrium[55] instead of the minimum necessary for stability. Later it was shown that no *n*th grade fluid with $n > 1$ has stable solutions.[56] After that, there was serious doubt that rational thermodynamics could be used for non-linear problems, and now it is accepted by most people that it cannot be so used.

In some ways this is a pity, because rational thermodynamics did employ some elegant and *rational* (sic!) arguments for the exploitation of the entropy inequality. These arguments are not lost, however, because they can be transferred to *extended thermodynamics* which we proceed to consider now, – after this:

Truesdell's outspoken partisanship of rational thermodynamics and his flamboyant style fuelled some lively controversies between adherents of TIP and the protagonists of rational thermodynamics, chiefly Truesdell himself. His attacks on *Onsagerism* were advanced with much satirical verve, that makes them fun to read for those who were not targeted. However, the defenders of TIP tried their best to pay Truesdell back in his own coin. Woods pointed out some awkward features of rational thermodynamics in a paper entitled "The bogus axioms of continuum mechanics."[57] And Ronald Samuel Rivlin (1915–2005) delighted a worldwide audience with a frequently repeated humorous lecture under the title "On red herrings and other sundry unidentified fish in modern continuum mechanics."

Truesdell was a consummate theoretician. He showed nothing but disdain for experiments, be they conducted in the laboratory or on the computer.

So, when Truesdell visited me in Berlin, on the first day he came and said: "Ingo, can I ask you for a favour?" And I, eager to please my visitor and one-time mentor, said: "Of course, Clifford, what can I do for you ?"

Truesdell: "Please, don't show me your lab."

Fig. 8.5. Clifford Ambrose Truesdell III

55 J.E. Dunn, R.L. Fosdick: "Thermodynamics, stability, and boundedness of fluids of complexity 2 and fluids of second grade." Archive for Rational Mechanics and Analysis 56 (1974).

56 D.D. Joseph: "Instability of the rest state of fluids of arbitrary grade greater than one." Archive for Rational Mechanics and Analysis. 75 (1981).

57 L.C. Woods: Bulletin of Mathematics and its Applications 1 (1981).

Extended Thermodynamics

Formal Structure

The objective of extended thermodynamics is the determination of n fields, synthetically denoted by $u_\alpha(\boldsymbol{x},t)$ ($\alpha = 1,2,\ldots n$). Invariably the first five of these fields are the densities of mass, momentum and energy. But in extended thermodynamics the number of fields is extended (sic) and it may contain the stress, the heat flux and more, see below.

We need n field equations and these are based upon n equations of balance

$$\frac{\partial u_\alpha}{\partial t} + \frac{\partial F_\alpha^{\ a}}{\partial x_a} = \Pi_\alpha \qquad (\alpha = 1,2,\ldots n).$$

The fields u_α may therefore be called densities, the $F_\alpha^{\ a}$ are called the corresponding fluxes and the Π_α are called productions. The first five productions are zero in accord with the conservation of mass, momentum, and energy. And all productions vanish in equilibrium.

In order to obtain field equations for the densities u_α, the balance equations must be supplemented by constitutive equations. Such constitutive equations relate the fluxes $F_\alpha^{\ a}$ and the productions Π_α to the densities in a materially dependent manner. In extended thermodynamics the constitutive relations have the forms

$$F_\alpha^{\ a} = \hat{F}_\alpha^{\ a}(u_\beta) \qquad \text{and} \qquad \Pi_\alpha = \hat{\Pi}_\alpha(u_\beta)$$

so that the fluxes $F_\alpha^{\ a}$ and the productions Π_α at a point and a time depend only on the densities u_α at that point and time. We may say that the constitutive equations are *local* in space-time.[59]

If the constitutive functions $\hat{F}_\alpha^{\ a}$ and $\hat{\Pi}_\alpha$ were known explicitly, we could eliminate $F_\alpha^{\ a}$ and Π_α from the equations of balance and obtain explicit field equations for the u_α's. They form a quasi-linear system of partial differential equations of first order. Every solution of this system is called a *thermodynamic process.*

[59] Thus no gradients or time derivatives do occur among the variables in the constitutive equations. In particular, there is no temperature gradient, and yet heat conduction is accounted for, because the heat flux is counted among the fields.

Symmetric Hyperbolic Systems

In reality, of course, the constitutive functions are not known and it is the task of the constitutive theory to determine those functions or, at least, to reduce their generality. The tools of the constitutive theory are certain universal physical principles which represent expectations based on long experience. The main principles are

• entropy inequality, • requirement of concavity, • principle of relativity.

The first two of these represent the entropy principle and, in particular, the second one guarantees thermodynamic stability and hyperbolicity of the field equations.

The entropy inequality is an additional balance law. We write

$$\frac{\partial h}{\partial t}+\frac{\partial h^a}{\partial x_a} = \qquad \geq 0 \qquad \text{for all thermodynamic processes.}$$

h is the entropy density, and h^a is the entropy flux. Σ is the entropy production. All three are constitutive quantities so that in extended thermodynamics we have

$$h = \hat{h}(u_\beta), \quad h^a = \hat{h}^a(u_\beta) \quad \text{and} \qquad = \qquad (u_\beta).$$

The requirement of concavity demands that h be a concave function of u_α:

$$\frac{\partial^2 h}{\partial u_\alpha \partial u_\beta} \quad - \quad \text{negative definite.}$$

The principle of relativity states that the field equations and the entropy inequality have the same form in all Galilei frames.[60]

The key to the exploitation of the entropy inequality lies in the observation that the inequality must hold for thermodynamic processes, i.e. solutions of the field equations. In a manner of speaking the field equations provide constraints for the fields that must satisfy the entropy inequality. A

60 In relativistic thermodynamics we require invariance of the equations under Lorentz transfomations, but this is not a subject of this book, although relativistic thermodynamics is an interesting application of extended thermodynamics. See: I-Shih Liu, I. Müller, T. Ruggeri: "Relativistic thermodynamics of gases." Annals of Physics 100 (1986). Also: I. Müller, T. Ruggeri: "Rational Extended Thermodynamics." Springer, New York (1998) 2nd edition.

lemma by Liu[61] proves that it is possible to use Lagrange multipliers Λ_α – functions of u_α – to eliminate such constraints. Indeed, the new inequality

$$\frac{\partial h}{\partial t}+\frac{\partial h^a}{\partial x_a}-\Lambda_\alpha\left(\frac{\partial u_\alpha}{\partial t}+\frac{\partial F_\alpha{}^a}{\partial x_a}-\Pi_\alpha\right)\geq 0 \text{ must hold for all fields } u_\alpha\,(x_i,t).$$

This implies

$$\mathrm{d}h=\Lambda_\alpha \mathrm{d}u_\alpha, \qquad \mathrm{d}h^a=\Lambda_\alpha \mathrm{d}F_\alpha{}^a, \qquad \text{and} \qquad \Lambda_\alpha \Pi_\alpha \geq 0,$$

so that in equilibrium all but the first five Lagrange multipliers vanish. The residual inequality $\Lambda_\alpha \Pi_\alpha \geq 0$ represents the entropy source or dissipation.

In order to appreciate the mathematical structure of the system of field equations we change variables from the densities u_α to the Lagrange multipliers Λ_α and obtain for the scalar and vector potentials $h' = -h + \Lambda_\alpha u_\alpha$ and $h^{a\prime} = -h^a + \Lambda_\alpha F_\alpha{}^a$

$$\frac{\partial h'}{\partial \Lambda_\alpha}=u_\alpha, \qquad \frac{\partial h'^a}{\partial \Lambda_\alpha}=F_\alpha{}^a \qquad \text{so that the field equations read}$$

$$\boxed{\frac{\partial^2 h'}{\partial \Lambda_\alpha \partial \Lambda_\beta}\frac{\partial \Lambda_\beta}{\partial t}+\frac{\partial^2 h'^a}{\partial \Lambda_\alpha \partial \Lambda_\beta}\frac{\partial \Lambda_\beta}{\partial x_a}=\Pi_\alpha \qquad (\alpha=1,2,\ldots n).}$$

All four matrices in this system are symmetric and the first one is negative definite.[62] Therefore the system of field equations – written in terms of the Lagrange multipliers – is a *symmetric hyperbolic system.*

Hyperbolicity guarantees finite speed of propagation and *symmetric* hyperbolic systems have convenient and desirable mathematical properties, namely well-posedness of Cauchy problems, i.e. existence, uniqueness and continuous dependence on the data. The desire for finite speeds of propagation was the primary original incentive for the formulation of extended thermodynamics, see below. There are n speeds of propagation and they may be calculated from the characteristic equation of the system of field equations, viz.

---

[61] I-Shih Liu: "Method of Lagrange multipliers for the exploitation of the entropy principle." Archive for Rational Mechanics and Analysis 46 (1972).

[62] This follows from the concavity of the entropy density in terms of the densities u_α, since $h' = -h + \Lambda_\alpha u_\alpha$ defines a Legendre transformation associated with the map $\Lambda_\alpha \leftrightarrow u_\alpha$..

$$\det\left(\frac{\partial^2 h'}{\partial \Lambda_\alpha \partial \Lambda_\beta} V - \frac{\partial^2 h'^a}{\partial \Lambda_\alpha \partial \Lambda_\beta} n_a\right) = 0 .$$

n_a and V denote direction and speed of propagation. Obviously, before any values for wave speeds can actually be calculated, the synthetic character of the equations of this section must be replaced by more concrete relations so that $h'(\Lambda_\beta)$ and $h'^a(\Lambda_\beta)$ can be identified. The most immediate concretization of the present formal framework is provided by extended thermodynamics of moments, see below.

Growth and Decay of Waves

Non-linear hyperbolic equations tend to evolve discontinuities in the fields, even if the initial data are smooth. On the other hand, steep gradients involve strong dissipation with a tendency to smooth out the solution. Thus there exists a competition between non-linearity and dissipation which may lead to smooth solutions for all times. This is important for a system of field equations to be realistic, since most phenomena that occur in the real world, are smooth. After all: *Natura non fecit saltus.*[63]

An instructive example for the competition of non-linearity and dissipation is the growth and decay of acceleration waves, i.e. moving singular surfaces along which $u_\alpha(\boldsymbol{x},t)$ ($\alpha = 1,2...n$) are continuous, but their gradients are not. As one moves with the wave, its amplitude A – representing the jumps of the gradients – obeys a Bernoulli equation, provided that the wave moves into an area of a homogeneous and time-independent equilibrium[64]

$$\frac{\delta A}{\delta t} - \underbrace{\frac{\partial V}{\partial u_\beta} d_\beta}_{a} A^2 - \underbrace{l_\alpha \frac{\partial \Pi_\alpha}{\partial u_\beta} d_\gamma}_{b} A = 0 .$$

[63] According to Aristoteles: "Historia animalium." Aristoteles said it in Greek, of course, and in quite a different context. The familiar quotation is often used in connection with the steep, but smooth structure of shock waves.

[64] An excellent review of waves – in particular acceleration waves – is given by P. Chen: "Growth and decay of waves in solids. Mechanics of Solids III" Handbuch der Physik 6A/3 Springer, Heidelberg (1973).
I believe that the first person to calculate the rate of change of the amplitude $A(t)$ of an acceleration wave was W.A. Green: "The growth of plane discontinuities propagating into a homogeneous deformed material." Archive for Rational Mechanics and Analysis 16 (1964).
The present compact form of the Bernoulli equation – with right and left eigenvectors – is due to G. Boillat: "La propagation des ondes." Gauthier-Villars, Paris (1965).

V is a characteristic speed and l_α and d_α are the left and right eigenvalues of the matrix $\frac{\partial F_\alpha^{\ 1}}{\partial u_\beta}$ in the one-dimensional field equations

$$\frac{\partial u_\alpha}{\partial t} + \frac{\partial F_\alpha^{\ 1}}{\partial x_1} = \Pi_\alpha \qquad (\alpha = 1,2,...n).$$

The solution of the Bernoulli equation reads

$$A(t) = \frac{A(0)e^{-bt}}{1 - A(0)\frac{a}{b}(e^{bt} - 1)}$$

so that $A(t)$ remains finite unless the initial amplitude $A(0)$ is large.

In general – for arbitrary solutions instead of merely acceleration waves – the condition for smooth solutions is not decisively known. There exists a sufficient condition for smoothness[65] which, however, is not necessary.

Characteristic Speeds in Monatomic Gases

We recall the generic equations of transfer in the kinetic theory of gases, cf. Chap. 4, and apply this to a polynomial in velocity components by setting $\psi = \mu c_{i_1} c_{i_2} ... c_{i_l}$. In this manner we obtain equations of balance for *moments* $u_{i_1 i_2 \cdots i_l} = \int \mu c_{i_1} c_{i_2} ... c_{i_l} f \mathrm{d}\mathbf{c}$ of the distribution function f which read

$$\frac{\partial u_{i_1 i_2 \cdots i_l}}{\partial t} + \frac{\partial u_{i_1 i_2 \cdots i_l a}}{\partial x_a} - \Pi_{i_1 i_2 \cdots i_l} \qquad (l = 0,1,2...N).$$

Since each index may assume the values 1,2,3, there are

$$n = \tfrac{1}{6}(N+1)(N+2)(N+3)$$

equations. These equations fit into the formal framework of extended thermodynamics, see above, but they are simpler. Indeed, on the left hand side there is only one flux, namely $u_{i_1 i_2 \cdots i_l a}$ – the last one – which is not explicitly related to the fields $u_{i_1 i_2 \cdots i_l}$ $(l = 1,..N)$.

---

[65] S. Kawashima: "Large-time behaviour of solutions to hyperbolic-parabolic systems of conservation laws and applications." Proceedings of the Royal Society of Edinburgh A 106 (1987).

Therefore the results of the previous sections may be carried over to the present case, in particular the exploitation of the entropy inequality. That inequality reads according to the kinetic theory of gases, cf. Chap. 4

$$\frac{\partial}{\partial t}\left(-k\int f\ln\frac{f}{\mathrm{e}Y}\mathrm{d}\mathbf{c}\right)+\frac{\partial}{\partial x_a}\left(-k\int c_a f\ln\frac{f}{\mathrm{e}Y}\mathrm{d}\mathbf{c}\right)\geq 0\,.$$

The exploitation makes use of the Lagrange multipliers $\Lambda_{i_1 i_2 \cdots i_l}$ $(l = 1,2,\ldots N)$ and the moment character of the densities and fluxes implies that the distribution function has the form

$$f = Y\exp\left(-\tfrac{1}{k}\sum\nolimits_{l=0}^{N}\Lambda_{i_1 i_2\cdots i_l}\,\mu c_{i_1}c_{i_2}\ldots c_{i_l}\right)$$

so that the scalar and vector potentials may be written as

$$h' = -kY\int \exp\left(-\tfrac{1}{k}\sum\nolimits_{l=0}^{N}\Lambda_{i_1 i_2\cdots i_l}\,\mu c_{i_1}c_{i_2}\ldots c_{i_l}\right)\mathrm{d}\mathbf{c}\quad\text{and}$$

$$h'^a = -kY\int c_a\exp\left(-\tfrac{1}{k}\sum\nolimits_{l=0}^{N}\Lambda_{i_1 i_2\cdots i_l}\,\mu c_{i_1}c_{i_2}\ldots c_{i_l}\right)\mathrm{d}\mathbf{c}\,.$$

Insertion into the characteristic equation for the calculation of wave speeds gives

$$\det\left(\int (c_a n_a - V)\; c_{i_1}..c_{i_l}c_{j_1}..c_{j_n} f_{equ}\mathrm{d}\mathbf{c}\right)=0$$

provided that the wave propagates into a region of equilibrium. f_{equ} is the Maxwell distribution, cf. Chap. 4.

Thus the calculation of characteristic speeds and, in particular, the maximal one, *the pulse speed* requires no more than simple quadratures and the solution of an nth order algebraic equation. It is true that the dimension of the determinant increases rapidly with N: For N = 10 we have 286 columns and rows, while for N = 43 we have 15180 of them. But then, the calculation of the elements of the determinant and the determination of $V_{\max}$ may be programmed into the computer and Wolf Weiss (1956–) has the values ready for any reasonable N at the touch of a button, see Fig. 8.6. We recognize that the pulse speed goes up with increasing N and it never

N	n	$V_{max}/\sqrt{\frac{5}{3}\frac{k}{\mu}T}$
10	286	4.018
20	1771	6.080
30	5456	7.663
40	12341	8.997

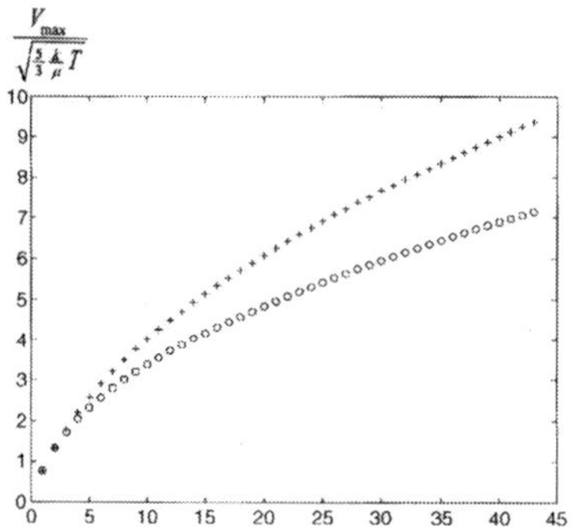

Fig. 8.6. Pulse speeds in relation to the normal speed of sound. Table and crosses: Calculations by Weiss[68]. Circles: Lower bound $\sqrt{\frac{6}{5}(N-\frac{1}{2})}$ by Boillat and Ruggeri[69]

stops.[66] Indeed, Guy Boillat (1937–) and Tommaso Ruggeri (1947–) have provided a lower bound for V_{max} which tends to infinity for $N \to \infty$.[67]

The fact that V_{max} is unbounded represents something of an anticlimax for extended thermodynamics, because the theory started out originally as an effort to find a *finite speed of heat conduction.* Let us consider this:

Carlo Cattaneo (1911–1979)

Fourier's equation of heat conduction is the prototypical parabolic equation and it predicts an infinite speed of propagation of disturbances in temperatures. This phenomenon became known as the *paradox of heat conduction.* Neither engineers nor physicists generally were much worried about the paradox. It is quantitatively unimportant in solids and liquids and even in gases under normal pressures and temperatures. And yet, the paradox represented an awkward feature of thermodynamics and in 1948 Carlo Cattaneo made an attempt to resolve it.

Upon reflection it was clear to Cattaneo that Fourier's law was to blame and he amended it. We refer to Fig. 8.7 and recall the *mechanism* of heat conduction in gases as described in the elementary kinetic theory. If there is a downward temperature gradient across a small volume element – of the dimensions of the mean free path – an atom moving upwards will, in the mean, carry more energy than an atom moving downwards. Therefore there

---

66 W. Weiss: "Zur Hierarchie der erweiterten Thermodynamik." [On the hierarchy of extended thermodynamics] Dissertation TU Berlin.
See also: I. Müller, T. Ruggeri: "Rational Extended Thermodynamics." loc.cit.

67 G. Boillat, T. Ruggeri: "Moment equations in the kinetic theory of gases and wave velocities." Continuum Mechanics and Thermodynamics 9 (1997).

68 W. Weiss: loc.cit.

69 G. Boillat, T. Ruggeri: "Moment equations ..." loc.cit.

is a net flux of energy upwards, i.e. opposite to the temperature gradient, associated with the passage of a pair of particles across the middle layer. That flux is obviously proportional to the temperature gradient, just as Fourier's law requires for the heat flux.

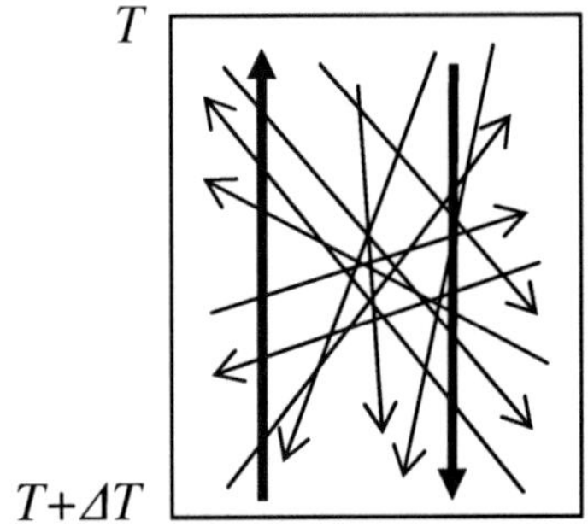

$$q_i + \tau \frac{\partial q_i}{\partial t} = -\kappa \frac{\partial T}{\partial x_i}$$

Fig. 8.7. Carlo Cattaneo. The Cattaneo equation

Cattaneo[70] changed that argument slightly. He argued that there is a time-lag between the start of the particles at their points of departures and the time of passage through the middle layer. If the temperature changes in time, it is clear that the heat flux at a certain time depends on the temperature gradient at a time τ earlier, where τ is of the order of magnitude of the mean time of free flight. Therefore it seems reasonable to write an *non-stationary Fourier law* in the form

$$q_i = -\kappa \left(\frac{\partial T}{\partial x_i} - \tau \frac{\partial}{\partial t} \frac{\partial T}{\partial x_i} \right) \quad \text{with } \tau > 0 .$$

Now, this equation is badly flawed, because it predicts that for $q_i = 0$ the temperature gradient tends exponentially toward infinity. Nor does this modified Fourier law lead to a finite speed, so that it does not resolve the paradox. Cattaneo must have known this – although he does not say so (!) – because he proceeded by converting his non-stationary Fourier law into something else in a sequence of three steps which deserve to be called *mathematically creative*.

[70] C. Cattaneo: "Sulla conduzione del calore." [On heat conduction] Atti del Seminario Matematico Fisico della Università di Modena, 3 (1948).

$$q_i = -\kappa\left(\frac{\partial T}{\partial x_i} - \tau\frac{\partial}{\partial t}\frac{\partial T}{\partial x_i}\right) \Rightarrow \frac{1}{1-\tau\frac{\partial}{\partial t}} q_i = -\kappa\frac{\partial T}{\partial x_i}$$

$$\Rightarrow \left(1+\tau\frac{\partial}{\partial t}\right) q_i = -\kappa\frac{\partial T}{\partial x_i}$$

$$\Rightarrow \quad q_i + \tau\frac{\partial q_i}{\partial t} = -\kappa\frac{\partial T}{\partial x_i}.$$

The end result, now usually called the *Cattaneo equation,* is acceptable. It provides a stable state of zero heat flux for $\frac{\partial T}{\partial x_i} = 0$ and, if combined with the energy equation, it leads to a telegraph equation and predicts a finite speed of propagation of disturbances of temperature.

So, however flawed Cattaneo's reasoning may have been, he is the author of the first hyperbolic equation for heat conduction. Let us quote him how he defends the transition from the non-stationary Fourier law to the Cattaneo equation:

> Nel risultato ottenuto approfitteremo della piccolezza del parametro τ per trascurare il termine che contiene a fattore il suo quadrato, conservando peraltro il termine in cui τ compare a primo grado. Naturalmente, per delimitare la portata delle conseguenze che stiamo per trarre, converrà precisare un po' meglio le condizioni in cui tale approssimazione è lecita. Allo scopo ammetteremo esplicitamente che il feno-meno di conduzione calorifica avvenga nell'intorno di uno stato stazionario o, in altri termini, che durante il suo svolgersi si mantengano abbastanza piccole le derivate temporali delle varie grandezze in giuoco.
>
> In the result we take advantage of the smallness of the parameter τ so that terms with squares of τ may be neglected. First order terms in τ are kept, however. Of course, in order to appreciate the effect on the consequences, which we are about to derive, it would be proper to investigate the conditions when that approximation is valid. For that purpose we stress that the heat conduction should remain nearly stationary. Or, in other words, that the time derivatives of the various quantities at play remain sufficiently small, while the stationary state changes slowly.

Well, if the truth were known, this is not a valid justification. How could it be, if it leads from an unstable equation to a stable one and from a parabolic to a hyperbolic equation.

Let me say at this point that Cattaneo's argument leading to the non-stationary Fourier law is the nut-shell-version of the first step in an iterative scheme that is often used in the kinetic theory of gases. In that field the objective is an improvement of the treatment of viscous, heat-conducting gases beyond what the

Navier-Stokes-Fourier theory can achieve. The iterative scheme is called the Chapman-Enskog method and its extensions are known as Burnett approximation and super Burnett. The scheme leads to inherently unstable equations and should be discarded. The reason why the fact was not recognized for decades is that the authors have all concentrated on stationary processes.[71] And the reason why it is still used is natural inertia and lack of imagination and initiative.

The situation is quite similar mathematically and psychologically to the one mentioned in the context of rational thermodynamics of unstable equilibria of *n*th grade fluids with $n > 1$, see above.

However, whatever the peculiarities of its derivation may have been, the Cattaneo equation on the paradox of heat conduction served as a stimulus. Müller[72] generalized Cattaneo's treatment within the framework of TIP, taking care – at the same time – of a related paradox of shear motion. And then, after rational thermodynamics appeared, Müller and I-Shih Liu (1943–)[73] formulated the first theory of rational extended thermodynamics, still restricted to 13 moments, but complete with a constitutive entropy flux – rather than the Clausius-Duhem expression – and with Lagrange multipliers.

Thus the subject was prepared for being joined to the mathematical theory of hyperbolic systems. Mathematicians had studied quasi-linear first order systems for their own purposes, – without being motivated by the paradoxon of infinite wave speeds. Godunov,[74] Friedrichs and Lax,[75] and Boillat[76] discovered that such systems may be reduced to a symmetric hyperbolic form, if they are compatible with a *convex extension*, i.e. an additional relation of the type of the entropy inequality. Ruggeri and

---

[71] The instabilities involved in the Chapman-Enskog iterative scheme have recently been reviewed by Henning Struchtrup (1956–). H. Struchtrup: "Macroscopic Transport Equations for Rarefied Gases – Approximation Methods in Kinetic Theory" Springer, Heidelberg (2005).

[72] I. Müller: "Zur Ausbreitungsgeschwindigkeit von Störungen in kontinuierlichen Medien." [On the speed of propagation in continuous bodies.]. Dissertation TH Aachen (1966). See also: I. Müller: "Zum Paradox der Wärmeleitungstheorie." [On the paradox of heat conduction]. Zeitschrift für Physik 198 (1967).

[73] I-Shih Liu, I. Müller: "Extended thermodynamics of classical and degenerate gases." Archive for Rational Mechanics and Analysis 46 (1983).

[74] S.K. Godunov: "An interesting class of quasi-linear systems." Soviet Mathematics 2 (1961).

[75] K.O. Friedrichs, P.D. Lax: "Systems of conservation equations with a convex extension." Proceeding of the National Academy of Science USA 68 (1971).

[76] Boillat: "Sur l´éxistence et la recherche d´équations de conservations supplémentaires pour les systèmes hyperbolique." [On the existence and investigation of supplementary conservation laws for hyperbolic systems] Comptes Rendues Académie des Sciences Paris. Ser5. A 278 (1974).

Strumia[77] recognized that the Lagrange multipliers – their *main field* – could be chosen as thermodynamic fields and, if they were, the field equations of extended thermodynamics were symmetric hyperbolic. The formal structure of the theory was refined by Boillat and Ruggeri,[78],[79] and eventually they proved that for infinitely many moments the pulse speed tends to infinity, although it is always finite for finitely many moments, see above.[80]

As mentioned before this phenomenon is a kind of anti-climax for a theory that had originally set out to calculate finite speeds. However, the infinite limiting case has its own appeal and anyway: Extended thermodynamics had by this time outgrown its original motivation and had become a predictive theory for processes with large rates of change and steep gradients, as they might occur in shock waves. Let us consider this:

Field Equations for Moments

Once the distribution function is known in terms of the Lagrange multipliers, see above, it is possible – in principle – to change back from the Lagrange multipliers $\Lambda_{i_1 i_2 \cdots i_l}$ to the moments $u_{i_1 i_2 \cdots i_l}$ by inverting the relation

$$u_{i_1 i_2 \cdots i_l} = \int \mu c_{i_1} \ldots c_{i_l} Y \exp\left(-\tfrac{1}{k}\sum_{l=0}^{N} \Lambda_{i_1 i_2 \cdots i_l} \mu c_{i_1} c_{i_2} \ldots c_{i_l}\right) \mathrm{d}c \, .$$

Once this is done, we may determine the *last flux*

$$u_{i_1 i_2 \cdots i_N a} = \int \mu c_{i_1} \ldots c_{i_N} c_a Y \exp\left(-\tfrac{1}{k}\sum_{l=0}^{N} \Lambda_{i_1 i_2 \cdots i_l} \mu c_{i_1} c_{i_2} \ldots c_{i_l}\right) \mathrm{d}c$$

in terms of $u_{i_1 i_2 \cdots i_l}$ $(l = 0.1,..N)$. Also in principle the productions may thus be calculated after we choose an appropriate model for the atomic interaction, e.g. the model of Maxwellian molecules, cf. Chap. 4.

---

77 T. Ruggeri, A. Strumia: "Main field and convex covariant density for quasi-linear hyperbolic systems. Relativistic fluid dynamics." Annales Institut Henri Poincaré 34 A (1981).

78 T. Ruggeri: "Galilean invariance and entropy principle for systems of balance laws. The structure of extended thermodynamics." Continuum Mechanics and Thermodynamics 1 (1989).

79 G. Boillat, T. Ruggeri: "Moment equations ..." loc.cit.

80 Incidentally, in the relativistic version of extended thermodynamics the maximal pulse speed for infinitely many moments is *c*, the speed of light.

In reality the calculations of the flux $u_{i_1 i_2 \cdots i_N a}$ and of the productions $\Pi_{i_1 i_2 \cdots i_l} (l = 6,7 \ldots N)$ [81] require somewhat precarious approximations, since integrals of the type occurring in the last equations cannot be solved analytically. However, when everything is said and done, one arrives at explicit field equations, e.g. those of Fig. 8.8, which are valid for $N = 3$ so that there are 20 individual equations. The equations written in the figure are linearized and the canonical notation has been introduced like ρ for u, ρv_i for u_i, $3 \rho \, {}^k/_\mu T$ for the trace u_{ii}, $t_{<ij>}$ for the deviatoric stress and q_i for the heat flux. The moment $u_{<ijk>}$ has no conventional name, – other than *trace-less third moment* – because it does not enter equations of mass, momentum and energy. But it does have to satisfy an explicit fields equation, see figure.

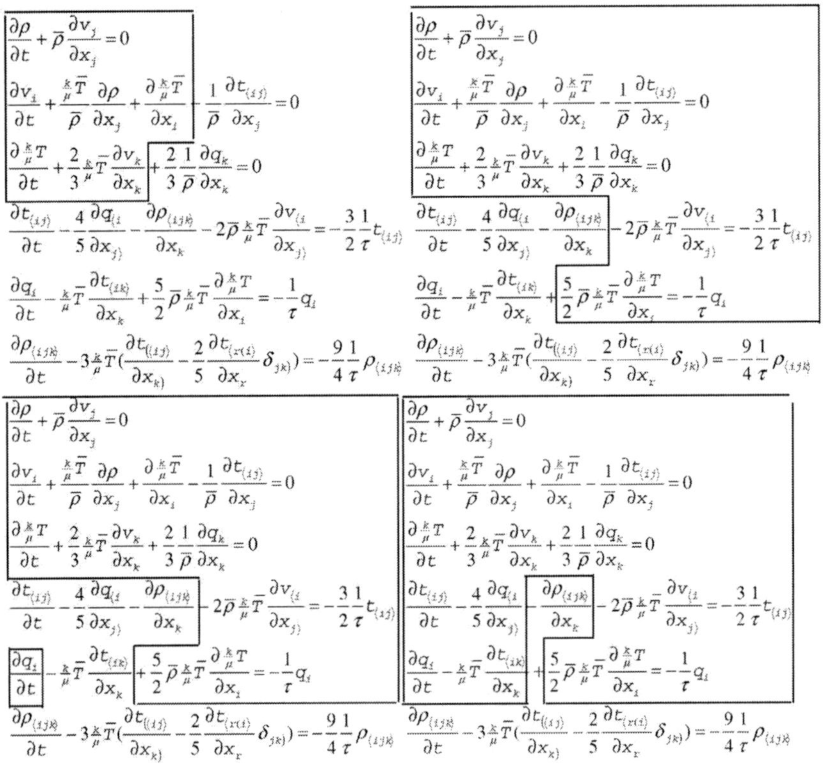

Fig. 8.8. 4 times field equations of extended thermodynamics for $N= 3$ *Top left*: Euler. *Top right*: Navier-Stokes. *Bottom left*: Cattaneo. *Bottom right*: 13 moment

[81] Recall that the first five productions are zero which reflects the conservation of mass, momentum and energy.

Figure. 8.8 shows the same set of 20 equations four times so as to make it possible to point out special cases within the different frames:

- On the upper left side we see the equations for the Euler fluid, which is entirely free of dissipation and thus without shear stresses and heat flux.
- The upper right box contains the Navier-Stokes-Fourier equations with the stress proportional to the velocity gradient and the heat flux proportional to the temperature gradient. This set identifies the only unspecified coefficient τ as being related to the shear viscosity η. We have $\eta = \frac{4}{3}\tau\rho\frac{k}{\mu}T$ so that η grows linearly with T as is expected for Maxwellian molecules, cf. Chap. 4.
- In the fifth equation of the third box I have highlighted the *Cattaneo equation* which has provided the stimulus for the formulation of extended thermodynamics, see above. The Cattaneo equation is essentially a Fourier equation, but it includes the rate of change of the heat flux as an additional term even though it ignores other terms.
- The fourth box exhibits the 13-moment equations. These are the ones best known among all equations of extended thermodynamics, because they contain no unconventional terms, – only the 13 moments familiar from the ordinary thermodynamics, viz. ρ, v_i, T, $t_{\langle ij \rangle}$, and q_i.

For interpretation we may focus on the upper right box in Fig. 8.8, the one that emphasizes the Navier-Stokes theory. In this way we see that some specific terms are left out of that theory, namely

$$\frac{\partial t_{\langle ij \rangle}}{\partial t} \quad \text{and} \quad \frac{\partial q_i}{\partial t} \quad \text{and} \quad \frac{\partial t_{\langle ik \rangle}}{\partial x_k} \quad \text{and} \quad \frac{\partial q_i}{\partial x_k}.$$

For rapid rates and steep gradients we may suspect that these terms do count and, indeed, they do, and we must go to the full set of 20 equations, or to equations with even more moments. Since rapid rates and steep gradients are measured in terms of mean times of free flight and mean free paths, we may suspect that extended thermodynamics becomes necessary for rarefied gases.

Shock Waves

Properly speaking shock waves do not exist, at least not as discontinuities in density, velocity, temperature, etc. What seems like shock waves turns out to be *shock structures* upon close experimental inspection, i.e. smooth but steep solutions of the field equations, which assume different equilibrium values at the two sides. Scientists and engineers are interested to calculate the exact form of the shock structures; and they have realized that the Navier-Stokes-Fourier theory fails to predict the observed thickness.[82] Since this is a case of steep gradients or rapid rates, it is appropriate, perhaps, to apply extended thermodynamics.

To be sure we cannot use the formulae of Fig. 8.8, because these are linearized. Their proper non-linear form is too complicated to be written here. Let it suffice therefore to say that, yes, extended thermodynamics does provide improved shock structures. But the work is hard, because even for rather weak shock – which move with a Mach number of 1.8 – the required number of moments goes into the hundreds as Wolf Weiss[83] and Jörg Au have shown.[84]

An interesting feature of that research – first noticed, but apparently not understood by Grad[85] – is the observation that, when the Mach number reaches the pulse speed and exceeds it, a sharp shock occurs within the shock structure. Obviously those Mach numbers are *truly supersonic* and not just bigger than 1. That is to say that the upstream region has no way of being warned about the onrushing wave, if that wave comes along faster than the pulse speed. For the mathematician this is a clear sign that he has over-extrapolated the theory: He should take more moments into account and, if he does, the sharp shocks disappear, or rather they are pushed to a higher Mach number appropriate to the bigger pulse speed of the more extended theory.

Boundary Conditions

Extended thermodynamics up to 1998 is summarized by Müller and Ruggeri.[86] Since the publication of that book boundary value problems have been at the focus of the research in the field, and some problems of the 13-moment theory have been solved:

- It has been shown for thermal non-equilibrium between two co-axial cylinders that the temperature measured by a contact thermometer is not

[82] This was decisively shown by D. Gilbarg, D. Paolucci: "The structure of shock waves in the continuum theory of fluids." Journal for Rational Mechanics and Analysis 2 (1953).

[83] W. Weiss: "Die Berechnung von kontinuierlichen Stoßstrukturen in der kinetischen Gastheorie." [Calculation of continuous shock structures in the kinetic theory of gases] Habilitation thesis TU Berlin (1997). See also: W. Weiss: Chapter 12 in: I. Müller, T. Ruggeri: "Rational Extended Thermodynamics" loc.cit.
W. Weiss: "Continuous shock structure in extended Thermodynamics." Physical Review E, Part A 52 (1995).

[84] Au: "Lösung nichtlinearer Probleme in der Erweiterten Thermodynamik." [Solution of non-linear problems in extended thermodynamics'']. Dissertation TU Berlin, Shaker Verlag (2001).

[85] H. Grad: "The profile of a steady plane shock wave." Communications of Pure and Applied Mathematics 5 Wiley, New York (1952).

[86] I. Müller, T. Ruggeri: "Rational Extended Thermodynamics." loc.cit.

equal to the kinetic temperature, a measure of the mean kinetic energy of the atoms,[87] cf. Inserts 8.2, and 8.3 and

- It has been shown that a gas cannot rotate rigidly, if it conducts heat.[88] Both results differ from those that are predicted by the Navier-Stokes-Fourier theory, indeed, they are *qualitatively* and *quantitatively* different.

Thus some extrapolations away from equilibrium, that we have grown fond of, must be revised in the light of extended thermodynamics. Notably this is true for the principle of local equilibrium and for the Clausius-Duhem inequality. Both lose their validity when non-equilibrium becomes severe.

The problem with more than 13 moments is, that there is no possibility to prescribe and control higher moments – like $u_{<ijk>}$, or u_{iijk}, etc. – initially or on the boundary. Thus we face the situation that we do have specific field equations for those moments, but that we are unable to use them for lack of initial and boundary values.

On the other hand, it can be shown that an arbitrary choice of boundary values of u_{iijk} (say) may affect the temperature field in a drastic – and totally unacceptable, since unobserved – manner. Therefore it seems to be inevitable to conclude that a gas *itself* adjusts the uncontrollable boundary values and the question is which criterion the gas employs. It has been suggested[89] that the boundary values *adjust themselves* so as to minimize the entropy production in some norm. Another suggestion is that the uncontrollable boundary values fluctuate with the thermal motion and that the gas reacts to their mean values.[90]

In all honesty, however, the problem of assigning data in extended thermodynamics must still be considered open so far. At the present time only such problems have been resolved by extended thermodynamics – with more than 13 moments – which do not need boundary and initial conditions or which possess trivial ones. These include *shock waves,* which have been treated with moderate success, see above, and *light scattering,* which has been dealt with very satisfactorily indeed, cf. Chap 9.

Minor intrinsic inconsistencies within extended thermodynamics have been removed by a cautious reformulation of the theory[91,92].

---

[87] I. Müller, T. Ruggeri: "Stationary heat conduction in radially symmetric situations – an application of extended thermodynamics." Journal of Non-Newtonian Fluid Mechanics 119 (2004).

[88] E. Barbera, I. Müller: "Inherent frame dependence of thermodynamic fields in a gas." Acta Mechanica, 184 (2006) pp. 205-216.

[89] H. Struchtrup, W. Weiss: "Maximum of the local entropy production becomes minimal in stationary processes." Physical Review Letters 80 (1998).

[90] E. Barbera, I. Müller, D. Reitebuch, N.R. Zhao: "Determination of boundary conditions in extended thermodynamics." Continuum Mechanics and Thermodynamics 16 (2004).

[91] I. Müller, D. Reitebuch, W. Weiss: "Extended thermodynamics – consistent in order of magnitude." Continuum Mechanics and Thermodynamics 15 (2003).

Heat conduction between circular cylinders.
Fourier theory and 13-moment theory [93]

For stationary heat conduction in a gas at rest between two concentric cylinders the BGK- version [94] of the 13-moment equations reads

$$\text{momentum balance}: \frac{\partial\left(p\delta_{ik} - t_{\langle ik\rangle}\right)}{\partial x_k} = 0, \ \text{energy balance}: \frac{\partial q_k}{\partial x_k} = 0,$$

$$t_{\langle ij\rangle} - \ \text{balance}: -\frac{2}{5}\left(\frac{\partial q_i}{\partial x_j} + \frac{\partial q_j}{\partial x_i}\right) = -\frac{1}{\tau} t_{\langle ij\rangle},$$

$$q_i - \text{balance}: \frac{\partial\left(5p\frac{k}{\mu}T\delta_{ik} - 7\frac{k}{\mu}Tt_{\langle ik\rangle}\right)}{\partial x_k} = -\frac{2}{\tau} q_i.$$

In the physical cylindrical coordinates appropriate to the problem the solution can easily be found

$$p \sim \ \text{homogeneous}, \qquad t_{\langle ij\rangle} = \begin{bmatrix} -\frac{4}{5}\tau\frac{c_1}{r^2} & 0 & 0 \\ 0 & +\frac{4}{5}\tau\frac{c_1}{r^2} & 0 \\ 0 & 0 & 0 \end{bmatrix},$$

$$q\langle i\rangle = \begin{bmatrix} \frac{c_1}{r} \\ 0 \\ 0 \end{bmatrix}, \qquad T = c_2 - \frac{c_1}{5\frac{k}{\mu}\tau p} \ln\left(\frac{28}{25}\frac{\tau}{p}c_1 + r^2\right).$$

---

[92] D. Reitebuch: "Konsistent geordnete Erweiterte Thermodynamik." [Consistently ordered extended thermodynamics] Dissertation TU Berlin (2004).

[93] I. Müller, T. Ruggeri: "Stationary heat conduction ..." loc. cit. (2004).

[94] P.L. Bhatnagar, E.P. Gross, M. Krook: "A model for collision processes in gases. I. Small amplitude processes in charge and neutral one-component systems." Physical Review 94 (1954).

The model approximates the collision term in the Boltzmann equation by $\frac{1}{\tau}(f_{equ} - f)$ with a constant relaxation time τ of the order of a mean time of free flight. The BGK model is popular for a quick check and qualitative results. In the present case it permits an analytical solution, which cannot be obtained by a more realistic collision term.

Figure 8.9 shows the comparison of the temperature fields in this solution and of the Navier-Stokes-Fourier solution in a rarefied gas – with $p = 1$mbar – for a boundary value problem as indicated in the figure

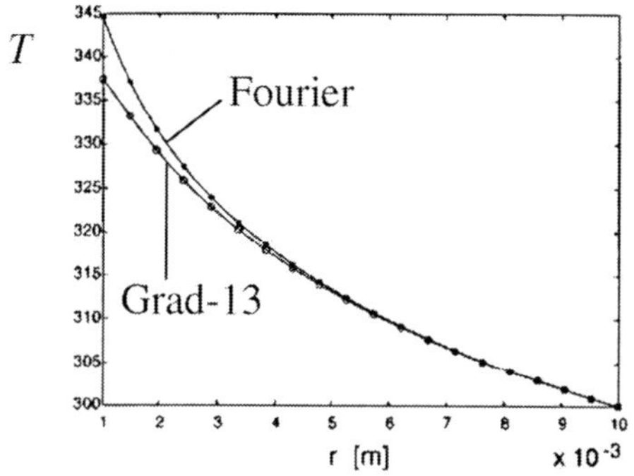

$r_{int} = 10^{-3}$m $\quad r_{ext} = 10^{-2}$m

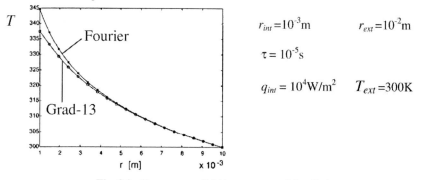

$\tau = 10^{-5}$s

$q_{int} = 10^4$W/m^2 $\quad T_{ext} = 300$K

Fig. 8.9. Temperature field between coaxial cylinders

As expected, the difference becomes noticeable where the temperature gradient is big. Note that the Fourier solution becomes singular for $r \to 0$, but the Grad solution remains finite.

Insert 8.2

Kinetic and thermodynamic temperatures[95,96]

We recall Insert 4.5 where the non-convective entropy flux Φ_i was calculated. It was unequal to q_i / T. In fact it was given by

$$\Phi_i = \frac{q_i}{T} + \frac{2}{5}\frac{t_{\langle ij\rangle} q_j}{pT},$$

so that T is not continuous at a diathermic, non-entropy-producing – i.e. thermometric – wall, where the normal components of the heat flux and the entropy flux are continuous.

In the case of heat conduction – treated in Insert 8.2 – there are only radial components of $\boldsymbol{\Phi}$ and $\boldsymbol{q}$ and we have

$$\Phi\langle 1\rangle = \underbrace{\frac{1}{T}\left(1 + \frac{2}{5}\frac{t\langle 11\rangle}{p}\right)}_{\frac{1}{\Theta}} q\langle 1\rangle.$$

[95] I. Müller, T. Ruggeri: "Stationary heat conduction ..." loc. cit (2004).

[96] I. Müller, P. Strehlow: "Kinetic temperature and thermodynamic temperature." In: Dean C. Ripple (ed.) "Temperature: Its Measurement and Control in Science and Industry." Vol. 7 American Institute of Physics (2003).

Thus Θ is the *thermodynamic temperature*, the temperature shown by a contact thermometer. Θ is not equal to T, the kinetic temperature, except in equilibrium, of course. Figure 8.10 shows the ratio of the two temperatures in a rarefied in the situation investigated in Insert. 8.2 for the Grad 13-moment theory.

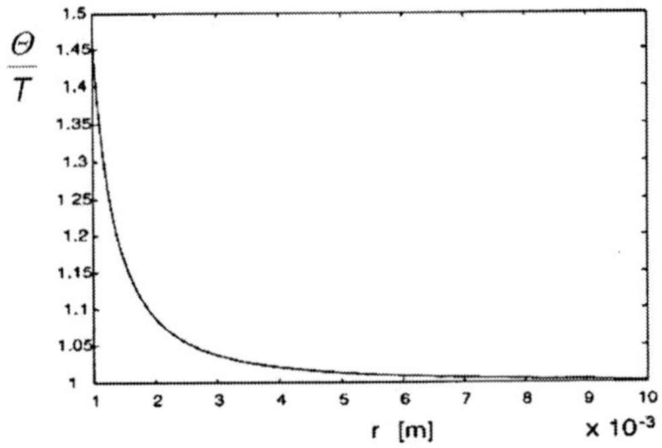

Fig. 8.10. The ratio of thermodynamic to kinetic temperature

Insert 8.3

9 Fluctuations

Fluctuations are random and therefore unpredictable, except in the mean, or on average. They are due to the irregular thermal motion of the atoms. An instructive example – and the first one to be described analytically – is the Brownian motion of nearly macroscopic particles suspended in a solution. The velocity of such a particle fluctuates around zero in an apparently irregular manner. Some regularity reveals itself, however, in the *mean regression* of the velocity fluctuations. In fact, in some approximation the mean regression is akin to the non-fluctuating velocity of a macroscopic ball thrown into the solution.

That observation has been extrapolated to arbitrary fluctuating quantities by Lars Onsager. Applied to the fluctuating density field in a gas, or a liquid, Onsager's mean-regression hypothesis furnishes the basis for the exploitation of light scattering experiments: The light scattered by a gas carries information about the transport coefficients of the gas, like the thermal conductivity and the viscosity, although the gas is macroscopically in equilibrium.

In a rarefied gas, where extended thermodynamics is appropriate, the Onsager hypothesis – if accepted – permits the prediction of the shape of the scattering spectrum. Experiments confirm that prediction.

Brownian Motion

Brownian motion is observed in suspensions of tiny particles which follow irregular, erratic paths visible under the microscope. The phenomenon was reported by Robert Brown (1773–1858) in 1828.[1] He was not the first person to observe this, but he was first to recognize that he was not seeing some kind of self-animated biological movement. He proved the point by observing suspensions of organic *and* inorganic particles. Among the latter category there were *ground-up fragments of the Sphinx,* surely a dead substance, if ever there was one. All samples showed the same behaviour

[1] R. Brown: "A brief account of microscopic observations made in the months of June, July and August 1827 on the particles contained in the pollen of plants; and on the general existence of active molecules in organic and inorganic bodies." Edinburgh New Philosophical Journal 5 (1828) p. 358.

and no convincing explanation or description could be given for nearly 80 years. According to Brush the phenomenon was mentioned in books on the microscope which gave warnings about Brownian motion, lest observers should mistake it for a manifestation of life and *attempt to build fantastic theories on it.*[2]

After the kinetic theory of gases was proposed and slowly accepted, the impression grew that the phenomenon provides a *beautiful and direct experimental demonstration of the fundamental principles of the mechanical theory of heat.*[3] That interpretation was supported by the observation that at higher temperatures the motion becomes more rapid. However, none of the protagonists of the field of kinetic theory addressed the problem, neither Clausius, nor Maxwell, nor Boltzmann. It may be that they did not wish to become involved in liquids.

A great difficulty was that the Brownian particles were about 10^8 times more massive than the molecules of the solvent so that it seemed inconceivable that they could be made to move appreciably by impacting molecules.

It was Poincaré – the mathematician who enriched the early history of thermodynamics on several occasions with his perspicacious remarks – who identified the mechanism of Brownian motion when he said:[4]

> Bodies too large, those, for example, which are a tenth of a millimetre, are hit from all sides by moving atoms, but they do not budge, because these shocks are very numerous and the law of chance makes them compensate each other; but the smaller particles receive too few shocks for this compensation to take place with certainty and are incessantly knocked about.

Also Poincaré noted that the existence of Brownian motion was in contradiction to the second law of thermodynamics when he said:

> … but we see under our eyes now motion transformed into heat by friction, now heat changed inversely into motion, and [all] that without loss, since the movement lasts forever. This is the contrary of the principle of Carnot.[5]

And indeed, the existence of Brownian motion demonstrates that the second law is a law of probabilities. It cannot be expected to be valid when few particles or few collisions are involved. If that is the case, there will be sizable fluctuations around equilibrium.

---

[2] S.G. Brush: "The kind of motion we call heat." loc.cit. p. 661.

[3] G. Cantoni: Reale Istituto Lombardo di Scienze e Lettere. (Milano) Rendiconti (2) 1, (1868) p. 56.

[4] J.H. Poincaré: In: "Congress of Arts and Science. Universal Exhibition Saint Louis 1904." Houghton, Miffin & Co. Boston and New York (1905).

[5] Ibidem.

Brownian Motion as a Stochastic Process

And so we come to the third one of Einstein's seminal papers of the *annus mirabilis*: "On the movement of small particles suspended in a stationary liquid demanded by the molecular-kinetic theory of heat."[6] After Poincaré's remarks the physical explanation of the Brownian motion was known, but what remained to be done was the mathematical description.

Actually Einstein claimed to have provided both: The physical explanation and the mathematical formulation. As a matter of fact, he even claimed to have foreseen the phenomenon on general grounds, without knowing of Brownian motion at all. Brush is sceptical. Says he:[7]

> … there is some doubt about the accuracy of these [claims]

and he reminds the reader of Einstein's own pronouncement quoted before, cf. Chap. 7:

> Every reminiscence is coloured by today's being what it is, and therefore by a deceptive point of view.[8]

People do have a way of treading lightly around Einstein's claims of priority, because there is a certain amount of hero-worship. The fact is, however, that in later life Einstein sometimes overreached himself; so when he claims to have developed statistical mechanics because he had no knowledge of Boltzmann and Gibbs's work in 1905.[9] In fact, however, he had quoted Boltzmann's book in an earlier paper published in 1902.[10]

Be that as it may. The fact remains that Einstein opened a new chapter of thermodynamics when he treated Brownian motion.

Obviously, after the insight provided by Poincaré, the Brownian motion had to be considered as stochastic, i.e. random, or determined by chance and probabilities. As far as I can tell, it was Einstein who invented a method

---

[6] A. Einstein: "Die von der molekularkinetischen Theorie der Wärme geforderte Bewegung von in ruhenden Flüssigkeiten suspendierten Teilchen." Annalen der Physik (4) 17 (1905) pp. 549–560.
All of Einstein's early papers on the Brownian motion were later edited by R. Fürth: "Untersuchungen über die Theorie der Brownschen Bewegungen." [Investigations on the theory of the Brownian movement] Akademische Verlagsgesellschaft, Leipzig (1922). This collection has been translated into English by A.D. Cowper and is available as a Dover booklet.

[7] S.G. Brush: "The kind of motion we call heat." loc. cit. p. 673.

[8] P.A. Schilpp (ed.): "Albert Einstein Philosopher-Scientist". New York. "Library of Living Philosophers" (1949).
The Schilpp collection contains an autobiographical note by A. Einstein from which the above quotation is taken.

[9] Schilpp collection. Autobiographical notes. loc.cit p. 17/18.

[10] A. Einstein: "Kinetische Theorie des Wärmegleichgewichtes und des zweiten Hauptsatzes der Thermodynamik." [Kinetic theory of heat equilibrium and of the second law of thermodynamics] Annalen der Physik (4) 9 (1902)pp. 417–433.

to deal with such a process.[11] We shall consider a one-dimensional and simplified version of his argument:

Let the x-axis be subdivided into equal intervals of length Δ and let a Brownian particle jump – right or left with equal probability, i.e. probability $^1/_2$ – to neighbouring intervals after each time interval τ. The jumps occur because the particle is hit by solvent molecules but no explicit account is given of the mechanics of the collisions.

From what has been said, the probability $w(x,t)$ of finding the particle at position x at time t must satisfy the difference equation

$$w(x,t) = \tfrac{1}{2} w(x-\Delta, t-\tau) + \tfrac{1}{2} w(x+\Delta, t-\tau) .$$

If Δ and τ are small, one may expand the right hand side into a Taylor series breaking off at the leading non-zero terms in Δ and τ. Thus one obtains the differential equation

$$\frac{\partial w}{\partial t} = \frac{\Delta^2}{2\tau} \frac{\partial^2 w}{\partial x^2} .$$

Einstein says: *This is the well-known diffusion equation and we recognize that $D = \Delta^2/2\tau$ is the coefficient of diffusion.*

Many solutions of this equation are known – primarily through Fourier's work, cf. Chap. 8. In particular, if at time $t = 0$ the particle was in the interval at X, its probability to be at position x at time t is given by

$$w(x,t) = \frac{1}{\sqrt{4\pi D t}} \exp\left(-\frac{(x-X)^2}{4Dt} \right)$$

and the root mean square distance λ from X comes out as

$$\lambda = \sqrt{\overline{(x-X)^2}} = \left(\int_{-\infty}^{\infty} (x-X)^2 w(x,t) dx \right)^{1/2} = \sqrt{2Dt} ,$$

so that it is determined by the diffusion coefficient. Thus by repeated careful observations of Brownian motion and averaging over the results one could determine D.

Einstein, however, favoured another application of the formula for λ. He had determined a relation between the unknown diffusion coefficient D – of a Brownian particle of radius r in a solvent – and the known viscosity η of the solvent, viz, cf. Insert 9.1

[11] A. Einstein: "Investigations ... " loc.cit. § 4.

$$D = \frac{kT}{6\pi\eta r} \quad \text{so that he could write} \quad \lambda = \sqrt{\frac{kT}{3\pi\eta r} t}\ .$$

Thus measurements of λ for known values of η and r could determine the value of the Boltzmann constant k. Therefore Einstein concludes his paper with the words: *It is to be hoped that some enquirer may succeed shortly in solving this problem* [the experimental determination of k]... *which is so important in connection with the theory of heat.*

This remark is obviously a reflection of the then still ongoing – albeit obsolete – discussion between Mach and Boltzmann in Vienna, where the former maintained that atoms were a fiction of imagination, since their properties could not be determined; [Mach ignored Loschmidt's rough and ready calculation of 1865, cf. Chap. 4.] The rest of the world watched this out-dated debate in amazement[12] but Einstein seems to have taken it seriously.[13]

I cannot help feeling that the importance and feasibility of measuring k in this manner is somewhat exaggerated here by Einstein. After all, this recipe would involve a cumbersome observation of the *mean* motion of a Brownian particle. No doubt that it can be done, but why should it be done? A good value of the Boltzmann constant was already known from the Rayleigh-Jeans formula, cf. Chap. 7, which was perfectly convincing and indubitably correct for low-frequency radiation.

Relation between diffusion coefficient D and viscosity η

When Brownian particles of mass μ, radius r, and with particle density $n(x,t)$ are suspended – macroscopically at rest – in a solvent of temperature T, they are denser at the bottom than on top, because they must satisfy the stationary momentum balance

[12] Robert Andrews Millikan (1868-1953) – the man who determined the elementary charge e – writes in his autobiography: *The amazing thing is that this question could be debated at all at that time* [1904] ... *and that even the brilliant philosopher Ernst Mach could at that epoch oppose atomic theories.*
R. A. Millikan: "The autobiography of Robert A. Millikan." Arno Press, New York (1980).
D. Lindley, the author of "Boltzmann's atom" loc.cit. writes: *To an audience of young New World scientists, this debate must have seemed an intrusion into their fresh universe from the Old World's attic.*

[13] Einstein writes: *If the movement discussed here can actually be observed ... an exact determination of actual atomic dimensions is then possible. On the other hand, had the prediction of this movement proved to be incorrect, a weighty argument would be provided against the molecular kinetic conception of heat.*

$$\frac{\partial p(n,T)}{\partial x} = -n\mu g \quad \text{or with } p = nkT, \text{ according to van't Hoff's law for dilute solutions, cf. Chap. 5:}$$

$$\frac{\partial n}{\partial x} = -\frac{\mu g}{kT}n \qquad \text{(barometric formula).}$$

We may think of the particles as being macroscopically at rest, because two flow velocities compensate each other:

a downward flow with $v_{\downarrow} = \frac{\mu g}{6\pi\eta r}$ according to Stokes´s law for a spherical particle under gravity, cf. Chap. 8 and

an upward flow $v_{\uparrow} = -D\frac{1}{n}\frac{\partial n}{\partial x}$ according to Fick´s law, cf. Chap. 8.

Hence follows with the barometric formula

$$\frac{\mu g}{6\pi\eta r} = D\frac{\mu g}{kT} \qquad \text{or} \qquad D = \frac{kT}{6\pi\eta r}.$$

This is Einstein's relation between D and η.

Insert 9.1

Einstein's paper carries the mark of genius in a positive and negative sense: The positive aspect is that the paper introduces stochastic arguments into Brownian motion and this made such arguments acceptable to thermodynamicists. But then the paper is also carelessly written, it shows a benign neglect of detail and direction that might – and did – throw people off the track. Thus Brush[14] complains about the muddled presentation. He says that

> Einstein did not emphasize very strongly the significance of his result that λ is proportional to the square root of time, and in fact it is quite probable that most early readers of the paper gave up in bewilderment before they got to the result.

Indeed, it makes no sense that the initial growth rate of λ is infinite as is implied by the result. And surely this prediction should have warranted a remark. It may in fact be understood as a shortcoming of the stochastic model by which the Brownian particle, – in executing its random jumps – is

[14] S.G. Brush: "The kind of motion we call heat." loc.cit. p. 681.

not ascribed an inertia. The physicist Paul Langevin (1872–1946)[15] looked into the argument and he came up with an improved equation of the form

$$\lambda^2 = 2D\left[t - \frac{\mu}{6\pi\eta r}\left(1 - \exp\left(-\tfrac{6\pi\eta r}{\mu}t\right)\right)\right]$$

by taking inertia into account. To be sure, for typical values of η, μ, and r the second term in the square brackets is usually negligible, so that Einstein's results holds approximately. But this is not so for small times.

Mean Regression of Fluctuations

In the Brownian motion we see a nearly macroscopic body – the Brownian particle – kicked around by the atoms or molecules in the manner envisaged by Poincaré, see above. The force $F(t)$ of impact by the molecules on the particle fluctuates, and it stands to reason that, averaged over a long time, or averaged – at one time – over many Brownian particles, the force is zero. Since the particle moves in a viscous fluid, its equation of motion reads

$$\dot{v} + \frac{6\pi\eta r}{\mu} v = \frac{1}{\mu} F(t).$$

This equation is known as the *Langevin equation.* On the basis of that equation Langevin was able to correct Einstein's result for the root mean square distance λ, see above.

If the mass μ of the particle is very big, its equation of motion is unaffected by the fluctuating force $F(t)$ and the velocity decays exponentially as a function of time

$$v(t) = v(t_0)\exp\left(-\tfrac{6\pi\eta r}{\mu}(t - t_0)\right).$$

I shall refer to this solution as the *macroscopic law of decay.* For the Brownian motion the decay is exponential, but this need not be so in other cases of fluctuating quantities; indeed, the decay may be a damped oscillation on other occasions.

On the other hand, when the particle has a small mass, the fluctuating force makes its velocity fluctuate as well about an average velocity zero as illustrated in the upper part of Fig. 9.1. The graph of this velocity fluctuation seems totally irregular, and certainly in no way related to the macroscopic law of decay. And yet, some regularity is hidden in the fluctuations; and that regularity is brought forth, if we construct the *mean regression of a fluctuation.*

[15] P. Langevin: Comptes Rendues Paris 146 (1908) p. 530.

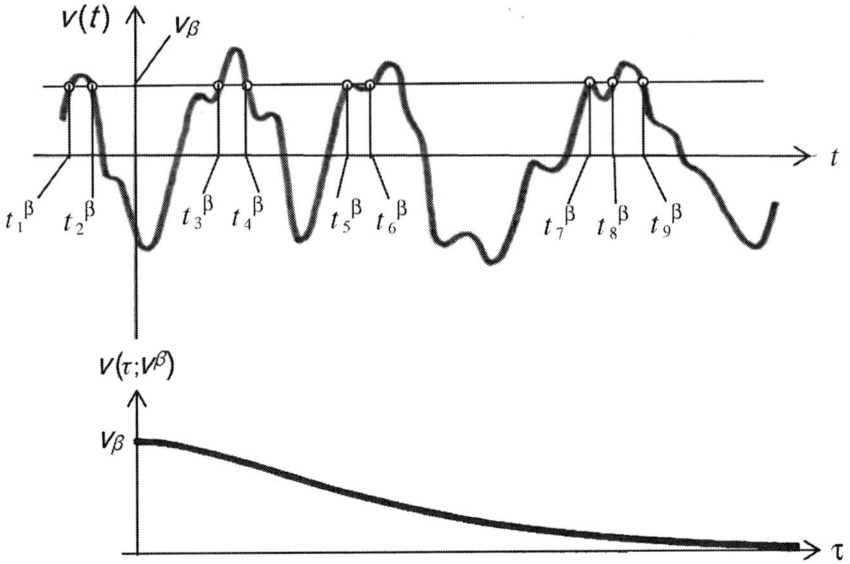

Fig. 9.1. *Top:* Velocity fluctuations of a Brownian particle. *Bottom:* Mean regression of a fluctuation

When we consider very many, say N, velocity fluctuations of a particular fixed size v_β which occur at the times t_α^β (α = 1,2,…N), we may ask for the sizes of the fluctuation at a later time $t_\alpha + \tau$. They are all different, of course, but upon averaging we obtain

$$\overline{v(\tau, v_\beta)} = \frac{1}{N} \Sigma_{\alpha=1}^{N} v(t_\alpha^{\ \beta} + \tau) .$$

This function of τ is the mean regression of the fluctuation v_β. We may draw $\overline{v(\tau; v_\beta)}$ as a graph of the type shown in the lower part of Fig. 9.1. According to Lars Onsager the mean regression is given by the same function as a macroscopic decay. This can be proved – after a fashion – for Brownian particles, see Insert 9.2.

The mean regression of fluctuation for a Brownian particle

The formal solution of the Langevin equation reads with $\tau_0 = \dfrac{\mu}{6\pi\eta\ r}$

$$v(t) = v(t_0) e^{-\frac{t-t_0}{\tau_0}} + \frac{1}{\mu} e^{-\frac{t}{\tau_0}} \int_{t_0}^{t} e^{\frac{t'}{\tau_0}} F(t') dt'$$

so that the mean regression of a fluctuation v_β comes out as

$$\overline{v(\tau;v_\beta)}=\frac{1}{N}\sum_{\alpha=1}^{N}\left(v(t_\alpha^{\ \beta})e^{-\frac{(t_\alpha^{\ \beta}+\tau)-t_\alpha^{\ \beta}}{\tau_0}}-\frac{1}{\mu}e^{-\frac{t_\alpha^{\ \beta}+\tau}{\tau_0}}\int_{t_\alpha^{\ \beta}}^{t_\alpha^{\ \beta}+\tau}e^{\frac{t'}{\tau_0}}F(t')dt'\right)$$

$$=e^{-\frac{\tau}{\tau_0}}\left(v_\beta-\frac{1}{\mu}\frac{1}{N}\sum_{\alpha=1}^{N}e^{-\frac{t_\alpha^{\ \beta}}{\tau_0}}\int_{t_\alpha^{\ \beta}}^{t_\alpha^{\ \beta}+\tau}e^{\frac{t'}{\tau_0}}F(t')dt'\right)$$

Given time the force $F(t)$ in the integrand is fluctuating between positive and negative values so that the integral itself may have a positive or negative value, but it is definitely finite. Therefore for large enough N the second term vanishes, so that the mean regression becomes

$$\overline{v(\tau;v_\beta)}=v_\beta e^{-\frac{\tau}{\tau_0}}$$

which is equal to the macroscopic decay. This may be considered proof of the Onsager hypothesis, at least for Brownian particles.

[The fallacy of this proof for small values of τ is obvious: Indeed, for small values of τ the force $F(t')$ in the interval $t_\alpha^\beta < t' < t_\alpha^\beta + \tau$ is most likely close to the force $F(t_\alpha^\beta)$, because the force does not really jump, although it may change quickly. Thus for small τ the Onsager hypothesis fails. Another way to see this is as follows: For small values of τ there are obviously equally many values $v(t_\alpha^{\ \beta}+\tau)$ bigger and smaller than $v(t_\alpha^{\ \beta})$ so that $\overline{v(\tau;v_\beta)}$ must start out horizontally, i.e. it cannot decay exponentially at the outset.]

Insert 9.2

Auto-correlation Function

The auto-correlation function – denoted by $\langle v(0)v(\tau)\rangle$ for the velocity of a Brownian particle – is a mean value over mean fluctuation regressions of all occurring initial values v_β; let their number be denoted by M. So as to avoid the trivial result zero for the mean value, the mean regressions are pre-multiplied by v_β before the mean value is taken. Thus the auto-correlation function is defined as

$$\langle v(0)v(\tau)\rangle=\frac{1}{M}\sum_{\beta=1}^{M}v_\beta\overline{v(\tau;v_\beta)}$$
$$=\frac{1}{MN}\sum_{\alpha,\beta=1}^{N,M}v(t_\alpha^{\ \beta})v(t_\alpha^{\ \beta}+\tau)\ .$$

Between all N values t_α^β and all M values with v_β the summation covers a coherent large time interval T so that one may write

$$\langle v(0)v(\tau)\rangle = \frac{1}{T}\int_0^T v(t)v(t+\tau)dt.$$

Since all mean fluctuation regressions are equal in their functional behaviour to the macroscopic law of decay, – according to the Onsager hypothesis – this is also true for their mean value, i.e. the auto-correlation function.

The auto-correlation function is often easier to calculate and to measure than the mean regression of a particular size of fluctuation. Therefore the Onsager hypothesis is most often pronounced by saying that *the auto-correlation function is equal to the macroscopic decay function.*

Extrapolation of Onsager's Hypothesis

Brownian particles provide the first fluctuating phenomenon that has been studied and they are simple enough to be amenable to intuitive argument *and to* calculation. Therefore they serve as prototypes for the treatment of fluctuation and for the proof of Onsager's hypothesis, see Insert 9.2.

The hypothesis is not restricted to Brownian particles, however. It is supposed to hold for all fluctuating systems. And it is usually called Onsager's *theorem.* Physicists have a way to quickly become very defensive of Onsager when challenged, probably because of the precariousness of the proof of the *theorem,* or because they do not understand it, or because Onsager has been canonized by the Nobel prize in 1968, see Fig. 9.2. There is some uneasiness, however. We have already quoted the popular textbook by de Groot and Mazur,[16] who give faint praise to Onsager by calling his hypothesis *not altogether unreasonable.*[17]

Light Scattering

While Brownian particles and their erratic motion can be seen, albeit only under the microscope, fluctuations of mass density, and velocity and temperature in air cannot be seen. And yet they are there, and they affect the transmission of light. Indeed, very tiny and very short-lived local

---

[16] S.R. de Groot, P. Mazur: loc.cit.

[17] And we have seen above why the proof of the hypothesis is flawed for small times even in Brownian motion. In the sequel I shall ignore that qualification. Physicists tell me that it is pedantic.

Onsager left his native Norway in 1928 and came to the United States. Later he held a chair of theoretical chemistry at Yale University, where he taught Statistical Mechanics I and II to chemistry students.

Among the students his course was known as Norwegian I and II.[18]

Fig. 9.2. Lars Onsager receiving the Nobel prize for chemistry in 1968

compressions and expansions of air – and gases generally – occur as a result of the random motion of molecules and atoms and they affect the dielectric constant, because it depends on the mass density.

Because of these fluctuations some light is scattered sideways, see Fig. 9.3. Most of the scattered light has the frequency $\omega^{(i)}$ of the incident mono-chromatic light, but neighbouring frequencies ω are also present. Typically the spectrum $S(\omega)$ of light – scattered in a gas and passed through an interferometer to a photo-multiplier – exhibits three peaks, if the gas is normally dense. In a moderately rarefied gas one sees a flatter curve with lateral *shoulders*, cf. Fig. 9.4

The blue frequencies in sunlight are 16 times more efficiently scattered than the red frequencies. Therefore the cloudless sky appears blue. It was John Tyndall – the admirer of Robert Mayer who recognized this phenomenon after studying Lord Rayleigh's work on electro-magnetic waves.

Sir James Dewar – the low temperature physicist – had thought erroneously that the blue sky is due to the oxygen content of the air; he knew that liquid oxygen has a blue colour.

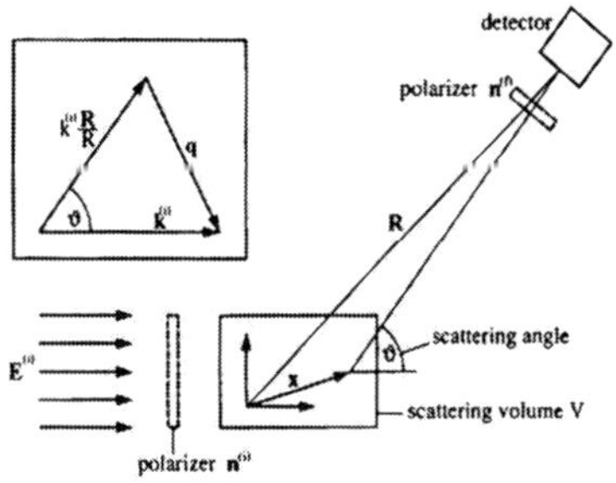

Fig. 9.3. Light scattering, schematic

[18] According to J. Meixner: "Chemie Nobelpreis 1968 für Lars Onsager." [Nobel prize 1968 for chemistry for Lars Onsager] Physikalische Blätter 2 (1969).

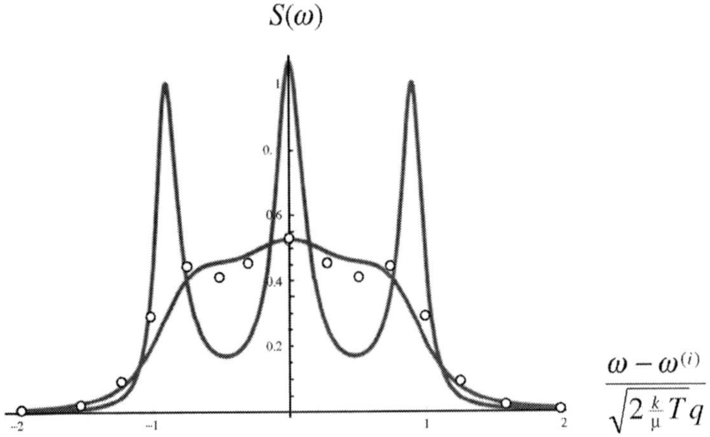

Fig. 9.4. Scattering spectrum $S(\omega)$ in a normally dense gas and in a moderately rarefied gas. *Dots:* Measurements by Clarke for rarefied gas.[19] *Lines:* Calculation from a Navier-Stokes-Fourier theory

If the Onsager hypothesis is accepted, $S(\omega)$ can also be *calculated* from the field equations of the gas, e.g. the Navier-Stokes equations. For dense gases the measured and calculated curves fit perfectly, and thus they support the hypothesis. For the rarefied gas, however, the fit is not good, cf. Fig. 9.4, and we may conclude that the discrepancy is due to the Navier-Stokes equations which, indeed, according to Chap. 8 are expected to fail in a rarefied gas.

So, this is a case where extended thermodynamics can prove its usefulness and practicality. Wolf Weiss[20] has applied the linearized field equations of 20, 35, 56, and 84 moments to the problem and has obtained the scattering spectra of Fig. 9.5 (*top*) for small pressures as in Fig. 9.4. They differ among themselves and none of them fits the experimental points well. Nor can we adjust parameters to obtain a better fit, because there are no adjustable parameters in the theories of extended thermodynamics. Or rather, one might say that the only parameter is the number of moments and moment equations. So Weiss went ahead to 120 through 286 moments and obtained *convergence as well as a perfect agreement with experimental results*, cf. Fig. 9.5 (*bottom*).

Here we have another instance where a result of thermodynamics is satisfactory, amazing and disappointing at the same time.[21]

---

[19] N.A. Clarke: "Inelastic light scattering from density fluctuations in dilute gases. The kinetic-hydrodynamic transition in a monatomic gas." Physical Review A 12 (1975).

[20] W. Weiss: "Zur Hierarchie der Erweiterten Thermodynamik." loc. cit.
See also: W. Weiss, I. Müller: "Light scattering and extended thermodynamics." Continuum Mechanics and Thermodynamics 7 (1995).

[21] Recall Schrödinger's comment on gas degeneracy in Chap. 6.

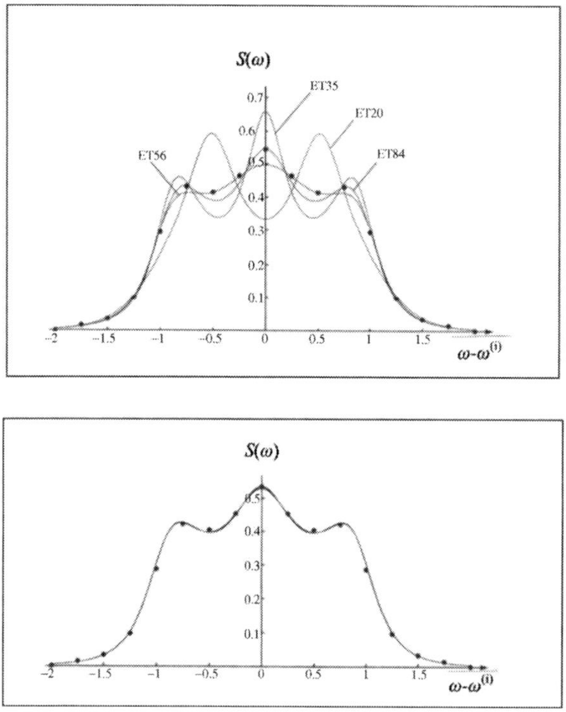

Fig. 9.5 . Light scattering spectra in a moderately rarefied gas. *Top:* Extended thermodynamics$N$= 20, 35, 56, 84. *Bottom*: Extended thermodynamics N= 120, 165, 220, 286. *Dots:* Experimental points measured by N.A. Clarke[22]

- Satisfaction comes from the fact that extended thermodynamics combined with the Onsager hypothesis *is capable* of representing light scattering in rarefied gases.
- The amazing feature is the convergence at some finite number of moments; that observation carries information about the range of validity of the theory, see below.
- Disappointment stems from the large number of moments needed to achieve convergence. We might have hoped that 13 or, perhaps, 14 or 20 moments could give good results. That would have given us a manageable system of equations. Instead, we need 120 of them, – at least for the low pressure to which the curves of Fig. 9.5 refer.

The convergence put in evidence by the plots of Fig. 9.5 permits us to conclude that extended thermodynamics determines its own range of applicability without any reference to experiments. This is something that is often said a theory cannot do. But then, extended thermodynamics is not a

[22] N.A. Clarke: loc.cit.

single theory, rather it is a *theory of theories,* one each for a given number of moments. So if – after an increase of that number – we obtain the same function $S(\omega)$ as before, in some norm, we have reached convergence and can fully trust the theory and predict the light spectrum, *without making a single experiment.*

More Information About Light Scattering

In the previous section we have used light scattering to advertise the validity and usefulness of the Onsager hypothesis in extended thermodynamics. This is only one rather special aspect and application of light scattering. There are others, more practical ones, and we give the briefest possible description about the method and its practical application.

The scattered electric field which arrives at the interferometer consists of the high frequency carrier wave – from a laser with $\omega^{(i)} = 4.7\ 10^{15}$ Hz (say) – modulated in its amplitude by a fluctuating spatial Fourier harmonic – of wave number q – of the density field. The value of q is determined by the position of the detector. The interferometer of type Fabry-Perot,[23] cf. Fig. 9.6, superposes light that was scattered at different times in the past. In that way it registers the auto-correlation function $\langle E(0)E(\tau)\rangle$ of the scattered field or, in fact, the temporal Fourier transform of that function, i.e. the spectral density $I(q,\omega)$, whose essential part is the scattering spectrum $S(\omega)$ discussed above.

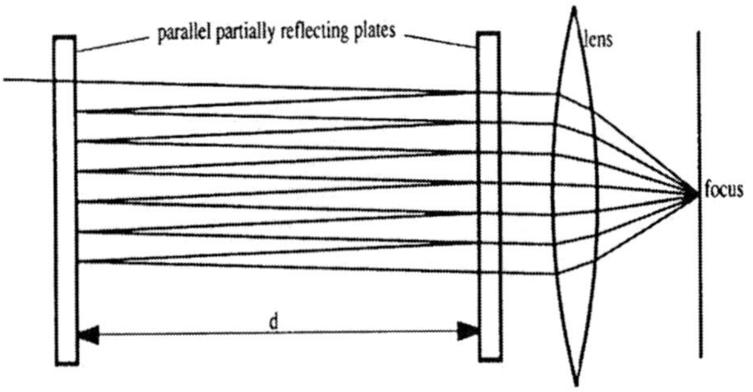

Fig. 9.6. Schematic view of a Fabry-Perot interferometer

In practical physics and engineering the scattering of light has by now become a powerful and elegant tool for the measurement of thermodynamic state functions and of transport coefficients. Let us consider this:

[23] A lucid description of this remarkable instrument is given by G. Simonsohn: "The role of the first order auto-correlation function in conventional grating spectroscopy." Optics Communications 5 (1972). See also: I. Müller, T. Ruggeri: "Rational Extended Thermodynamics." loc.cit. pp. 233–236.

The Onsager hypothesis permits the calculation of the scattering spectrum from the field equations of a gas (say). In particular, in a normally dense gas, where the Navier Stokes-Fourier theory is applicable we obtain three well-developed peaks, cf. Fig. 9.4. The heights and widths and distance of the peaks permit the determination of the constitutive properties of the gas as listed in Table 9.1. Thus it is possible to read off specific heats, sound speed, and transport coefficients like thermal conductivity κ, and viscosity η from the properties of light scattered by a gas *in equilibrium*.

Table 9.1. Constitutive data determine the shape of the scattering spectrum

	Central Peak	Lateral Peaks
site	$\omega = 0$	$\omega = \pm\sqrt{\frac{c_p}{c_v}\left(\frac{\partial p}{\partial \rho}\right)_T}\, q$
height	$\frac{c_p - c_v}{c_p}\frac{\rho c_p}{\kappa}\frac{1}{q^2}$	$\frac{c_v}{c_p}\left(\frac{c_p - c_v}{c_v}\frac{\kappa}{\rho c_v} + \frac{4}{3}\frac{\eta}{\rho}\right)^{-1}\frac{1}{q^2}$
half width	$\frac{\kappa}{\rho c_p} q^2$	$\frac{1}{2}\left(\frac{c_p - c_v}{c_v}\frac{\kappa}{\rho c_v} + \frac{4}{3}\frac{\eta}{\rho}\right) q^2$

10 Relativistic Thermodynamics

The theory of relativity must have implications in thermodynamics on two counts. Firstly, hot bodies are heavier than cold ones, because their atoms, or molecules have a bigger speed and therefore more mass. And secondly, since no particle can move faster than the speed of light, the velocity distribution of the particles must reflect the fact.

To be sure, both effects are minuscule and it takes extraordinary conditions – extraordinarily high temperatures – to make relativistic corrections of classical formulae relevant numerically; the conditions inside the sun are *not* sufficiently extreme, despite a temperature of millions of K in the solar centre. In fact it seems that white dwarfs are the only bodies for which relativity matters, and where thermodynamic arguments may still be employed without entering the realm of science fiction. For white dwarfs the relativistic effects are intermingled with quantum effects, because the density of the stars is so great that the de Broglie wave lengths of the free electrons overlap.

In the beginning of this book I have given much space to the idiosyncrasies of the early authors in the field of thermodynamics. One must not think, however, that wild ideas, oversimplifications and shallow answers are a privilege of physicists of the 19th century. They do occur at all times and among the most distinguished people. As a case in point I describe – briefly – what has become known as the Ott-Planck imbroglio.

Ferencz Jüttner

Although Planck was slow to accept his own theory of quantization as true, he was quick to trust Einstein's theory of relativity. It was therefore soon obvious to him that the Maxwell distribution had to be revised in order to accommodate the upper bound on the speeds of atoms. We recall from Chap. 2 that no mass can be accelerated beyond the speed of light c. Planck suggested the problem to Ferencz Jüttner, who says in his paper:

> It is a pleasant duty for me to express my warmest thanks to Hrn. Geheimrat[1] Planck for the kind suggestion of this work and for his benevolent advise.

Jüttner solved the problem in a satisfactory manner and published the result in 1911.[2] What he did was basically very simple. In effect he obtains the equilibrium distribution by maximizing the entropy

$$S = k \ln W \text{ with } W = \frac{N!}{\prod_{x,p} N_{xp}!},$$

under the constraints of a fixed number N of atoms and fixed energy cP^0 and momentum P^a

$$N = \sum_{x,p} N_{xp} \quad \text{and} \quad P^A = \sum_{x,p} p^A N_{xp},$$

where N_{xp} is the number of atoms at place $\boldsymbol{x}$ with momentum $\boldsymbol{p}$.

Once again I apologize for a somewhat anachronistic presentation because, indeed, Jüttner did not employ the elegant four-dimensional notation of relativistic formulae which became standard later. Capital indices run from 0 to 3 such that $x^0 = ct$ represents time, while x^a are spatial coordinates. The four-momentum of an atom of velocity q^a and mass $\mu = \mu' / \sqrt{1 - \frac{q^2}{c^2}}$ combines its energy $cp^0 = \quad c^2$ and its momentum $p^a = \quad q^a$ ($a = 1,2,3$) in one four-vector p^A. In that notation we have to distinguish between co- and contravariant components of a generic vector V_A and V^A respectively. Both are related through $V_A - g_{AB}V^B$ by the tensor g_{AB} which in Lorentz frames is given by

$$g_{AB} = \begin{bmatrix} 1 & 0 & 0 & 0 \\ 0 & -1 & 0 & 0 \\ 0 & 0 & -1 & 0 \\ 0 & 0 & 0 & -1 \end{bmatrix}, \text{ so that } p_A p^A = \mu'^2 c^2.$$

---

[1] Privy Councillor. This is a honorific bestowed on eminent German – and Austrian – scientists in pre-WWI-times.

[2] F. Jüttner: "Das Maxwellsche Gesetz der Geschwindigkeitsverteilung in der Relativtheorie." [Maxwell's law of the velocity distribution in the theory of relativity] Annalen der Physik 84 (1911) pp. 856–882.
F. Jüttner: "Die Dynamik eines bewegten Gases in der Relativtheorie." [Dynamics of a moving gas in the theory of relativity] Annalen der Physik 35 (1911) pp. 145–161.
The first paper deals with a gas at rest, while the second one deals with a moving gas. The second paper is much influenced by Planck's erroneous opinion that temperature should be transformed between two Lorentz frames, see below. In my account I present a streamlined modern version.

The maximization of entropy follows the usual steps known from the corresponding non-relativistic arguments. The result is known as the *Maxwell-Jüttner distribution*

$$N_{xp}^{equ} = a \cdot \exp\left(-\frac{U_A p^A}{kT} \right) \qquad \text{with} \quad N = a \sum_{xp} \exp\left(-\frac{U_A p^A}{kT} \right) \quad ,$$

where U^A is the four-vector of velocity v^a of the gas

$$U^A = \left(\frac{c}{\sqrt{1 - v^2/c^2}}, \frac{v^a}{\sqrt{1 - v^2/c^2}} \right),$$

and T is its temperature, a scalar quantity with respect to Lorentz transformations. a is a Lagrange multiplier and it must be calculated as a function of N and T from the constraint on N. That calculation is best done in the rest frame of the gas, where $U^A = (c,0,0,0)$ holds. In the general case the summation – or integration – leads to Hankel functions which makes the expressions cumbersome although, of course, Hankel functions have been calculated numerically and are tabulated. So, they are available, if needed.

More instructive, however, than the full solutions in terms of Hankel functions are expansions in terms of what may be called the *relativistic coldness* $\frac{\mu' c^2}{kT}$, which represents the ratio of the total energy $\mu' c^2$ of the rest mass to the thermal energy kT. This is obviously a large number for *normal* temperatures. The thermal and caloric equations of state may be given by such an expansion. Somewhat miraculously the thermal equation of state is unaffected by relativity; it still reads $p = nkT$, as it did for Mariotte and Avogadro. But the caloric equation of state becomes more complex, namely

$$u = \mu' c^2 \left(1 + \frac{3}{2}\left(\frac{kT}{\mu' c^2} \right) + \frac{15}{8}\left(\frac{kT}{\mu' c^2} \right)^2 - \frac{15}{8}\left(\frac{kT}{\mu' c^2} \right)^3 + \ldots \right).$$

Thus the internal energy is still only a function of T, but its derivative with respect to T – the specific heat c_v – is no longer constant and universal as it is in the classical first order term. Rather it depends on T *and on* μ', so that the equipartition of energy is violated in a mixture of gases.

Despite the successful completion of his task, Jüttner is rather despondent about observability and applicability of all this. He calculates the value of the relativistic coldness, and for helium he finds it equal to $\frac{\mu' c^2}{kT} = \frac{4.32 \cdot 10^{13}}{T/\mathrm{K}}$. He comments that result by saying:

> We recognize that for all temperatures amenable to experiment the parameter has a very high value for *all* monatomic gases: Even when we

consider the temperatures of some stars, which have been calculated as 20,000K, the parameter would not sink below1 billion.[3] for any gas.

Maybe Jüttner would have been less discouraged, had he known that the centre of the sun has a temperature of 20 million K. But even so, the relativistic coldness would still be roughly one million, so that no noticeable relativistic effect can be expected in the sun.

And yet, Jüttner's work was to achieve some relevance in the end, although he had to wait for it.

Seventeen years after the work of 1911, the phenomenon of quantum degeneration was brought to Jüttner's attention. He studied the works of Einstein,[4] Fermi,[5] and Dirac,[6] in which Bose's new method of counting realizations of a state were employed – and in which the difference between Fermions and Bosons was recognized. Jüttner incorporated these modifications of classical, i.e. non-quantum physics into his relativistic formula and obtained [7]

$$N_{xp}^{equ} = \frac{1}{\frac{1}{a}\exp\left(\frac{U_A p^A}{kT}\right) \mp 1} \quad \text{with} \quad N = \sum_{xp} \frac{1}{\frac{1}{a}\exp\left(\frac{U_A p^A}{kT}\right) \mp 1} \quad \begin{matrix}\text{bosons}\\ \text{fermions}\end{matrix} .$$

The modification introduces more complex special functions even than Hankel functions into the equations of state, and the results are of little suggestive value to the non-expert. General results are listed in the literature on relativistic thermodynamics, e.g.[8, 9, 10]. More instructive are the limiting expressions of the equilibrium distribution function for either small or large relativistic coldness, or small or large quantum mechanical degeneration. Some of these are exhibited in table 10.1.

As for relevance under physically realistic circumstances Jüttner was still pessimistic. He says:

> The significance of both generalized gas theories [relativistic only, and relativistic plus quantum corrections] is, however, essentially theoretical. One has to consider that deviations of the relativistic from the Newtonian mechanics can only occur at such high temperatures that the speeds of the

---

[3] I am using the American nomenclature here: What Wall Street calls a billion is a milliard, i.e. 10^9, in the rest of the world.

[4] A. Einstein: Sitzungsberichte (1924) loc. cit.

[5] E. Fermi: Zeitschrift für Physik (1926) loc.cit.

[6] P.A.M. Dirac: Proceedings of the Royal Society (1927) loc.cit.

[7] F. Jüttner: "Die relativistische Quantentheorie des idealen Gases." [The relativistic quantum theory of the ideal gas] Zeitschrift für Physik 47 (1964), pp. 542–566.

[8] S.R. de Groot, W.A. van Leeuven, Ch.G. van Weert: "Relativistic Kinetic Theory." North Holland Publishers Amsterdam (1980).

[9] I. Müller, T. Ruggeri: "Rational Extended Thermodynamics." loc.cit. (1998).

[10] C. Cercignani, G.M. Kremer: "The Relativistic Boltzmann Equation. Theory and Applications." Birkhäuser Verlag, Basel (2002).

particles become comparable with the speed of light. On the other hand, the quantization of the translational energy makes itself felt as gas degeneration only at small temperatures. Therefore the full theory could only be checked at intermediate temperatures by measurements conducted with extreme accuracy, and only, if the van der Waals corrections were taken into account properly.

In other words, Jüttner did not believe that his formulae had any actual relevance anywhere. In that pessimistic evaluation he was wrong, however, as we shall see now.

White Dwarfs

The first white dwarf was detected by the eminent astronomer Friedrich Wilhelm Bessel (1784–1846) in 1844, although the star was not actually *seen* by Bessel; it was only *conjectured* from the observation of the wavy line of the proper motion of the bright star Sirius. Therefore, ironically, the first white dwarf entered the literature as the *dark companion* of Sirius, also called Sirius B. The first person to *see* it – in 1862 – as a dim spot was the astronomer Alvan Graham Clark (1832-1897). And in 1914, Walter Sidney Adams (1876-1956) succeeded to measure the spectrum of Sirius B, and could thus conclude that its surface temperature is about 10000K, making it white hot, considerably hotter than the sun. Given that fact, the star had to be quite small in order to appear dim. Thus it came to be called a *white dwarf*. From the known distance the diameter could be estimated as $2.7 \cdot 10^7$m, or 4% of the solar diameter. And yet, according to Eddington's mass-luminosity relation, cf. Chap. 7, Sirius A was twice as massive as the sun and, in order to be forced into the observed orbit by the companion, the companion had to have about the same mass as the sun. This meant that the average density had to be a fantastic 140000 times that of the sun, or 200000 times that of water. One cm^3 has a mass of 200kg!

Any scepticism about such numbers was quickly silenced when – at the suggestion of Eddington – Adams re-examined his spectroscopic data and found the relativistic red shift of spectral lines which must be expected for the intense gravitational field of a massive and compact white dwarf. More white dwarfs were discovered as time went on – despite their dimness – and some are considerably denser and hotter even than Sirius B.

Therefore, it was clear that no atoms could exist in a white dwarf, only nuclei and electrons. And the nuclei must be fairly heavy nuclei, because astronomers have good reasons to believe that white dwarfs are old stars, which have essentially burned up their light-weight-fuel, the protons. Now they consist of many electrons and few nuclei of atoms of intermediate mass, like iron, which cannot serve as fuel for further combustion. If that is so, the large majority of all particles are electrons and the mean relative molecular mass is $\mu/\mu_0 = 2$, cf. Chap. 7.

Table 10.1 Equilibrium distribution function in a gas at rest, i.e. with $U_A=(c,0,0,0)$ for a degenerate relativistic gas and limit values for weak and strong degeneration and for non-relativistic and ultra-relativistic case

	Non-relativistic $\frac{\mu' c^2}{kT} >> 1$	Relativistic	Ultra-relativistic $\frac{\mu' c^2}{kT} << 1$
non-degenerate ln$a$<<1	$a\exp(-\frac{\mu' c^2}{kT})\exp(-\frac{p^2}{2\mu' kT})$ Maxwell distribution	$a\exp(-\frac{\mu' c^2}{kT}\sqrt{1+\frac{p^2}{(\mu' c)^2}})$	$a\exp(-\frac{cp}{kT})$
degenerate	$\frac{1}{\frac{1}{a}\exp(\frac{\mu' c^2}{kT})\exp(\frac{p^2}{2\mu' kT})\mp 1}$	$\frac{1}{\frac{1}{a}\exp(\frac{\mu' c^2}{kT}\sqrt{1+\frac{p^2}{\mu' c^2}})\mp 1}$ Maxwell-Jüttner distribution	$\frac{1}{\frac{1}{a}\exp(\frac{cp}{kT})\mp 1}$
strongly degenerate Fermi $\ln a-\frac{\mu' c^2}{kT} >> 1$	1 for $0 \le \sqrt{2\mu kT(\ln a-\frac{\mu' c^2}{kT})}$ 0 else	1 for $0 \le \frac{p}{\mu' c} \le \sqrt{\left[\frac{\ln a}{\frac{\mu' c^2}{kT}}\right]^2-1}$ 0 else	1 for $0 \le p \le \frac{kT}{c}\ln a$ 0 else
strongly degenerate Bose $\ln a-\frac{\mu' c^2}{kT} \le 0$	$\frac{1}{\exp(\frac{p^2}{2\mu' kT})-1}$ $p \ne 0$	$\frac{1}{\exp[\frac{\mu' c^2}{kT}(\sqrt{1+\frac{p^2}{(\mu' c^2)}})-1)]-1}$ $p \ne 0$	$\frac{1}{\exp\frac{cp}{kT}-1}$ $p \ne 0$ Planck distribution for $p = h\nu/c$

The only remaining source of energy for a white dwarf is gravitational contraction, – Helmholtz fashion. That keeps the star hot in the centre, perhaps hot enough – a thousand times as hot as the sun – that it must be considered a relativistic gas. Note that the small electronic mass helps in this respect, because the relativistic coldness $\frac{\mu' c^2}{kT}$ is more than 10^3 times smaller for electrons than for nuclei, or atoms at the same temperature. Now, large speeds make for small de Broglie wave lengths, so that quantum effects should be small. However, the large gravitational pressure compresses the star to such a degree that even the small de Broglie wave lengths interfere and thus produce quantum degeneration. Therefore in a white dwarf the electron gas can perhaps be both: a relativistic gas and a quantum gas. Chandrasekhar adopted this assumption as the basis for his theory of white dwarfs. In this way he provided an application for Jüttner's formulae.

Thermal equation of state inside a white dwarf

In relativistic thermodynamics the conservation of mass is replaced by the conservation of the number of particles, and momentum and energy conservation are combined in a vector equation. We have

$$N^A{}_{,A} = 0 \quad \text{and} \quad T^{AB}{}_{,B} = 0, \quad \text{where}$$

N^A is the particle flux vector and T^{AB} is the energy-momentum tensor. The equilibrium quantities n, e, and p are related to N^A and T^{AB} as shown in the following table.

number density	energy density	pressure
$n = \frac{1}{c^2} U_A N^A$	$e = \frac{1}{c^2} U_A U_B T^{AB}$	$p - \frac{1}{3}(\frac{1}{c^2} U_A U_B - g_{AB}) T^{AB}$

In a gas in equilibrium N^A and T^{AB} are moments of Jüttner's equilibrium distribution

$$F = \frac{Y}{\frac{1}{a}\exp\left(-\frac{U_A p^A}{kT}\right) \mp 1}$$

so that we have

$$N^A = \int p^A F \frac{dp^1 dp^2 dp^3}{p_o} \quad \text{and} \quad T^{AB} = c \int p^A p^B F \frac{dp^1 dp^2 dp^3}{p_o}.$$

$\frac{dp^1 dp^2 dp^3}{p_o}$ with $p_o = \mu' c \sqrt{1 + \frac{p^2}{\mu'^2 c^2}}$ is the scalar element of momentum space, and $1/Y$ – or h^3 – determines the cell of the phase space.

For a strongly degenerate Fermi gas we thus have, cf. Table 10.1

$$n = 4\pi(\mu' c)^3 Y \int_o^x z^2 dz \quad \text{and} \quad p = \frac{1}{3} c 4\pi\, (\mu' c)^4 Y \int_o^x \frac{z^4}{\sqrt{1+z^2}} dz\,,$$

where $x = \sqrt{(kT \ln a)^2 - 1}$. It follows that p depends only on n, not on T ! An explicit form of the relation – the thermal equation of state – can be obtained, if the integrals are evaluated, so that x can be eliminated.

If relativistic effects were ignored, the square root in the integrand for p would be absent.

Insert. 10.1

Subramanyan Chandrasekhar (1910–1995)

Chandrasekhar was an astrophysicist with a particular interest in white dwarfs. As Eddington did for normal stars, he argued that inside a white dwarf the atoms are broken down into nuclei and electrons, so that there is a lot of space for the particles to move in freely, -- even when the densities are as big as described above: If the total mass of the star is big enough, however, the free space between the particles can be squeezed out, as it were. The electrons are then pushed together and the resulting compact cluster of electrons resists the gravitational pull. That equilibrium can persist even when the white dwarf cools and becomes a red dwarf and eventually, a black one. But not all stars can follow that course as we shall now see.

In part of his work Chandrasekhar assumed that the electron gas is a strongly degenerate relativistic Fermi gas.[11] In that case it was fairly easy to consider the limit of the *ultimate white dwarf* characterized by an infinite mass density at the centre and zero radius. Surely no other star could be denser and, presumably, have more mass. That ultimate white dwarf came

[11] S. Chandrasekhar: "The maximum mass of ideal white dwarfs." Astrophysical Journal 74 (1931) p. 81.
S. Chandrasekhar: "The highly collapsed configurations of a stellar mass, I and II." Monthly Notices of the Royal Astronomical Society 91 (1931) p. 456 and 95 (1935) p. 207.
See also: S. Chandrasekhar: "An Introduction to the Study of Stellar Structure" University of Chicago Press (1939). This book is available in a Dover edition, first published in 1957.

out to have a mass of approximately 1.4 solar masses, cf. Insert 10.2. This limiting mass for white dwarfs became known as the *Chandrasekhar limit.* It was confirmed by observation in the sense that no white dwarf was ever seen that has more than Chandrasekhar's limit mass.

The Chandrasekhar limit

Since the mean value of the relative molecular mass is 2, by Insert 10.1 the mass density and the pressure are given by

$$\rho = Ax^3 \qquad \text{with} \qquad A = 2\mu_o \frac{4\pi}{3}(\mu' c)^3 Y \qquad \text{and}$$

$$p = B\int_0^x \frac{z^4}{\sqrt{1+z^2}}\mathrm{d}z \qquad \text{with} \qquad B = \frac{4\pi}{3}c(\mu' c)^4 Y.$$

Therefore the momentum balance reads, see Chap. 7

$$\frac{\mathrm{d}p}{\mathrm{d}r} = -\rho G \frac{M_r}{r^2} \qquad \text{where} \qquad M_r = 4\pi\int_0^r \rho(r')r'^2\mathrm{d}r'.$$

Differentiation with respect to r and the use of the thermal equation of state, cf. Insert 10.1, provides

$$\frac{1}{r^2}\frac{\mathrm{d}}{\mathrm{d}r}\left(r^2\frac{d\sqrt{1+(\rho/A)^{2/3}}}{dr}\right) = -\underbrace{\frac{4\pi GA^2}{B}}_{1/L^2}\left(\sqrt{1+(\rho/A)^{2/3}}^{\,2} - 1\right)^{3/2}.$$

Non-dimensionalization with the unknown central value ρ_c of ρ provides

$$\frac{1}{\eta^2}\frac{\mathrm{d}}{\mathrm{d}\eta}\left(\eta^2\frac{\mathrm{d}}{\mathrm{d}\eta}\underbrace{\sqrt{\frac{1+(\rho/A)^{2/3}}{1+(\rho_c/A)^{2/3}}}}_{\Phi(\eta)}\right) = -\left(\underbrace{\sqrt{\frac{1+(\rho/A)^{2/3}}{1+(\rho_c/A)^{2/3}}}^{\,2}}_{\Phi^2(\eta)} - \frac{1}{1+(\rho_c/A)^{2/3}}\right)^{3/2},$$

where $\eta = \sqrt{(1+(\rho_c/A)^{2/3})}\,\frac{r}{L}$ is the dimensionless radius.

We investigate the case that ρ_c is infinite. Presumably that assumption characterizes the *ultimate white dwarf* in the sense that no other one could be denser and have more mass. In that case it is easy to solve – numerically – the differential equation for the central values $\Phi(0) = 1$ and $\Phi'(0) = 0$ and one obtains the graph shown in Fig. 10.1. On the surface of the star, at $r = R$, we must have $\rho = 0$, hence $\Phi = 0$. According to the figure, that value occurs for $\eta = 6.9$, so that R is zero, but the mass is not. It can be calculated as follows:

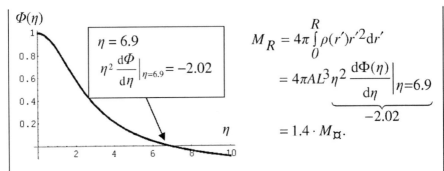

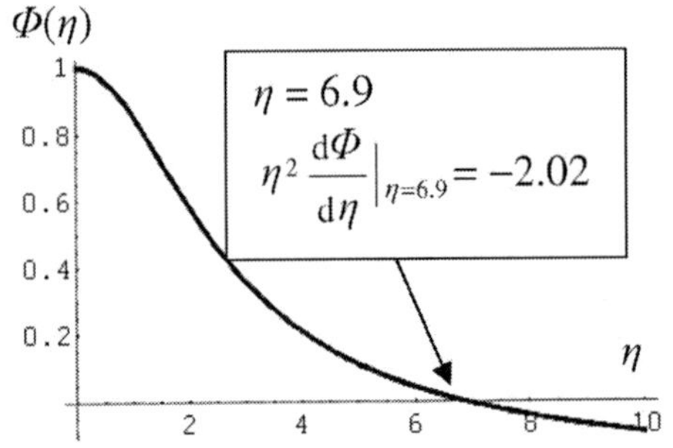

$$M_R = 4\pi \int_0^R \rho(r')r'^2 \mathrm{d}r' = 4\pi A L^3 \eta^2 \underbrace{\left.\frac{\mathrm{d}\Phi(\eta)}{\mathrm{d}\eta}\right|_{\eta=6.9}}_{-2.02} = 1.4 \cdot M_{☼}.$$

Fig. 10.1. A kind of density distribution in the ultimate white dwarf

The last step makes use of the differential equation in the form

$$\rho = -AL^2 \frac{1}{r^2}\frac{\mathrm{d}}{\mathrm{d}r}\left(r^2 \frac{\mathrm{d}\sqrt{1+(\rho/A)^{2/3}}}{\mathrm{d}r} \right).$$

Obviously, degeneration of the electron gas has played a decisive role in the forgoing analysis. It is less clear that the relativistic square root in the equation for p is essential for the result. However, it is! Without that relativistic contribution there is no mass limit.

Insert 10.2

The usual interpretation of the Chandrasekhar limit is that the electron gas cannot withstand the gravitational pull of bigger masses than 1.4 $M_{☼}$. It is assumed that under great pressure the electrons are pushed into the protons of the iron nuclei to form neutrons. The star thus becomes a *neutron star*, with a truly enormous mass density: 10^{15} times the already large density of a white dwarf. Neutron stars have their own mass limit $-3.2\ M_{☼}$ – according to a theory presented by J. Robert Oppenheimer (1904–1967) in 1939. If a star is bigger than that, – and does not get rid of the excess mass by nova- or supernova-explosions – it collapses into a black hole, at least according to current wisdom. There seems to be no conceivable mechanism to stop the collapse. It is tempting to pursue the matter further in this book. However, there is a touch of science fiction in the subject and I desist, – with regret.

Chandrasekhar has left his mark in several fields of physics. In his autobiography he says that he was … *motivated, principally, by a quest after perspectives…compatible with my taste, abilities and temperament.* Stellar dynamics was the subject of only the first such quest. Others followed: • Brownian motion, • radiative transfer, • hydrodynamic stability, • relativistic astrophysics, • mathematical theory of black holes. Whenever he found that he understood the subject, he published one of his highly readable books, – in his words: *a coherent account with order, form, and structure.* Thus he has left behind an admirable library of monographs for students and teachers alike. His work on white dwarfs, but also his lifelong exemplary dedication to science, was rewarded with the Nobel prize in physics in 1983, fifty years after he discovered the Chandrasekhar limit.

The maximal mass of a white dwarf is not alone in having been named after Chandrasekhar. There is also the NASA X-ray observatory which is called Chandrasekhar observatory, and a *minor planet,*– one of about 15000 – which was named Chandra in 1958.

Fig. 10.2. Subrahmanyan Chandrasekhar

Maximum Characteristic Speed

After Jüttner there was a period of stagnation in the development of relativistic thermodynamics. To be sure, there was *some* interest, and in 1957 John Lighton Synge (1897–1995) streamlined Jüttner's results in a neat small book[12] which, however, did not significantly add to previous results.

Also Eckart provided a relativistic version of thermodynamics of irreversible processes,[13] in which he improved Fourier's law of heat conduction by accounting for the inertia of energy, cf. Chap. 8. However, his differential equation for temperature was still parabolic so that the *paradox of heat conduction* persisted. Understandably that paradox has irritated relativists more than it did non-relativistic physicists. After all, if no atom, or molecule can move faster than the speed of light, heat conduction should

[12] J.L Synge: "The Relativistic Gas." North Holland, Amsterdam (1957).

[13] C. Eckart: "The thermodynamic of irreversible processes III: Relativistic theory of the simple fluid." loc. cit.

not be infinitely fast. This problem was the original motive for Müller to develop extended thermodynamics, cf. Chap. 8, and its relativistic version.[14] Shortly afterwards, Israel[15] published a very similar theory and, eventually, it was shown by Boillat and Ruggeri[16] that extended thermodynamics of infinitely many moments predicts the speed of light for heat conduction. Thus the paradox was resolved; the field is conclusively explained by Müller in a recent review article.[17]

A decisive step forward in the general theory was done by N.A. Chernikov in 1964 [18] when he formulated a relativistic Boltzmann equation. Let us consider this now.

Boltzmann-Chernikov Equation

I have already mentioned the elegant four-dimensional formulation which is now standard in relativity. It was introduced by Hermann Minkowski (1864–1909). Minkowski had taught Einstein in Zürich and later he became the most eager student of Einstein's paper on special relativity. He suggested that the theory of relativity makes it possible to take time into account as a kind of fourth dimension and he introduced the *distance* $\mathrm{d}s$ between two events at different places and *different times*[19]

$$\mathrm{d}s^2 = \underbrace{g'_{AB}\mathrm{d}x'^A\mathrm{d}x'^B = c^2\mathrm{d}t'^2 - (\mathrm{d}x'^1)^2 - (\mathrm{d}x'^2)^2 - (\mathrm{d}x'^3)^2}_{\substack{\text{in a Lorentz frame} \\ \text{with coordinates } ct', x'^a}}.$$

---

[14] I. Müller: "Zur Ausbreitungsgeschwindigkeit ..." Dissertation (1966) loc. cit.
A streamlined version of relativistic extended thermodynamics may be found in:
I-Shih Liu, I. Müller, T. Ruggeri: "Relativistic thermodynamics of gases." Annals of Physics 169 (1986).

[15] W. Israel: "Nonstationary irreversible thermodynamics: A causal relativistic theory." Annals of Physics 100 (1976).

[16] G. Boillat, T. Ruggeri: "Moment equations in the kinetic theory of gases and wave velocities." (1997) loc.cit.

[17] I. Müller: "Speeds of propagation in classical and relativistic extended thermodynamics." http:/www.livingreviews.org/Articles/Volume2/1999-1mueller.

[18] N.A. Chernikov: "The relativistic gas in the gravitational field." Acta Physica Polonica 23 (1964).
N.A. Chernikov: "Equilibrium distribution of the relativistic gas." Acto Physica Polonica 26 (1964).
N.A. Chernikov: "Microscopic foundation of relativistic hydrodynamics." Acta Physica Polonica 27 (1964).

[19] H. Minkowski: "Raum und Zeit." [Space and time] Address delivered at the 80th Assembly of German Natural Scientists and Physicists, at Cologne. September 21st, 1908.
The address has been translated into English and is reprinted in "The Principle of Relativity. A collection of original memoirs on the special and general theory of relativity." Dover Publications pp. 75–91

In this manner the tensor g'_{AB}, whose invariance defines the Lorentz frames, may be interpreted as a *metric tensor* of space-time. Its components in a arbitrary frame $x^A = x^A(x'^B)$ can be calculated from

$$g_{AB} = \frac{\partial x'^C}{\partial x^A}\frac{\partial x'^D}{\partial x^B} g'_{CD}.$$

In particular, for a rotating frame – on a carousel (say) – with coordinates (ct, r, θ, z) given by

$$t' = t, \;\; x'^1 = r\cos(\theta + \omega t), \; x'^2 = r\sin(\theta + \omega t), \;\; x'^3 = z$$

the metric tensor reads

$$g_{AB} = \begin{pmatrix} 1 - \frac{\omega^2 r^2}{c^2} & 0 & -\frac{\omega r}{c} & 0 \\ 0 & -1 & 0 & 0 \\ -\frac{\omega r}{c} & 0 & -r^2 & 0 \\ 0 & 0 & 0 & -1 \end{pmatrix}.$$

The metric tensor has some significance, because it allows us to write the equation of motion of a free particle, whose orbit is parametrized by τ, in the form

$$\frac{d^2 x^B}{d\tau^2} = -\Gamma^B_{AC}\frac{dx^A}{d\tau}\frac{dx^B}{d\tau}, \qquad \text{where} \quad \Gamma^B_{AC} = \frac{1}{2} g^{BD}\left(\frac{\partial g_{DA}}{\partial x^C} + \frac{\partial g_{DC}}{\partial x^A} - \frac{\partial g_{AC}}{\partial x^D}\right).$$

Indeed, in a Lorentz frame, with $\Gamma^B_{AC} = 0$, the solution of this equation is a motion in a straight line with constant velocity, which is the defining feature of an inertial frame. The parameter τ is usually chosen as the *proper time* of the moving particle, i.e. the time read off from a clock in the momentarily co-moving Lorentz frame. With that, the equation of motion may be written in the form

$$\frac{dp^B}{d\tau} = -\frac{1}{\mu'}\Gamma^B_{AC} p^A p^B, \quad \text{where} \quad p^A = \frac{dx^A}{d\tau}$$

is the four-momentum of the particle as before.

The equation of motion represents the equation of a geodesic in space-time. This is a nice feature, much beloved by theoretical physicists, because it supports their predilection for a specious geometrical interpretation of the theory of relativity. The notion was useful for Einstein, when he developed the theory of general relativity; but most often it is used to confuse laymen with talk about *curved space*, etc.

Of course, nobody will try to solve the equation of the geodesic in its general form in order to calculate the orbit of a free particle. It is so much easier to solve it in a Lorentz frame and transform the straight line obtained there to an arbitrary frame.

The relativistic – non-quantum – formulation of the Boltzmann equation was derived in a series of three remarkable papers by N.A. Chernikov. It is an integro-differential equation for the relativistic distribution function $F(x^A, p^a)$ which reads

$$p^A \frac{\partial F}{\partial x^A} - \Gamma^d_{AB} p^A p^B \frac{\partial F}{\partial p^d} = \int (F(p'^C)F(q'^C) - F(p^C)F(q^C) h < pq > \mathrm{d}e\mathrm{d}Q.$$

Comparison with the classical Boltzmann equation, cf. Chap. 4, easily identifies the individual terms. I do not go into that, other than saying that

- the term with Γ represents the acceleration of a particle between two collisions,[20] and
- the collision term on the right hand side vanishes for the Maxwell-Jüttner distribution because of conservation of the energy and momentum vector p^A in the collision.

Chernikov uses the equation for the formulation of equations of transfer for moments of the distribution function and he concentrates on 13 moments, which is rather artificial for a relativistic theory; it is more appropriate to include the dynamic pressure and thus come up with a theory of 14 moments.[21] But we shall not pursue this question here, because so far – apart from the finite characteristic speeds – the multi-moment theory has not provided any suggestive results that go beyond Eckart's reformulation of the Fourier law, see Chap. 8. Let us concentrate on equilibrium instead:

Seeing that the collision term vanishes for the Maxwell-Jüttner distribution, we must ask whether the Boltzmann-Chernikov equation is satisfied by that distribution, or what conditions on the fields $a(x^B)$, $T(x^B)$, and $U^A(x^B)$ are required by the equation. Insertion of the distribution leads to the requirements

$$\frac{\partial a}{\partial x^A} = 0 \quad \text{and} \quad \left(\frac{U_B}{kT}\right)_{;A} + \left(\frac{U_A}{kT}\right)_{;B} = 0,$$

where the semi-colon denotes covariant derivatives.

---

[20] The possibility of such a term was ignored in Chap. 4, because I wished to be brief. The term is only present in a non-inertial frame.

[21] See: I. Müller, T. Ruggeri: "Rational Extended Thermodynamics." loc. cit.

Since a is a function of n and T, it follows that a temperature gradient must exist in equilibrium, if there is a density gradient. That conclusion may be made more concrete by exploiting the second condition for the special case of a gas at rest on a carousel. We obtain

$$\frac{T}{\sqrt{g_{00}}} \sim \text{homogeneous or, see above: } \frac{T(r)}{\sqrt{1-\frac{\omega^2 r^2}{c^2}}} \sim \text{homogeneous.}$$

This result is eminently plausible, because it reflects the inertia of the thermal energy in the field of the centrifugal potential $\omega^2 r^2$. Indeed, if energy has mass – and weight – it should be subject to sedimentation, as it were, by centrifugation.

Einstein has postulated – in his general theory of relativity – that inertial forces and gravitational forces are equivalent. Accordingly non-homogeneous temperature fields are also created by gravitational fields – not only by centrifugal fields – because they lead to stratification of mass density. I have already commented on that aspect in the context of Eckart's relativistic paper.

In view of the following argument, I should like to stress that the last relation does *not* imply a transformation formula for the temperature. It represents a property of the scalar temperature field as a solution of the energy balance equation in a centrifugal force field.

Ott-Planck Imbroglio

In 1907 the theory of relativity was new. A fundamental change had occurred in mechanics, and physics in the immediate aftermath was in a state of flux. The extension of the new concepts to thermodynamics was clearly desirable. Everything seemed possible and so Planck[22] came up with the idea to modify the Gibbs equation. Einstein[23] elaborated on that idea and introduced a working term $-q\mathrm{d}G$ into the heating of a body moving with the

---

[22] M. Planck: "Zur Dynamik bewegter Systeme." [On the dynamics of moving systems] Sitzungsberichte der königlichen preußischen Akademie der Wissenschaften (1907).

Printed version: Annalen der Physik 26 (1908) p. 1.

[23] A.. Einstein: "Über das Relativitätsprinzip und die aus demselben gezogenen Folgerungen." [On the principle of relativity and the conclusions drawn from it] Jahrbuch der Radioaktivität und Elektronik 4 (1907) pp. 411–462. Reprinted in: "Albert Einstein, die grundlegenden Arbeiten." [Albert Einstein, the basic works] K.v. Meyenn (ed) Vieweg Verlag (1990).

In the reprinting the modified Gibbs equation is misprinted: It says dQ instead of dG.

speed q. G is the momentum; it includes a relativistically small term, because the mass is $m' + \frac{U}{c^2}$. The modified Gibbs relation thus reads

$$TdS = dQ = dU + pdV - qdG\,.$$

The transformation of dU, p,dV, and dG between the moving body and the body at rest were known and thus Einstein produced the relation

$$dQ = \sqrt{1 - \frac{q^2}{c^2}}\, dQ_0$$

between the heating of the moving body and the heating of the body at rest.

Now Planck had already argued that the entropy of the body should be unaffected by motion, and therefore the second law written as dq = TdS seemed to require

$$T = \sqrt{1 - \frac{q^2}{c^2}}\, T_0\,.$$

That relation was later rephrased by epigones of the argument in the words: *A moving body is cold.*

On the surface the argument appears plausible. It does ignore, however, the fact that the Gibbs relation is a relation for a body *at rest*: The heating consists of the *non-convective* part of the energy flux and the internal energy is the *non-convective* part of the energy. The power, or working of the force dG has no place in the Gibbs equation therefore, or it should not have.

Also, the heating of a body in the Gibbs equation is the integral of the heat flux over the surface. And relativistically the heat flux forms three components of the energy-momentum tensor. It is that fact which determines the transformation of the heating, not its position in the Gibbs equation.

None of the serious physicists in the following years and decades followed Planck and Einstein in this precarious thermodynamic argument, neither Eckart, nor Synge, nor Chernikov. Consequently one might have thought that the argument was discarded as a valiant, perhaps, though erroneous early attempt on relativistic thermodynamics.

Not so, however! In 1962, H. Ott[24] revisited the argument on a slightly different basis involving Joule heating, and he came to the conclusion that

[24] H. Ott: "Lorentz-Transformation der Wärme und der Temperatur." [Lorentz transformation of heat and temperature] Zeitschrift für Physik 175 (1963) 70–104.

$$\mathrm{d}Q = \frac{\mathrm{d}Q_0}{\sqrt{1-\frac{q^2}{c^2}}} \quad \text{holds, hence} \quad T = \frac{T_0}{\sqrt{1-\frac{q^2}{c^2}}},$$

such that: *A moving body is hot.*

Serious people in the field ignore the subject, which was appropriately termed the *Ott-Planck imbroglio* by Israel and Stewart.[25] However, the farce continues and Peter Thomas Landsberg[26] – himself an enthusiastic contributor to the imbroglio – cites papers on the subject of temperature transformation in special relativity as recent as the late 1990's.[27]

---

[25] W. Israel, J.M. Stewart: "On transient relativistic thermodynamics and kinetic theory II." Proceeding of the Royal Society London Ser. A 365 (1979).

[26] www.maths.soton.ac.uk/staff/Landsberg

[27] I have a personal memory of all this: Ott's paper was in the process of publication when he died. So the proof sheets – already adorned with the multi-coloured marks of the copy-editor of the pre-computer era – where sent to Josef Meixner for his evaluation. Meixner was my advisor at the time and he gave the paper to me, his most junior assistant. Naturally, perhaps, I thought that my opinion was being requested. And so – having already studied Jüttner's papers and Synge's booklet – I put my precocious and very junior thumb down on the paper. But Ott had been an important member of the German Physical Society, and he was not to be embarrassed, not even posthumously, and certainly not by the Zeitschrift für Physik. So the paper was published, and the imbroglio took another turn.

11 Metabolism

If the truth were known, thermodynamics would be seen as explaining little about the details of life functions in animals and plants, at least compared to what there is to be explained. This is no different than with engines: Thermodynamics cannot provide a recipe for their construction, or give information about where and how to arrange seals and boreholes for lubrication, and how to operate the valves and where to install them. What thermodynamics *can* do about engines is to give an account of the balance of in- and effluxes of mass, momentum, energy and entropy, and that is essentially what it can also do about life. For the engine that task has been done satisfactorily; for animals and plants maybe there remains something to be done.

Having said this, I hasten to stress that, what thermodynamics *is* able to provide, is good enough to refute esoteric theories, and to convince people with an open mind that nothing unnatural occurs in the living body: No *vitalistic force* of old, nor Niels Bohr's *complimentarity of life and physics,* akin to the wave-particle dualism of quantum mechanics.[1]

I have previously – cf. Chap. 4 – warned against an over-interpretation of entropy as a measure of disorder and I stress that caution again. To be sure, an animal definitely seems more ordered than the sum of its atoms, loosely distributed, and it does probably have a lower entropy. But then, what *is* the entropy of an animal? Or let us ask the easier question: What is the entropy of a molecule like hemoglobin, one of the simpler proteins with only about 500 amino acids? Maybe molecular biologists can come up with an answer; if so, I do not know about it. But I do know that surely it must be a case of simplism when Schrödinger says[2] that animals maintain their highly ordered state, because they eat highly ordered food. Indeed, before the animal body makes use of the food in any way, – and sets about to create order – it breaks the food down to much less ordered fragments than those which it ingests.

---

[1] In his later years Bohr expressed doubts that life functions can be reduced to physics and chemistry. See: N. Bohr: "Atomphysik und menschliche Erkenntnis." [Atomic physics and human knowledge] Vieweg Verlag, Braunschweig (1985).

[2] E. Schrödinger: "What is life? The physical aspect of the living cell" Cambridge: At the University Press, New York: The Macmillan Company (1945) p. 75.

In writing this chapter on metabolism I disregard Schrödinger's warning that a *scientist is usually expected not to write on any topic of which he is not a master.*[3] But then, Schrödinger did not heed that warning himself. And the subject *is* interesting, and it seems to be replete with unsolved problems of a quantitative nature. Therefore it is easy to yield to the temptation to write about it, albeit in a layman's manner.

Carbon Cycle

One of the truly mind-expanding discoveries of all times, concerning life and life functions, was the observation that carbon, hydrogen and oxygen cycle through living organisms, driven by solar radiation: Plants use water from the soil and carbon dioxide from the air to produce their tissue and they release oxygen. Animals on the other hand breathe oxygen and use it to break down plant tissue. In the process they release carbon dioxide and water. The plants perform their task only in the light.

Jan Baptista van Helmont (1577–1644) was an alchemist on the verge of becoming a chemist or, perhaps, a biochemist. On the one hand he claimed to have seen and used the *philosopher's stone* – the hypothetical ultimate tool of alchemy – but on the other hand he was keen enough as an experimenter to see that water was essential for plant growth, while soil was not, or not to the same degree. Helmont did not recognize the importance of carbon dioxide for plants, even though he actually discovered that gas, which he called *gas sylvestre*, i.e. wood gas, because he had found that it was released by burning wood. It took another hundred years before the significance of that observation was recognized by Stephen Hales (1677–1761). Carbon dioxide has originally entered the wood from the air surrounding the leaves of a plant, thus furnishing the second component – after water – that is essential for plant growth.

Another generation later Joseph Priestley (1733–1804), one of the discoverers of oxygen, noticed that oxygen was used up in the air by breathing and that, plants can restore the freshness of used-up air, obviously by giving off oxygen. These observation were all couched in the language of the phlogiston theory, – even then obsolete[4] –, but Jan Ingenhousz (1730–1799) was able to penetrate the verbiage and to see a broad scheme of balance in nature: Plants consume the carbon dioxide of the air and

---

[3] Ibidem. p. vi.

[4] The phlogiston theory is the forerunner of Lavoisier's caloric theory, see Chap. 2. In the 18th century a weightless fluid called phlogiston was supposed to flow from a body when that body burns, or rusts, or is just cooling. As far as burning and rusting was concerned, Lavoisier refuted the concept, because he showed that both phenomena are due to the combination of a body with oxygen. Heating or cooling was another matter. Lavoisier maintained that heat was indeed a weightless fluid which he called caloric.

release oxygen, while animals breathe oxygen and give off carbon dioxide. In this manner there is a stable balance. Ingenhousz showed that the plants need light in order to build up their tissue. That is why we now call the process *photosynthesis.*

Ingenhousz, who was first to discover this grand scheme, is not very much known nowadays, but he was a celebrity in his time. Being a physician, he became an early expert on inoculation, particularly smallpox inoculation, and he travelled all over Europe serving the members of royal families with smallpox, as it were, – in small doses!

Respiratory Quotient

It was the eminent chemist Berzelius, cf. Chap. 4, who introduced the distinction of organic and inorganic substances in 1807. The former were the substances of life, and – in Berzelius's view – they called for a separate type of chemistry from the chemistry of elements and of their simple stoichiometric compounds that were the stock-in-trade of his own work and everybody else's at the time. There were vague notions that a *vis viva*, a vitalistic force, was involved in living bodies, *a spark of life*. Berzelius himself and his followers even conceived of a strict barrier between the chemistries of life and non-life.

Seeing and appreciating the difference between rock and lizard, as it were, one must admit that there is a certain plausibility to the idea and it took at least half a century to refute it. This required an improved knowledge of the life functions, and exact measurements. The first organic process to be thoroughly investigated was respiration. Even Lavoisier and Henry Cavendish (1731–1810) had understood that respiration supported a kind of combustion in the body of animals by which the oxygen of the air was partially consumed and converted to carbon-dioxide and water. Obviously therefore, whatever substance, or substances fed the combustion had to contain carbon and hydrogen. Beyond that, the substances were unknown chemically, so that no quantitative conclusions could be drawn. However, it stood to reason that, whatever it was that *burned* had to be supplied to the animal – or man – with the food.

Early in the 19th century it became clear upon analysis of the food of animals that there were three main types

- carbohydrates • lipids • proteins.

The carbohydrates form the chief components of cereals, and of fruit and vegetables. They are of different types but closely related and, for the moment, we take sugar – more precisely glucose – as their representative.

The chemical formula is $C_6H_{12}O_6$, so that Gay-Lussac – one of the discoverers of the thermal equation of state of ideal gases – could assume that glucose consisted of 6 carbon atoms strung together and a water molecule attached to each one in the manner of hydrates. The structure is more complex, as we know now, see Fig. 11.1, but Gay-Lussac's concept led to the misnomer *carbohydrate,* which is here to stay. Actually, what we eat is not glucose itself, but rather something like starch or other substances which are built up from several or many glucose molecules. The large molecules are held together by glycoside bonds, having shedded water molecules in a process that is called *condensation* – obviously because it produces liquid water.

Again lipids, or fats are of varied types. Their pioneer was Michel Eugène Chevreul (1786–1889). Fats are used in manufacturing soap and as a young man Chevreul was involved in that business. He was able to isolate different insoluble organic acids – also called carbonic acids, or fatty acids – like stearic acid, palmitic acid and oleic acid. Lipids themselves result from the carbonic acids by esterification with glycerol $C_3H_8O_3$, giving off water, i.e. undergoing *condensation* cf. Fig. 11.1. A typical representative is oleine $C_{57}H_{104}O_6$, an ingredient of olive oil, or also of blubber, i.e. whale oil.

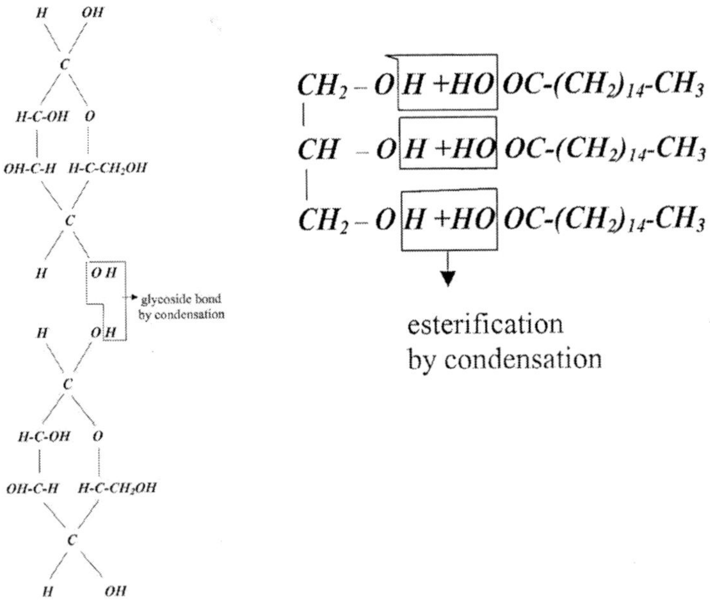

Fig. 11.1. *Left*: Two glucose molecules combining by a glycoside bond. *Right*: Olein. Glycerol combining with oleic acids

While carbohydrates and lipids contain only carbon, hydrogen and oxygen, the third type of food-stuff – of which egg-white is the best-known representative – also contains nitrogen, a little sulphur and, sometimes, still less phosphorus. The molecules are polymers formed from amino-acids which are bound together by a peptide link, again a bond formed by *condensation*. The detailed structure is too complex and varied to be easily characterized. In 1838 Gerardus Johannes Mulder devised a model molecule of 88 individual atoms which he hoped might be used to build up other *albuminous* substances. The word albuminous is derived from albus = white in Latin; it is sometimes used as a generic name for substances like egg-white.[5] More often these substances are called proteins in English, because Mulder called his model molecule *Protein,* from Greek, meaning *of first importance.* Otherwise the model sank into oblivion; it was too simple.

Now, if indeed food was involved in a combustion inside animals, and if CO_2 and H_2O were the reaction products, the reactions for carbohydrates and lipids had to obey the stoichiometric formulae

$$\tfrac{1}{6}\; C_6H_{12}O_6 + O_2 \rightarrow CO_2 + H_2O$$
$$\tfrac{1}{80} C_{57}H_{104}O_6 + O_2 \rightarrow \tfrac{57}{80} CO_2\, \tfrac{52}{80} H_2O \,.$$

The volume ratio of exhaled CO_2 to inhaled O_2 is called the *respiratory quotient,* abbreviated as RQ. Thus the stoichiometric formulae imply

RQ = 1	for the carbohydrate
RQ = 0.71	for the lipids,

since both CO_2 and O_2 are ideal gases. The value for proteins lies in-between, at roughly RQ = 0.8.

So, if chemistry is involved in respiration, the RQ should lie between 0.7 and 1. And indeed, the chemist Henri Victor Regnault[6] put animals in a cage and carefully measured the oxygen input and the carbon-dioxide output and found the ratio to be right. What is more, if he fed the animals a diet of carbohydrates, the RQ tended to one, while on a fat-rich diet it tended to 0.7. This was later confirmed for a *man* in a cage by the chemist Max von Pettenkofer (1818–1901) – the founder of scientific hygiene. All of this provided strong evidence that there was no *vis viva* involved, at least not in respiration.

[5] Actually, in German proteins are called "Eiweisse" [egg whites].

[6] We have met him before in connection with his 700 page-long memoir of careful measurements of vapour properties, cf. Chap. 3.

Metabolic Rates

So what about the energy to be gained from food? Was the first law satisfied, or did the intervention of a *vitalistic force* render thermodynamic laws invalid in the field of nutrition?

If sugar and fat and the mix of proteins normally eaten by an animal are burned in a calorimeter they provide heats of reaction as follows[7]

$$\begin{matrix} \text{sugar} \\ \text{proteins} \\ \text{lipids} \end{matrix} \qquad \Delta h_R = \begin{cases} 17.1\cdot 10^3 \\ 23.6\cdot 10^3 \\ 39.5\cdot 10^3 \end{cases} \frac{\text{kJ}}{\text{g}}.$$

The question is whether these values are also relevant when food is consumed by eating.

The experimental investigation was infinitely more difficult than the determination of the respiratory quotient. First of all, it requires calorimetric studies which are notoriously difficult even in the best of circumstances. Secondly, here the feces had to be analysed in order to find out which proportion of the ingested food remained unconsumed by the body. And a quantitative urine analysis had to determine the urea content, which is the substance by which the body gets rid of the nitrogen ingested with proteins. Naturally the RQ was also part of the investigation.

The person who did all this carefully was the physiologist Max Rubner (1854–1932). He presented his findings in a report[8] in which he came to the conclusion that the law of conservation of energy was maintained in nutrition just as punctiliously as in ordinary combustion. By now scientists were ready to believe that physical laws govern both: life and non-life.

Once this was understood, the distinction between organic and inorganic chemistry began to lose its original meaning. Organic chemistry became the branch that deals with carbon compounds.

The chemical changes that take place in animals and humans are called *metabolism*; from Greek: to *rearrange*. The metabolic rate may be measured in Watt – just like the power of a heat engine. The maximal metabolic rate that a person can achieve is approximately 700W, but that can only be sustained for a few minutes. So what is the minimum, the *basal metabolic rate*?

The basal metabolic rate is abbreviated as BMR; it can be achieved by a person lying down in a comfortably warm room, having fasted for some

---

[7] We are now back from the mol to ordinary mass units. The use of the mol in organic chemistry with it huge molecules would be totally impractical. Not so, however, for the glucose synthesis and the glucose decomposition, see below.

[8] M. Rubner: "Gesetze des Energieverbrauchs bei der Ernährung." [Laws of energy consumption in nutrition] (1902).

time and being mentally relaxed. In that case we measure a BMR of 50W for a typical adult man; that is the rate at which we need to feed him to keep him alive. A normally active person may need approximately twice that amount. And he or she emit this power as heat, which is why a crowded room needs no heating.

Digestive Catabolism

So far so good. But the fact remains that there is a lot of difference when food is burned in a fire or when it is consumed in the body. And indeed, the direct reaction between the sugar (say) and oxygen involves so large an *activation energy* that it takes an open flame to start it. This is not feasible in the body, of course. In the body the energy barrier must be *bypassed* by suitable catalysers rather than overcome by brute force, i.e. heat in this case. The catalysers were originally called *ferments.* Later – when their nature became clearer – they were called *enzymes*; and they are proteins. Reading about biochemistry, one gets the impression that we do not know much about body thermodynamics, when all we do know is that carbohydrates, or lipids, or proteins *burn* to give CO_2 and H_2O. The real question is how the body goes about this, and that makes biochemistry a science of enzymatic catalysis. Having said this, I hasten to add that in the sequel, although we shall always be dealing with enzyme-catalysed reactions, we shall largely ignore the enzymes; and we are able to do that, since presumably – or by definition – the catalysers do not contribute to the energies and entropies of the reactants and resultants.

The most evident difference between the burning in an open fire and burning inside animal bodies is that the latter occurs slowly and at body temperature. In fact, it is common knowledge that human life is severely jeopardized when a person *has a temperature* beyond 42°C. The reason for this high sensitivity of organic material against heat was discovered by Linus Carl Pauling (1901–1994), who suggested in 1936 that the proper functioning of proteins (say) depended to a large extent on weak *hydrogen bonds.* Such bonds provide a precarious stability to organic macromolecules when they are folded in a particular fashion. Pauling even envisaged helical protein molecules and thus became a forerunner of the biochemistry of the genetic code.[9]

As we eat them, starch, lipids and proteins have no chance to arrive where we need their structural units, the glucose, fatty acids and amino acids: We do *not* need them in the digestive tract but rather inside the body tissue, – in the blood, the liver, etc. The large molecules of food must be broken down before they can be absorbed by the tissue, and that break-

[9] The notion of molecular helicity helped Francis Harry Compton Crick (1916–2004) and James Dewey Watson (1928–) to uncover the shape of nucleic acids (DNA).

down happens during the digestive *catabolism*. Catabolism is the Greek word for break-down. Let us take starch as an example, which is essentially a long chain of glucose molecules.

Of course, it is common knowledge that the stomach contains acid juices, and they might go a long way to break up the starch into glucose. The study of gastric digestion begins in the Wild West in the year 1819 where William Beaumont (1785–1853) was surgeon of a border post in northern Michigan. One of his patients had received a bullet wound that left him with a fistula – an opening – leading to the stomach. Thus Beaumont was able to study the changes which the food undergoes in the stomach, and he did so with so much enthusiasm that the patient eventually ran away from him. That was a wise decision on the part of the patient, because away from his doctor he lived to the old age of 82 years,[10] always with the fistula.

Later, and in a different part of the world, the physiologist Claude Bernard (1813–1878) created fistulae artificially in different parts of the digestive tract of animals. He was heavily attacked for this by the anti-vivisectionists of the day, including his own wife, who left him over the issue. However, Bernard was able to discover that digestion does not exclusively happen in the stomach. By inserting foodstuffs into the small intestine he showed that the major part of the digestion takes place there, under the influence of the secretions of the pancreas, the large gland situated below the stomach.

As time went on, the *enzymes* were discovered and their nature as proteins with very specific capacities to catalyse reactions. Digestive enzyme activity begins actually in the mouth, where the saliva contains the enzyme *amylase* which breaks up starch, – or helps water to break up the glycoside links between the glucose molecules that form starch. This is why bread, if kept in the mouth long enough, develops a distinctive sweet taste. Further down the digestive track other enzymes pitch in, so that, when the small intestine is left, the food is largely split into its structural units: Not only starch into glucose, but also lipids into fatty acids, and proteins into amino acids. Whatever is not broken up at that point is excreted. Chemically speaking the break-up occurs through enzyme-assisted *hydrolysis*, the insertion of water molecules between the structural units of the macromolecules, or the reverse of *condensation*. Hydrolysis breaks up the glycoside- and ester- and peptide-bonds in the food. These are exothermic processes, although the heats of reaction are small.

That is the first step of *catabolism*, the food break-down. Now, the small break-down products, viz. glucose, fatty acids and amino acids are able to pass the intestinal membranes out of the digestive tract and into the body tissue itself, where they are decomposed further; remember that we must end up with CO_2 and H_2O – and urea.

---

[10] I. Asimov: "Biographies..." loc.cit. p. 268.

Tissue Respiration

The discovery of the modes of break-down of glucose inside the body tissue occurred in the first half of the 20th century. To a non-chemist like myself it represents the successful assembling of the most amazing inventive puzzle, based on the flimsiest evidence. In the beginning it was known that glucose (say) enters the tissue through the intestinal walls and that oxygen enters the blood through the lungs and is carried to the body cells by hemoglobin, the stuff that gives blood its red colour. But how do those two components come together in order to react and liberate the energy and consume entropy according to the stoichiometric equation, see above

$$C_6H_{12}O_6 + 6O_2 \rightarrow 6CO_2 + 6H_2O \qquad \begin{aligned} \Delta h_R &= -2798 \tfrac{\text{kJ}}{\text{mol}} \\ \Delta s_R &= 241 \tfrac{\text{J}}{\text{molK}} \end{aligned}$$

such that the Gibbs free energy – which is the essential quantity – decreases by 2873 $\frac{\text{kJ}}{\text{mol}}$, if the reaction occurs at the body temperature of 37°C.

Actually it turns out that the glucose molecule is first decomposed into two lactic acid molecules $C_3H_6O_3$ before the interesting things happen. Therefore we rephrase the above question and ask how lactic acid reacts with oxygen to form CO_2 and H_2O.

The problem was approached from opposite ends: The consumption of oxygen and the lactic acid oxidation. Both occur separately so that lactic acid and oxygen never get together directly chemically. The early champions of the discovery were the chemists Heinrich Otto Wieland (1877–1957) and Otto Heinrich Warburg (1883–1970) and both engaged in a fruitful scientific controversy.

Warburg had invented a manometer that could be used to measure the uptake of oxygen by tissue and he observed that the oxygen combined with heme enzymes. He did not know what the oxygen was doing there, but his insight and experimental acumen were rewarded with the 1931 Nobel prize. Wieland on the other hand recognized that the oxidation of lactic acid proceeds by dehydrogenation, i.e. the splitting-off of two hydrogen atoms from the organic molecule. Subsequently the two bonds left free in lactic acid – by the departure of the hydrogen atoms – join to form a double bond $C = O$ inside the molecule – a keto group – which, with water, is converted to a CO_2 molecule plus another pair of hydrogen atoms. There remains acetic acid CH_3COOH as the organic compound to be broken down further.

After Wieland, one of Warburg's students, Hans Adolf Krebs (1900–1981) – Sir Hans Adolf after 1958 – took up the matter of dehydrogenation and invented the *Krebs cycle* which can attach an acetic acid molecule to an enzyme and grind it down to individual *H*-atoms and CO_2 and then return and be ready to accept the next acetic acid molecule for grinding down, etc. The overall formula – starting from lactic acid – reads

$$C_3H_6O_3 + 3H_2O \rightarrow 3CO_2 + 12H \ .$$

The six pairs of hydrogen atoms are handed down a sequence of enzymes with which they build tighter and tighter bonds, before they reach oxygen and form water. The energetic downward steps are such that each hydrogen pair *activates* three adenosine tri-phosphate molecules. These so-called ATP's are the molecular energy carriers and we shall describe them and discuss their action in a short while.

Fig.11.2. Wieland, Warburg and Krebs, pioneers of intermediary metabolism

Before that, however, let it be said that the Krebs cycle is not only involved in glycolysis, the breaking up of sugar, but also in the catabolism of fatty acids and of amino acids. Fatty acids and amino acids are first broken down to acetic acid which can then enter the Krebs cycle just as the acetic acid originating from lactic acid does. The catabolism of fatty acids is particularly productive of new ATP's, which we shall now proceed to discuss.

Anabolism

Obviously the energy – or enthalpy – of reactions in the tissue does not all appear as heat, as it does in a flame. Indeed, an animal and man are able to exert power, and they *must do so,* at least to the extent of the basal metabolic rate. Also animals grow, and they are able – in their bodies – to produce fat even if they ingest primarily carbohydrates. So they are building up complex molecules from the simpler ones that have entered their tissue. The process is called *anabolism* from Greek: *to build up*.

A first case of anabolism was discovered as early as 1856 by Bernard, the vivisectionist. He noticed that glucose is converted into glycogen, a starch-like substance in the liver. And he also saw that glycogen regulates the sugar content of the blood: If the blood is swamped with glucose, glycogen is formed , and if there is too little glucose in the blood, glycogen falls back to sugar. Diabetes happens, if that balance fails to function. Therefore, obviously, the liver is capable of forming starch from glucose, just the opposite of what the digestive track achieves.

Two things are interesting about the balancing act between glucose and glycogen: Firstly, that it proceeds through sugar phosphate, albeit only as an intermediate,[11] and secondly that adenosine tri-phosphate is involved, an organic compound – invariably abbreviated as ATP – which was discovered in 1929 by the biochemist K. Lohmann. He found that phosphoric acid H_3PO_4, which had been thought to belong firmly to inorganic chemistry, played an important role in muscle action.

ATP results from phosphoric acid by condensation of three phosphor acid molecules and an adenosine molecule which we may write as $R-OH$, since its exact form does not concern us. Thus ATP has the structural formula

$$\begin{array}{ccc} O & O & O \\ \| & \downarrow\| & \downarrow\| \\ R-O-P-O-P-O-P-OH \\ | & | & | \\ OH & OH & OH \end{array}$$

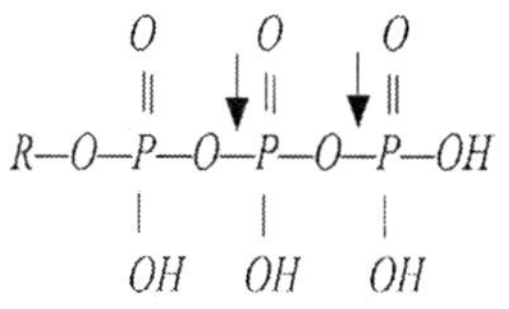

The biochemist Fritz Albert Lipman (1899–1986) noticed that the two phosphate ester bonds marked by an arrow can be more easily hydrolized than the bond near the adenosine, and his interpretation was that those two bonds lie at a higher level of free energy. Quantitatively it seems that there is about $30 \frac{kJ}{mol}$ to be gained from a reaction involving a high energy bond, twice as much as from the low energy one.

Now, back to the glucose–glycogen balance. This will help us to understand what ATP does with its high energy bonds. If we characterize a glucose molecule by $OH-\langle\ \underline{\ }\ \rangle-OH$, the glycogen molecule may be written in the form

$$OH-\langle\ \underline{\ }\ \rangle-O-\langle\ \underline{\ }\ \rangle-O-\langle\ \underline{\ }\ \rangle-O-\cdots-\langle\ \underline{\ }\ \rangle-OH$$

and one might assume that this chain results from a direct multiple condensation of glucose. However, this is not so. Indeed, in the 1930's Carl Ferdinand Cori (1896–1984) and his wife Gerty Theresa Radnitz Cori (1896–1957) found that the formation of glycogen proceeds in two steps as follows.

---

[11] The metabolic reactions inside the body tissue are called intermediary metabolism., because it is the intermediates that play the most decisive role.

Step (I): Formation of glucose phosphate and ADP from glucose and ATP

$$OH\text{-}\langle\rangle\text{-}OH + R-O-\underset{OH}{\overset{\overset{O}{\|}}{P}}-O-\underset{OH}{\overset{\overset{O}{\|}}{P}}-O-\underset{OH}{\overset{\overset{O}{\|}}{P}}-OH \rightarrow$$

$$\rightarrow OH\text{-}\langle\rangle\text{-}O-\underset{OH}{\overset{\overset{O}{\|}}{P}}-OH + R-O-\underset{OH}{\overset{\overset{O}{\|}}{P}}-O-\underset{OH}{\overset{\overset{O}{\|}}{P}}-OH$$

$$\underbrace{\phantom{OH\text{-}\langle\rangle\text{-}O-P-OH}}_{\text{Glucose phosphate}} \quad \underbrace{}_{\text{adenosine di-phosphate}}$$

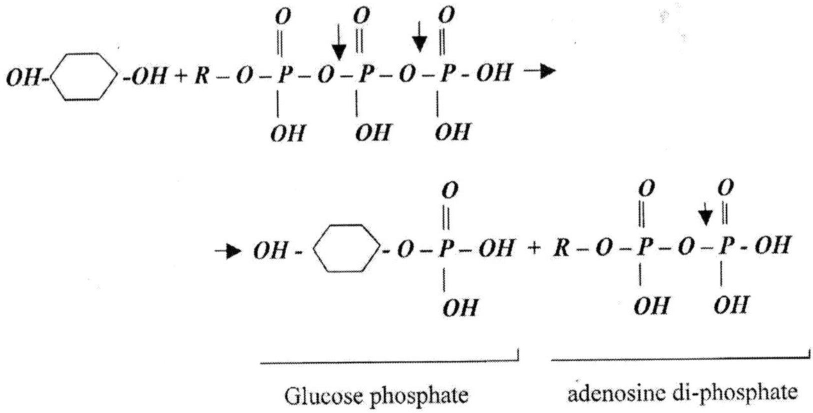

Step (II): Shedding of phosphorous acid:

$$n \times OH\text{-}\langle\rangle\text{-}O-\underset{OH}{\overset{\overset{O}{\|}}{P}}-OH \rightarrow$$

$$\rightarrow OH\text{-}\langle\rangle\text{-}O\cdots\cdots O\text{-}\langle\rangle\text{-}OH + n \times OH-\underset{OH}{\overset{\overset{O}{\|}}{P}}-OH$$

$$\underbrace{\phantom{OH\text{-}\langle\rangle\text{-}O\cdots O\text{-}\langle\rangle\text{-}OH}}_{\text{glycogen}} \quad \underbrace{}_{\text{phosphoric acid}}$$

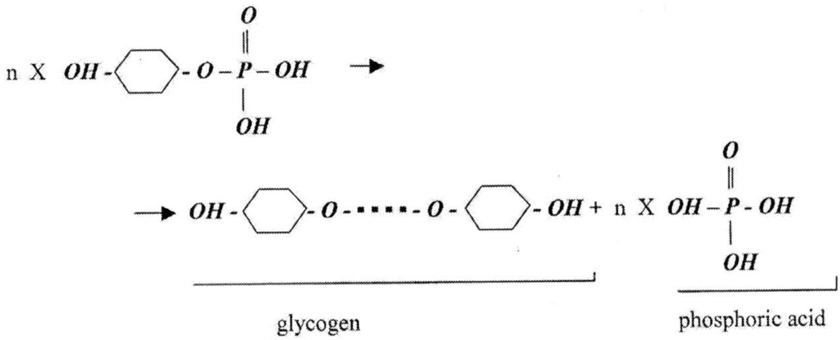

The energy-consuming step is the first one and the energy needed for the formation of glucose phosphate results from the *de-activation* of one of the high energy bonds of ATP which sinks down energetically to become ADP, i.e. adenosine di-phosphate with only *one* high energy bond.

Thinking mechanically we may say that the high energy bonds are like compressed springs. In that visualization, step (I) of the above reaction releases the spring and allows the subsequent uncoiling to lift the emerging compound glucose phosphate to its high level of energy. Actually, after Lipman's discovery, ATP has been found in body chemistry at all points where energy is needed. One may say that the large amount of energy contained in food is broken down – by tissue respiration as explained above – into energetic *small change* appropriate to *pay* for molecular reactions in the course of anabolism. Thus reactions with ATP allow a compound to move uphill energetically.

On Thermodynamics of Metabolism

One often hears it said that the functions of life create order and should therefore decrease entropy, cf. Chap. 4. Such a statement must be qualified, at least as far as animal life is concerned.[12] Indeed, one of the functions of life is the decomposition of glucose and that *increases* entropy as we saw above. Doubtless the decompositions of fatty acids and amino acids are the same in that respect, although I lack numbers for those cases.

It is true, however, that the decomposition of glucose in the tissue is accompanied by anabolism, which is also a function of life. Like when glucose builds glucose phosphate and then glycogen. We have seen that the assiduous ATP's carry their energy to the site of construction of glucose phosphate and we have implied that glycogen and glucose phosphate are energetically on the same level. Thus the two reactions involved may be written as

$$\text{Glucose} + \text{ATP} \rightarrow \text{glucose phosphate} + \text{ADP} + \Delta h_R^{(I)} \text{ with } \Delta h_R^{(I)} < 0$$
$$n \times \text{glucose phosphate} \rightarrow \text{glycogen} + n \times \text{phosphoric acid} + \Delta h_R^{(II)}$$
$$\text{with } \Delta h_R^{(II)} = 0.$$

Of course, one may ask why step (I) and step (II) occur at all. Why is the glucose ↔ glycogen balance not simply maintained by mass action via hydrolysis and condensation? And what about the entropy change of the reaction? It seems likely that entropy decreases – because *order is created* by the build-up of the long glycogen chain – but again I lack numbers.[13] If indeed entropy decreases, it must be that $\Delta h_R^{(I)}$ has a sufficiently large negative value, – i.e. the reaction (I) is exothermic to a large degree – in order to offset the entropy drop so that the free energy can decrease, as it must.

It seems to me that it might be worthwhile to study the thermodynamics of anabolism with an eye on the energies of reaction *and the entropies of reaction.* This may not actually teach us more about the reactions than we already know; but it may explain why a particular reaction occurs rather than another, seemingly simpler one.

On the other hand, the people, who disentangled the complex workings of intermediary metabolism, were probably not much concerned with thermodynamic questions. Even without that concern it must be admitted that they did an excellent job. Nor did they go unrecognised. Nearly all of those biochemists whom I have mentioned received the Nobel prize: Wieland, Warburg, Krebs, Lipman, Pauling,[14] and the Coris. The Germans

---

[12] We shall come to plant life in a short while.

[13] Those books which I have consulted for the writing of this chapter do not give entropies for such molecules as glucose phosphate and glycogen chains.

[14] Pauling is one of only two persons who received two Nobel prizes, – one for peace, because of his commitment against nuclear armament. The other person with two prizes is Marie Sklodowska Curie (1867–1934).

among them all had some difficulties with Adolf Hitler, or he with them.[15] Most of them emigrated, and life was not made easy for those who stayed.

What is Life?

Another emigrant was the eminent quantum-physicist Erwin Schrödinger (1887–1961) who found a fairly comfortable temporary home at the School of Advanced Studies in Dublin, Ireland. There, in 1943, he gave a course of public lectures entitled "What is life?" which was afterwards published as a booklet.[16]

To the modern reader – well-informed by newspapers about DNA and the human genome – the book is somewhat obsolete, but it is still worshipped by theoretical physicists. Schrödinger expounds the idea that the gene must be a molecule lest it be subject to constant change by the thermal motion. He observes that a gene seems to be stable over many generations as put in evidence by the persistence of the well-documented *Habsburg lower lip*, a slight deformity of the lip in the members of that illustrious and oft-portrayed family. Also, in order to account for mutations, Schrödinger emphasizes the need for meta-stable states in the gene, i.e. energetic minima separated from other, conceivably lower minima by barriers. He sees thermal motion, or possibly X-rays, or cosmic rays as the only means to overcome such barriers. This seems to offer a satisfactory explanation for the fact that a mutation is a rare event because, after all, we are not very often exposed to X-rays, and the temperature must not be increased much beyond 37°C.

The closest Schrödinger comes to answer his self-imposed question about life is, when he says [17]

> What is the characteristic feature of life? When is a piece of matter said to be alive? When it goes on "doing something", moving, exchanging material with its environment, and so forth, and that for a much longer period than we would expect an inanimate piece of matter to "keep going" under similar circumstances.

---

[15] In the 1930's German scientists were not allowed to accept the Nobel prize, because Hitler was angered when the 1935-peace prize was given to Carl von Ossietzky (1889–1938), a well-known pacifist who, at the time of the award, was kept in a concentration camp where he later died.

[16] E. Schrödinger: "What is life? The physical aspect of the living cell." Cambridge: At the University Press. New York: Macmillan Company (1945).

[17] E. Schrödinger: "What is life? ..." loc.cit. p. 70.

In 1918, after World War I, Schrödinger made up his mind to abandon physics for philosophy, but the city where he had hoped to obtain a university post was lost to Austria in the peace treaties. Therefore Schrödinger remained a physicist.[18]

Fig. 11.3. Erwin Schrödinger (1887–1961)

Well, that answer seems to be begging the question. Maybe we cannot have a better – and shorter – answer than that. But I, for one, would certainly wish to. Since we are on the subject, let me also quote Asimov,[19] himself a biochemist

> A living organism is characterized by the ability to effect a temporary and local decrease in entropy by means of enzyme-catalysed chemical reactions.

So, thus we are back on the subject of entropy. Schrödinger devotes the last part of his booklet to it. He says that *a living organism feeds on negative entropy,* meaning that an animal maintains a high level of order because it feeds on plants which have themselves a high degree of orderliness, i.e. a low entropy. He says:

> Indeed, in the case of higher animals we know the kind of orderliness they feed upon well enough, viz. the extremely well-ordered state of matter in more or less complicated organic compounds, which serve them as foodstuffs.

I do wonder though whether that argument is not a trifle superficial. After all, we have seen that, before the animal does anything constructive with the foodstuffs, it breaks them down to material of lesser order in the digestive process. So, at least we can say that the organism does not make the most of the order that is offered to it.

---

[18] I. Asimov: "Biographies ..." loc.cit. p. 621.

[19] I. Asimov: "Life and Energy." Avon Publishers of Bard, New York (1972).
I owe much of the information presented in this chapter to the study of that book by Asimov. On these pages I have often quoted Asimovs book of biographies and occasionally other science essays by the author. Indeed, if the truth were known, I do admire Asimov's way of writing, – except when he writes Science Fiction.

However, the metaphor *feeding on negative entropy* – the term was quickly changed into *negentropy* – has fired the imagination of physicists of the more esoteric type and theologians. I am told that Teilhard de Chardin, a Jesuit palaeontologist and anthropologist – who attempted to reconcile the theory of evolution with the teachings of the catholic church – was inspired by negentropy. Schrödinger cannot be blamed, perhaps. After all he was giving a public lecture to a mixed audience. And he lived to regret his simplified presentation. Indeed, in the German translation of his book in 1951 he says[20]

> My remarks about *negative entropy* have been criticized by experts in physics. I have to say to them that I should have used the word *free energy,* if I had spoken to them.

Let us look at plants next, surely one of the sources of negentropy for animals. This is an interesting subject in itself. What happens in a plant is the *photosynthesis* of glucose from the CO_2 of the air and from H_2O of the soil and the release of oxygen. We rewrite the stoichiometric formula from before, except in reverse order as is appropriate for the synthesis of glucose rather than its decomposition.

$$6CO_2 + 6H_2O \rightarrow C_6H_{12}O_6 + 6O_2 \qquad \begin{aligned} \Delta\overline{h}_R &= 2798\tfrac{\mathrm{kJ}}{\mathrm{mol}} \\ \Delta\overline{s}_R &= -241\tfrac{\mathrm{J}}{\mathrm{molK}} \end{aligned} .$$

In some way this is the worst possible case for a chemical reaction: The energy – or enthalpy – *increases* and the entropy *decreases.* Since the process occurs at constant pressure $p_R = 1\mathrm{atm}$ and at the normal temperature, roughly $T_R = 298\mathrm{K}$, the first law requires that we provide heat and the second law demands that we withdraw heat. Indeed we have

$$\overline{q} = \Delta\overline{h}_R > 0 \quad \text{by the first law,} \quad \text{and} \quad \overline{q} \le T_R\Delta\overline{s}_R < 0 \quad \text{by the second law.}$$

This is a clear contradiction and, if we did not know better, we could now come to the conclusion that the process is impossible.

Another way to emphasize the contradiction is to calculate the change of Gibbs free energy

$$\Delta\overline{g}_R = \Delta\overline{h}_R - T_R\Delta\overline{s}_R = 2870\tfrac{\mathrm{kJ}}{\mathrm{mol}} > 0 .$$

Thus the free energy grows, when we know very well that it should decrease according to Gibbs, Helmholtz and every other thermodynamicist since their time.

---

[20] E. Schrödinger: "Was ist Leben? Die lebende Zelle mit den Augen des Physikers betrachtet." 2nd edition. A. Francke Verlag, Bern and Leo Lehnen Verlag, München (1951).

So there is a dilemma! The only way out seems to be to conclude that the reaction cannot occur *by itself.* Apart from a supply of energy there must be an accompanying process which *increases* the entropy far enough to offset the negative entropy of reaction. In fact, the increase of entropy must even be big enough to effect an overall decrease of *Gibbs free energy.*

At first sight the supply of energy does not seem to present a problem, since the sun sends 1341W toward every square meter on the earth that is held perpendicular to the incoming radiation.[21] 75% of that radiative power reaches the earth's surface and a plant leaf absorbs 65% of that, primarily the red and yellow part of the spectrum, which is why the leaves are green.

So, a leaf receives 650W/m^2, and it emits the radiative power $\frac{c}{4}aT^4$ [22] appropriate to its temperature T. According to plant physiologists,[23] if the leaf works well at photosynthesis, each m^2 may produce 1g, or 1/180 mol glucose in one hour. Thus the energy balance reads

$$650\frac{\mathrm{W}}{\mathrm{m}^2}-\frac{c}{4}aT^4=2789\frac{\mathrm{kJ}}{\mathrm{mol}}\cdot\frac{1}{180}\mathrm{mol}\frac{1}{3600\,\mathrm{sm}^2}=4.3\frac{\mathrm{W}}{\mathrm{m}^2}.$$

Hence follows T = 327K, or 54°C, a temperature which is high enough to let the leaf wilt and die. Moreover, plant physiologists inform us that photosynthesis does not work anymore beyond a temperature of 35°C. Therefore, there is a problem even with respect to the first law. A possible key to the solution is known to farmers, gardeners and house-wives, who all know that a plant requires more water – much more, x-times more (say) – than dictated by the stoichiometric formula. The plant absorbs all that water in the roots, passes it upwards to the leaves and evaporates it there. Thus a plant cools its leaves in the same manner as animals cool their skins: By evaporation of water.[24] It is easy to calculate the value of x when we require that the temperature stay at 298K. We obtain $x \approx 500$ so that, for each gram of water that helps to build up glucose, the plant needs to evaporate 500grams to keep itself cool.

---

[21] We need to know, of course, the chemical "mechanism" by which the plant makes use of the radiative energy. Biophysicist are working hard on that question and I am told that they have not uncovered all parts of the reaction yet, although they are getting close.

[22] $\frac{c}{4}a$ equals $5.67\cdot10^{-8}$W/m^{2}K^4, cf. Chap. 7.

[23] E.g. see W. Larcher: "Ökophysiologie der Pflanzen. Leben, Leistung und Stressbewältigung der Pflanzen in ihrer Umwelt" [Ecophysiology of plants. Life, performance and stress management of plants in their environment] 5.Auflage, Verlag Eugen Ulmer Stuttgart (1994).

[24] As far as I know, this idea was first presented by myself and A . Klippel in the paper: A. Klippel, I. Müller: "Plant growth – a thermodynamicist's view." Continuum Mechanics and Thermodynamics 9, (1997).

Therefore evaporation is a process accompanying photosynthesis, and that process indeed increases entropy. However, that fact does not help with respect to the Gibbs free energy balance, since the free energy does not change upon evaporation. It is true that entropy grows, but the enthalpy also grows, such that the free energy *h-Ts* remains equal. So we are still looking for the entropy-producing accompanying process that could set the free energy balance right. Schrödinger made his task easy by saying:

> These [the plants] of course have their most powerful supply of *negative entropy* in the sunlight.

Let us see whether we can make sense out of this statement. We refer to Chap. 7 and recall that absorption and emission of radiation by one m^2 of leaf surface produces entropy at the rate of 1.7 W/K. That amount is far bigger than the entropy increase needed for the process accompanying the photosynthesis, which is only 0.014 W/K according to the numbers given above. Thus, as far as pure numbers are concerned, Schrödinger's suggestion about the *negative entropy* of the sunlight could be right. And yet, there remains a feeling that his answer is too pat: One does not see how the leaf incorporates all that entropy – or even part of it – into the chemical process. As far as I know this question has never been addressed.[25]

All of this does not really help to answer the question "What is life?" and, although so many eminent people have failed, I should like to try myself: *Life is the indefinite working of a complex machinery.* Thus even the steam engine, or a locomotive show traces of life. Obviously the question is: How complex is complex? Surely the locomotive is too simple to be called alive; its mechanism is too easy to understand. One is tempted to draw the analogy with art. Someone has said: *If I can do it, it is not art.* And so: *If I can understand it, it is not life.*

Eventually, of course, we shall understand the working of animals and plants as well as we now understand the working of the locomotive. The biophysicists and biochemists are quite successful in clearing up the living mechanisms better and better. To be sure, they will not find *life*, just as little as an engineer finds steam, when he disassembles a steam engine.

---

[25] I have suggested an alternative accompanying process, – leaving out entropic radiation altogether –, namely the mixing of the water evaporated by the leaf with the surrounding air. A. Klippel, I. Müller: "Plant growth ..." loc.cit.

Name Index

Breinigsville, PA USA
27 May 2010
238823BV00005B/117/P

9 783540 462262